T0289636

Biotechnology and Crop Improvement

Biotechnology and Crop Improvement

Edited by Aila Williams

SYRAWOOD
PUBLISHING HOUSE

New York

Published by Syrawood Publishing House,
750 Third Avenue, 9th Floor,
New York, NY 10017, USA
www.syrawoodpublishinghouse.com

Biotechnology and Crop Improvement
Edited by Aila Williams

International Standard Book Number: 978-1-64740-415-4 (Hardback)

Cataloging-in-publication Data

Biotechnology and crop improvement / edited by Aila Williams.
    p. cm.
Includes bibliographical references and index.
ISBN 978-1-64740-415-4
1. Agricultural biotechnology. 2. Crop improvement. 3. Biotechnology. 4. Crops. I. Williams, Aila.
S494.5.B563 B56 2023
338.16--dc23

TABLE OF CONTENTS

Preface..VII

Chapter 1   **Selection of the Root Endophyte *Pseudomonas brassicacearum* CDVBN10 as Plant Growth Promoter for *Brassica napus* L. Crops**..1
Alejandro Jiménez-Gómez, Zaki Saati-Santamaría, Martin Kostovcik, Raúl Rivas, Encarna Velázquez, Pedro F. Mateos, Esther Menéndez and Paula García-Fraile

Chapter 2   **Evaluating Biochar-Microbe Synergies for Improved Growth, Yield of Maize and Post-Harvest Soil Characteristics in a Semi-Arid Climate**........................37
Maqshoof Ahmad, Xiukang Wang, Thomas H. Hilger, Muhammad Luqman, Farheen Nazli, Azhar Hussain, Zahir Ahmad Zahir, Muhammad Latif, Qudsia Saeed, Hina Ahmed Malik and Adnan Mustafa

Chapter 3   **Large Scale Screening of Rhizospheric Allelopathic Bacteria and their Potential for the Biocontrol of Wheat-Associated Weeds**..56
Tasawar Abbas, Zahir Ahmad Zahir, Muhammad Naveed, Sana Abbas, Mona S. Alwahibi, Mohamed Soliman Elshikh and Adnan Mustafa

Chapter 4   **Agricultural Utilization of Unused Resources: Liquid Food Waste Material as a New Source of Plant Growth-Promoting Microbes**..72
Waleed Asghar, Shiho Kondo, Riho Iguchi, Ahmad Mahmood and Ryota Kataoka

Chapter 5   **Co-Inoculation of Rhizobacteria and Biochar Application Improves Growth and Nutrients in Soybean and Enriches Soil Nutrients and Enzymes**........................86
Dilfuza Jabborova, Stephan Wirth, Annapurna Kannepalli, Abdujalil Narimanov, Said Desouky, Kakhramon Davranov, Riyaz Z. Sayyed, Hesham El Enshasy, Roslinda Abd Malek, Asad Syed and Ali H. Bahkali

Chapter 6   **Volatile Organic Compounds from Rhizobacteria Increase the Biosynthesis of Secondary Metabolites and Improve the Antioxidant Status in *Mentha piperita* L. Grown under Salt Stress**..101
Lorena del Rosario Cappellari, Julieta Chiappero, Tamara Belén Palermo, Walter Giordano and Erika Banchio

Chapter 7   **Salicylic Acid Improves Boron Toxicity Tolerance by Modulating the Physio-Biochemical Characteristics of Maize (*Zea mays* L.) at an Early Growth Stage**..117
Muhammad Nawaz, Sabtain Ishaq, Hasnain Ishaq, Naeem Khan, Naeem Iqbal, Shafaqat Ali, Muhammad Rizwan, Abdulaziz Abdullah Alsahli and Mohammed Nasser Alyemeni

Chapter 8   **α-Tocopherol Foliar Spray and Translocation Mediates Growth, Photosynthetic Pigments, Nutrient Uptake and Oxidative Defense in Maize (*Zea mays* L.) under Drought Stress** 131
Qasim Ali, Muhammad Tariq Javed, Muhammad Zulqurnain Haider, Noman Habib, Muhammad Rizwan, Rashida Perveen, Shafaqat Ali, Mohammed Nasser Alyemeni, Hamed A. El-Serehy and Fahad A. Al-Misned

Chapter 9   **Rhizobacteria Isolated from Saline Soil Induce Systemic Tolerance in Wheat (*Triticum aestivum* L.) against Salinity Stress** 158
Noshin Ilyas, Roomina Mazhar, Humaira Yasmin, Wajiha Khan, Sumera Iqbal, Hesham El Enshasy and Daniel Joe Dailin

Chapter 10   **PGPR Modulation of Secondary Metabolites in Tomato Infested with *Spodoptera litura*** 177
Bani Kousar, Asghari Bano and Naeem Khan

Chapter 11   **Nematicidal Evaluation and Active Compounds Isolation of *Aspergillus japonicus* ZW1 against Root-Knot Nematodes *Meloidogyne incognita*** 198
Qiong He, Dongya Wang, Bingxue Li, Ambreen Maqsood and Haiyan Wu

Chapter 12   **Relevance of Plant Growth Promoting Microorganisms and their Derived Compounds, in the Face of Climate Change** 214
Judith Naamala and Donald L. Smith

Permissions

List of Contributors

Index

PREFACE

Biotechnology is the application of biology to the creation of new products, organisms and methods with the goal of benefiting the society as well as improving human health. It has a wide range of applications most notably in agriculture and medicine. Modern biotechnology offers innovative products and technologies to lessen environmental footprint, use cleaner and less energy, combat debilitating and uncommon diseases, and have cleaner, safer and more effective industrial manufacturing processes. Biotechnology in agriculture not only aids in increasing productivity but also improves quality of crop production. Another advantage of biotechnology is the creation of crops that are resistant to diseases and pests. Gene editing technology enables targeted and high-precision restructuring of plant genomes which decreases the cost of product development. This book explores all the important aspects of biotechnology and its applications in crop improvement. It consists of contributions made by international experts. This book will serve as a reference to a broad spectrum of readers.

This book is a result of research of several months to collate the most relevant data in the field.

When I was approached with the idea of this book and the proposal to edit it, I was overwhelmed. It gave me an opportunity to reach out to all those who share a common interest with me in this field. I had 3 main parameters for editing this text:

1. Accuracy – The data and information provided in this book should be up-to-date and valuable to the readers.

2. Structure – The data must be presented in a structured format for easy understanding and better grasping of the readers.

3. Universal Approach – This book not only targets students but also experts and innovators in the field, thus my aim was to present topics which are of use to all.

Thus, it took me a couple of months to finish the editing of this book.

I would like to make a special mention of my publisher who considered me worthy of this opportunity and also supported me throughout the editing process. I would also like to thank the editing team at the back-end who extended their help whenever required.

Editor

Selection of the Root Endophyte *Pseudomonas brassicacearum* CDVBN10 as Plant Growth Promoter for *Brassica napus* L. Crops

Alejandro Jiménez-Gómez [1,2,†], Zaki Saati-Santamaría [1,2], Martin Kostovcik [3,4], Raúl Rivas [1,2,5], Encarna Velázquez [1,2,5], Pedro F. Mateos [1,2,5], Esther Menéndez [1,2,5,6,*] and Paula García-Fraile [1,2,5,*]

[1] Microbiology and Genetics Department, University of Salamanca, 37007 Salamanca, Spain; alexjg@usal.es (A.J.-G.); zakisaati@usal.es (Z.S.-S.); raulrg@usal.es (R.R.); evp@usal.es (E.V.); pfmg@usal.es (P.F.M.)
[2] Spanish-Portuguese Institute for Agricultural Research (CIALE), Villamayor, 37185 Salamanca, Spain
[3] Department of Genetics and Microbiology, Faculty of Science, Charles University, 12844 Prague, Czech Republic; kostovci@biomed.cas.cz
[4] BIOCEV, Institute of Microbiology, the Czech Academy of Sciences, 25242 Vestec, Czech Republic
[5] Associated R&D Unit, USAL-CSIC (IRNASA), Villamayor, 37185 Salamanca, Spain
[6] MED—Mediterranean Institute for Agriculture, Environment and Development, Institute for Advanced Studies and Research (IIFA), Universidade de Évora, Pólo da Mitra, Ap. 94, 7006-554 Évora, Portugal
[*] Correspondence: esthermenendez@uevora.pt (E.M.); paulagf81@usal.es (P.G.-F.)
[†] Present address: School of Humanities and Social Sciences, University Isabel I, 09003 Burgos, Spain.

Abstract: Rapeseed (*Brassica napus* L.) is an important crop worldwide, due to its multiple uses, such as a human food, animal feed and a bioenergetic crop. Traditionally, its cultivation is based on the use of chemical fertilizers, known to lead to several negative effects on human health and the environment. Plant growth-promoting bacteria may be used to reduce the need for chemical fertilizers, but efficient bacteria in controlled conditions frequently fail when applied to the fields. Bacterial endophytes, protected from the rhizospheric competitors and extreme environmental conditions, could overcome those problems and successfully promote the crops under field conditions. Here, we present a screening process among rapeseed bacterial endophytes to search for an efficient bacterial strain, which could be developed as an inoculant to biofertilize rapeseed crops. Based on in vitro, in planta, and in silico tests, we selected the strain *Pseudomonas brassicacearum* CDVBN10 as a promising candidate; this strain produces siderophores, solubilizes P, synthesizes cellulose and promotes plant height in 5 and 15 days-post-inoculation seedlings. The inoculation of strain CDVBN10 in a field trial with no addition of fertilizers showed significant improvements in pod numbers, pod dry weight and shoot dry weight. In addition, metagenome analysis of root endophytic bacterial communities of plants from this field trial indicated no alteration of the plant root bacterial microbiome; considering that the root microbiome plays an important role in plant fitness and development, we suggest this maintenance of the plant and its bacterial microbiome homeostasis as a positive result. Thus, *Pseudomonas brassicacearum* CDVBN10 seems to be a good biofertilizer to improve canola crops with no addition of chemical fertilizers; this the first study in which a plant growth-promoting (PGP) inoculant specifically designed for rapeseed crops significantly improves this crop's yields in field conditions.

Keywords: *Pseudomonas*; PGPB; bioinoculants; endophytes; bacterial microbiome; culturome; genome sequencing; *Brassica napus*

1. Introduction

The FAO estimates that there will be 2.3 billion more people on the Earth in 2050, in a world already struggling to combat poverty and hunger. Thus, we need to increase our capability to produce food using the limited natural resources of our planet more efficiently while fighting climate change. Chemical fertilizers increase crop yields, but they have negative effects for human and animal health, contaminate soils and water, and their fabrication, which requires huge amounts of energy, contributes to resource depletion and global warming [1]. Moreover, the excessive or repetitive use of chemical fertilizers usually presents low efficiency in their use by plants because the soil biogeochemical cycles is often altered [2].

Alternatively, plant growth-promoting bacteria (PGPB), which are naturally occurring microbes that modulate plant growth due to their metabolic activities, can enhance crop yields when applied as biofertilizers [3–5]. PGPB can fix atmospheric nitrogen, produce siderophores and/or phytohormones, solubilize phosphorous and/or potassium and inhibit the growth of pathogenic microorganisms [6]. Within PGPB, endophytes are particularly interesting because, once inside the plant, they do not need to compete with the dense population of bacteria in the rhizosphere and they are protected from extreme abiotic conditions, so they have more chances to succeed when applied in the fields [7,8].

Endophytes are part of the plant microbiome and play essential roles for its fitness and survival [9]. Many of these microorganisms are non-cultivable in routine laboratory conditions and thus, culture independent methods allow us to unravel the complete microbial diversity living within the plants. These endophytic microbiomes, as occurs in animals, interact with their host in essential functions [10–12]; hence, plant microbiome research highlights the importance of indigenous microbial communities for host phenotypes such as growth and health [13].

Brassica napus L. (rapeseed, canola) is an important crop due to its cultivation not only as a food resource (human and animal), but also for biodiesel production, being one of the most significant oilseed crops in temperate climates [14]. In Europe, rapeseed seeds are the primary source of oil for biodiesel production, its by-product being a high protein source for animal feeding [15]. However, rapeseed cultivation requires important amounts of chemical fertilizers [16], and therefore, alternatives that enable the reduction in chemical fertilization for a more sustainable crop are very desirable. This implies the use of biofertilizers, which include endophytic PGPB.

Thus, the design of an efficient bacterial endophytic inoculant for rapeseed crops which could increase rapeseed crop yields with no addition of chemical fertilizers is very desirable. For that purpose, it is necessary to study the members of the bacterial endophytic population, those members of the endophytic community which can be artificially cultured and thus biotechnologically produced and formulated.

In terms of plant growth-promoting (PGP) functionality, in vitro PGP mechanisms have been analyzed in just a few rhizospheric [17] or endophytic bacteria associated with *B. napus* plants [18,19]. In addition, the information about the effects of PGPB in rapeseed plants is scarce [18,20–23]. Taking advantage of next generation sequencing, PGPB genome sequence annotation and analysis allow in silico studies of the genetic potential of a bacterium to promote plant growth, including the discovery of specific PGP traits and/or pathways, such as tolerance to different biotic and abiotic stresses, heavy metal detoxifying activity or biological control potential [24].

These massive parallel sequencing techniques are becoming even more interesting when applied to elucidate the taxonomic composition and biological functions of the plant and soil microbiome when plants grow under field conditions, where they can be used to recreate the microbial communities' dynamics [25].

Based on the hypothesis that bacterial endophytes can be efficient biofertilizers when applied as inoculants in the fields, the aim of this work was to isolate and select a rapeseed bacterial endophyte with the potential to promote rapeseed growth and yields. For that, we obtained a collection of rapeseed endophytic bacteria and analyzed the potential of our isolates as plant growth promoters, through a screening of a few in vitro PGP mechanisms followed by the analysis of the in planta effect with several selected isolates, evaluating their capability to promote rapeseed seedling growth. Once we had selected the best-performing strains in planta, we obtained the genome sequences of the best PGP endophytic strains to deepen the study of their molecular machinery implicated in plant colonization and growth promotion. The in silico and in vivo assays allowed us to select one particular strain, which was inoculated in a field trial, showing for the first time a significant increase in rapeseed yields using a PGP bacterium inoculum. As a novelty, we analyzed the impact of the inoculation of the strain not only in the plant development and crop yields, but also on the root endophytic community.

2. Materials and Methods

2.1. Isolation and Identification of Bacterial Isolates

Rapeseed plants (*B. napus* cv rescator) in the phenological stage of rosette were collected in February 2017 from two agricultural soils located in the municipalities of Castellanos de Villiquera (CDV) (province of Salamanca) and Peleas de Arriba (PDA) (province of Zamora), both in Spain. Plants were extracted from the soils, kept refrigerated and shipped to the laboratory, where they were processed within two hours from the time of extraction.

To isolate rapeseed root bacterial endophytes, roots were excised carefully and washed in sterile Petri dishes containing sterile distilled water (× 10 times) and then surface-disinfected by immersion in sodium hypochlorite (2%) for 2 min. After that, surface-disinfected roots were washed 5 times in sterile distilled water and dried with sterile filter paper. An aliquot of water from the last washing step of each sample after the disinfection protocol and a few entire disinfected roots were plated as disinfection controls. No bacterial growth was observed in those plates.

Surface-disinfected roots were smashed in a sterile mortar and the content was serially diluted with sterile distilled water. Then, 100 µL of the 10^{-2}, 10^{-3} and 10^{-4} dilutions were plated onto Petri dishes containing different media to target the isolation of a wider biodiversity: Tryptic Soy Agar (TSA; BD Difco, Franklin Lakes, NJ, USA), YMA (Laboratorios Microkit, Madrid, Spain), 869 medium (Tryptone (10 g/L), yeast extract (5 g/L), NaCl (5 g/L), D-glucose (1 g/L), $CaCl_2$ (0.345 g/L), and agar (20 g/L)) and ten times diluted 869 medium.

Plates were incubated at 28 °C for 21 days. The emerging bacterial colonies were regularly isolated to get pure cultures. Their names were composed by CDV or PDA, depending on the sampling origin, followed by BN, from *Brassica napus* and a correlative number. Then, isolated strains were stored in a sterile 20% glycerol solution at −80 °C for long-term storage.

For bacterial strain identification, DNA was obtained using the REDExtract-N-Amp™ PCR Ready Mix (Sigma-Aldrich Co. LLC), following the instructions given by the manufacturer. Then, strains were grouped at species or subspecies level based on their 879F-RAPD fingerprints, obtained as detailed by Igual et al. [26] and grouped by means of the UPGMA algorithm (unweighted pair grouping with mathematic average) using the software package BioNumerics version 4.5 (Applied Maths NV, Sint-Martens-Latem, Belgium), with a threshold of 75% similarity. To identify a representative bacterial isolate of each 879F-RAPD group, 16S rRNA gene sequences were amplified as described in Rivas et al. [27] and processed as described in Poveda et al. [28]. Nearly complete (~1500 bp) sequences were compared with those from type strains deposited in GenBank using BLASTn program [29] and EzTaxon tool [30].

In the case of those bacterial strains selected for genome sequences, housekeeping gene sequences (*gyr*B and *rpo*B) were retrieved from the genome and compared to those available in the GenBank database using BLASTn for a more accurate taxonomic identification.

In the case of the strain inoculated in the field trials, a phylogenetic analysis of the 16S rRNA gene sequence of the strain and those of the closely related species was done as detailed in Jiménez-Gómez et al. [31].

2.2. In Vitro Analyses of Plant Growth-Promoting Mechanisms and Biosynthesis of Polysaccharides

Bacterial siderophore production and solubilization of non-assimilable phosphates were evaluated as detailed in Jiménez-Gómez et al. [31]. Briefly, siderophore production was evaluated by inoculating in M9-CAS-agar medium plates [32] modified according to the suggestions given by Alexander and Zuberer [33]. The solubilization of non-soluble phosphates into soluble assimilable ions was analyzed in Pikovskaya medium plates [34], which contain bicalcium phosphate ($CaHPO_4$) or tricalcium phosphate $[Ca_3(PO_4)_2]$ as the P source. Polysaccharide (cellulose and cellulose-like polymers) biosynthesis ability of each isolate was determined as described by Robledo et al. [35]. All plates were incubated for up to 21 days at 28 °C, recording the results every week. Nitrogen fixation was assayed in liquid medium as detailed in Poveda et al. [28]. The method shows the ability of strains to grow in a N-free minimal liquid medium. Indole acetic acid (IAA)-like compound production was measured by the colorimetric method described in Khalid et al. [36].

2.3. Effects of Bacterial Isolates on Rapeseed Seedlings

Rapeseed seeds (cv rescator) were surface-disinfected with 70% ethanol for two min, followed by soaking in an aqueous 5% sodium hypochlorite solution for ten minutes. Then, they were washed five times with sterile water and pre-germinated on water-agar plates (1.5%) for 24 h. To prepare the inoculum, bacteria were grown in their respective isolation media for 3 days at 28 °C; afterwards, the Petri dishes were flooded with saline buffer (0.9% NaCl) in order to obtain the cell suspensions, which were adjusted to an O.D. (600 nm) of 0.5, corresponding to final concentrations of ~10^8–10^9 CFU/mL (this concentration was determined after counting the number of viable cells using the serial decimal dilution method). After the pre-germination and inoculum preparation steps, Petri dishes containing the seedlings were inoculated. Twelve plates per treatment with five seeds per plate were prepared for the in vitro analyses. Thirty seedlings per treatment were collected at five and fifteen days post-inoculation, respectively. Values of seedling height and root length were recorded at each collection time.

2.4. Draft Genome Sequencing and Annotation

The genome sequence was obtained from selected strains after the plant growth promotion tests on rapeseed seedlings. For genome sequencing, the DNA was obtained from selected bacteria after two days of growth at 28 °C using the Quick DNA Fungal/Bacterial Miniprep kit (Zymo Research, Irvine, CA, USA) following the procedure described by the manufacturer.

The draft genome of selected isolates was sequenced on an Illumina MiSeq platform as described by Saati-Santamaría et al. [37]. The sequence data were assembled using Velvet (v1.12.10) [38]. Gene calling, annotation, and search for genes related to plant growth promotion- and colonization-related capabilities was performed using RAST (v2.0) pipeline [39] and then re-checked by BLASTp against known conserved proteins from phylogenetically related or closest relatives *Pseudomonas* strains. The Genome Shotgun project for strains CDVBN10 and CDVBN20 has been deposited at DDBJ/EMBL/GenBank under the accessions VDLV00000000 and VDLW00000000, respectively. The versions described in this paper are versions VDLV01000000 and VDLW01000000.

2.5. Field Experiment

The most promising PGP bacterium according to *in vitro, in vivo,* and *in silico* experiments, strain *Pseudomonas brassicacearum* CDVBN10, was assayed in field conditions as a rapeseed biofertilizer.

The field trial was performed between September 2018 and May 2019 in the locality of Cañizal (Zamora; NS/EW coordinates: 41.152627/-5.356508). The field has a crop history of rotations between sunflower and barley. No rapeseed crops were sown previously in this soil. The soil is a non-saline soil with loamy-sandy texture, with a good organic matter content (5.6%), showing a very slightly basic pH (7.87) and a low EC (EC1:2 0.096 dS/m). The electrical conductivity and the pH were measured according to Dellavalle [40]. The mineral content of the soil is as follows: total N < 0.045%; assimilable

P 15 mg/kg, K 0.25%, Zn 21.9 mg/Kg, Fe 1.2%. The number of colony formation units (CFU) per gram of soil (counted in Plate Counting Agar (PCA; Sigma-Aldrich Co. LLC, St. Louis, MO, USA) plates incubated at 28 °C for 7 days) is 1.2×10^7.

The experimental field was divided in six rows with 5 m length by 2 m width ($10 \ m^2$) with a 0.5 m buffer non-cropped area between them to avoid the transfer of bacteria between plots. Plants were grown in a density of 12 plants per linear meter in each row and they were rainfed.

The experiment was arranged in a randomized block design with three replicates per treatment. No chemical fertilization was applied to the soil. One month after the seeding, once the seedlings had emerged, a bacterial suspension with a cell density of 10^9 CFU/mL was prepared on sterile saline buffer (0.9% NaCl), using 3 day old bacterial cultures grown at 28 °C in TSA. A total of 5 mL of the bacterial suspension was added to each plant. For uninoculated control, equal volume of sterile saline buffer was added per plant. Fifteen days after the inoculation of the plants, the application was repeated.

The rapeseed plants were collected at seed maturity stage, approximately 8 months after seeding. Thirty randomly selected plants of each plot were harvested and kept separately. Plants were quickly taken to the lab on ice, where we separated roots from shoots carefully. Roots were excised for further amplicon sequencing. From each plant, grain yield and total shoot dry biomass (oven-dried at 60 °C) were recorded. Dry plants were also used for the analysis of N, C, Fe, P, K at the Ionomics Service at CEBAS-CSIC (Murcia, Spain), using an Elemental Analyst model TruSpec CN628 equipment (Leco, St Joseph, MI, USA) for the N analysis, and ICP THERMO ICAP 6500DUO equipment (Thermo Fisher, Waltham, MA, USA) for the analysis of the remaining elements.

2.6. Amplicon Sequencing and Sequence Analysis

Total genomic DNA was obtained from rapeseed roots collected as explained in the previous section using the DNeasy Power Plant Pro Kit (Qiagen®, Venlo, Netherlands), following the instructions given by the manufacturer. For each location, DNA from roots of three different plants of each treatment was pooled and amplicons of the complete bacterial 16S rRNA gene (V1-V9 regions) were sequenced on a PacBio Sequel system using a SMRT Cell 1M V3 LR. PacBio circular consensus sequences (CCS) were used to obtain sequences with a low error rate in the consensus sequence resulting from the alignment between all the subreads from the same molecule.

Sequences with lengths ≥800 nt to ≤1600 nt were filtered using SEED2 software package [41]. QIIME (v1.9) software [42] was used for amplicon data analysis. The sequences were aligned and taxonomically classified (97% threshold) using the Greengenes 16S rRNA sequence database, release 13.8.97 [43] with an open-reference picking method for the OTU (Operational Taxonomic Units) clustering, using the default settings of the UCLUST algorithm. Chimeric sequences were removed using UCHIME (v6.1.544) [44]. Lineages belonging to chloroplast and mitochondria were removed with QIIME scripts. PacBio reads were deposited in NCBI under the SRA accession PRJNA601164.

Comparison between control and bacteria-treated samples and plots summarizing taxa were made following QIIME scripts. The alpha diversity was measured with the Phylogenetic Diversity (PD), Chao1, Shannon's, Simpson's and Good's coverage indexes. Comparisons between treatments were made using the Kruskal–Wallis statistic test [45] applying the Benjamini–Hochberg false discovery rate (FDR) procedure for multiple comparisons [46]. OTU tables were rarefied using the lower sequence count among all samples as maximum rarefaction depth. The beta diversity of the samples was measured using weighted and unweighted UniFrac distances. Beta diversity comparison of treatments was made through nonparametric p-values with the Bonferroni correction [47], calculated after 999 Monte Carlo permutations. A value of $p > 0.05$ was used as a threshold for statistical significance of OTU correlation to a control or treated samples.

2.7. Statistical Analysis of Plant Parameters

Statistical comparisons of plant growth assays, including parameters recorded of the plants collected from the field assay, were carried out using the StatView 5.0 (SAS Institute, Inc., Cary,

NC, USA) [48] and performed using one-way analysis of variance (ANOVA). P values of 0.05 or less ($p \leq 0.05$) were considered statistically significant. Fisher's protected least significant differences (LSD) test was used as post hoc test.

3. Results

3.1. Bacterial Culturome Shows the High Diversity of B. Napus Associated Endophytic Bacteria

Using a combination of rich and minimal media to target the isolation of a wider biodiversity, we obtained 112 bacterial isolates from surface-disinfected rapeseed roots collected in the same Spanish locations previously mentioned. From them, 31 strains were isolated from plants collected in PDA and 81 from plants collected in CDV (Table 1).

We used 879F-RAPD fingerprints to group the strains at infraspecific level in order to select representative strains for their identification. The 31 strains from the location of PDA (Zamora) clustered into 20 different 879F-RAPD groups, while the 81 bacterial isolates from the locality of CDV (Salamanca) clustered into 56 different groups (Table 1).

Afterwards, we chose a representative strain (marked in Table 1 with an asterisk) from each 879F-RAPD group to obtain its 16S rRNA gene sequence. Then, we compared the obtained sequences with those of the type strains of described species. The closest related species to each isolate is shown in Table 1. The bacterial community analysis of the culturable bacterial endophytes of the rapeseed roots of plant collected in the two agricultural lands of this study revealed the presence of 39 different species within 27 different genera (Table 1).

The dominant genera were *Pseudomonas*, *Pseudoarthrobacter* and *Bacillus*, with 49, 12 and 10 strains belonging to 29, 4 and 6 different 879F-RAPD groups, respectively. In addition, strains belonging to these three genera were found in plants cultivated in both locations of this study, while all the other genera were location specific.

3.2. In Vitro Analyses of Plant Growth-Promoting Mechanisms

The in vitro tests of PGP potential include the analyses of phosphorous (P) solubilization, siderophores production and cellulose biosynthesis.

The results of the in vitro analyses of the PGP traits performed in this study are summarized in Table 1. A total of 77.4% and 67.9% of the isolates associated with plants from PDA and CDV, respectively, solubilize phosphate. Concerning siderophores, 38.7% of the strains isolated from PDA showed siderophore production, whereas 55.5% of the bacterial isolates from CDV produced these iron-chelating molecules. Finally, more than half of the strains from this study showed capability to synthesize cellulose or cellulose-like polymers.

Regarding PGP traits of the bacteria selected for the in planta experiments, all strains but one synthesized IAA-like molecules, all but one solubilized tricalcium phosphate and only *Bacillus simplex* CDVBN6 was able to grow with no addition of a nitrogen source in the medium.

3.3. Plant Growth Promotion in Rapeseed Seedlings under Controlled Conditions and Additional PGP Traits

Those strains showing the best results in the in vitro test of PGP traits (grey-highlighted name in Table 1) were used to evaluate their PGP capability in planta, using rapeseed seedlings. These strains were also assayed for IAA-like production, nitrogen fixation and $Ca_3(PO_4)_2$ solubilization. The results for the PGP ability of these strains are summarized in Table 2.

Table 1. Identification of strains isolated in this study and in vitro plant growth-promoting mechanisms.

Strain	Bacterial Growth Medium	879F *	Most Closely Related Type Strain Based on the 16S rRNA Gene	% Similarity with the Most Closely Related Type Strain (16S rRNA)	Taxonomy	Siderophores	Cellulose	P Solub
CDVBN92A	869 1/10	I	*Pseudarthrobacter oxydans* ATCC 14358^T	-	Actinobacteria, Actinobacteria, Micrococcales, Micrococcaceae			
CDVBN98 *	869 1/10	I	*Pseudarthrobacter oxydans* ATCC 14358^T	99.58	Actinobacteria, Actinobacteria, Micrococcales, Micrococcaceae			
CDVBN100 *	869 1/10	II	*Isoptericola nanjingensis* H17^T	97.42	Actinobacteria, Actinobacteria, Micrococcales, Promicromonosporaceae			
PDABN24A *	YMA	IV	*Dermacoccus nishinomiyaensis* DSM 20448^T	99.35	Actinobacteria, Actinobacteria, Micrococcales, Dermacoccaceae			
CDVBN92B *	869 1/10	V	*Agromyces ramosus* DSM 43045^T	99.45	Actinobacteria, Actinobacteria, Micrococcales, Microbacteriaceae			
CDVBN29 *	YMA	VI	*Clavibacter capsici* LMG 29047^T	99.93	Actinobacteria, Actinobacteria, Micrococcales, Microbacteriaceae			
CDVBN34	TSA	VI	*Clavibacter capsici* LMG 29047^T	-	Actinobacteria, Actinobacteria, Micrococcales, Microbacteriaceae			
CDVBN89*	869 1/10	VII	*Microbacterium yannicii* DSM 23203^T	98.95	Actinobacteria, Actinobacteria, Micrococcales, Microbacteriaceae			

Table 1. *Cont.*

Strain	Bacterial Growth Medium	879F*	Most Closely Related Type Strain Based on the 16S rRNA Gene	% Similarity with the Most Closely Related Type Strain (16S rRNA)	Taxonomy	Siderophores	Cellulose	P Solub
CDVBN50*	869 1/10	VIII	*Microbacterium yannicii* G72^T	100	Actinobacteria, Actinobacteria, Micrococcales, Microbacteriaceae			
CDVBN46A	869 1/10	IX	*Arthrobacter humicola* KV-653^T	-	Actinobacteria, Actinobacteria, Micrococcales, Micrococcaceae			
CDVBN60*	869 1/10	IX	*Arthrobacter humicola* KV-653^T	99.71	Actinobacteria, Actinobacteria, Micrococcales, Micrococcaceae			
CDVBN84*	TSA	X	*Arthrobacter pascens* DSM 20545^T	98.73	Actinobacteria, Actinobacteria, Micrococcales, Micrococcaceae			
PDABN28*	869 1/10	XI	*Micrococcus yunnanensis* YIM 65004^T	99.57	Actinobacteria, Actinobacteria, Micrococcales, Micrococcaceae			
CDVBN49*	869 1/10	XII	*Pseudarthrobacter oxydans* ATCC 14358^T	99.58	Actinobacteria, Actinobacteria, Micrococcales, Micrococcaceae			
CDVBN42*	869 1/10	XIII	*Pseudarthrobacter oxydans* ATCC 14358^T	99.58	Actinobacteria, Actinobacteria, Micrococcales, Micrococcaceae			
CDVBN43	869 1/10	XIII	*Pseudarthrobacter oxydans* ATCC 14358^T	-	Actinobacteria, Actinobacteria, Micrococcales, Micrococcaceae			

Table 1. *Cont.*

Strain	Bacterial Growth Medium	879F *	Most Closely Related Type Strain Based on the 16S rRNA Gene	% Similarity with the Most Closely Related Type Strain (16S rRNA)	Taxonomy	Siderophores	Cellulose	P Solub
CDVBN44	869 1/10	XIII	*Pseudarthrobacter oxydans* ATCC 14358^{T}	-	Actinobacteria, Actinobacteria, Micrococcales, Micrococcaceae			
CDVBN53 *	869 1/10	XIV	*Pseudarthrobacter oxydans* ATCC 14358^{T}	99.58	Actinobacteria, Actinobacteria, Micrococcales, Micrococcaceae			
CDVBN73	869 1/10	XIV	*Pseudarthrobacter oxydans* ATCC 14358^{T}	-	Actinobacteria, Actinobacteria, Micrococcales, Micrococcaceae			
CDVBN57 *	869 1/10	XV	*Pseudarthrobacter oxydans* ATCC 14358^{T}	99.58	Actinobacteria, Actinobacteria, Micrococcales, Micrococcaceae			
CDVBN61	869 1/10	XV	*Pseudarthrobacter oxydans* ATCC 14358^{T}	-	Actinobacteria, Actinobacteria, Micrococcales, Micrococcaceae			
CDVBN51 *	869 1/10	XVI	*Pseudarthrobacter oxydans* ATCC 14358^{T}	99.58	Actinobacteria, Actinobacteria, Micrococcales, Micrococcaceae			
CDVBN33 *	TSA	XVII	*Pseudarthrobacter siccitolerans* LMG 27359^{T}	99.44	Actinobacteria, Actinobacteria, Micrococcales, Micrococcaceae			
CDVBN72 *	869 1/10	XVIII	*Nocardioides cavernae* YIM A1136^{T}	99.36	Actinobacteria, Actinobacteria, Propionibacteriales, Nocardioidaceae			

Table 1. *Cont.*

Strain	Bacterial Growth Medium	879F*	Most Closely Related Type Strain Based on the 16S rRNA Gene	% Similarity with the Most Closely Related Type Strain (16S rRNA)	Taxonomy	Siderophores	Cellulose	P Solub
CDVBN90 *	869 1/10	XIX	*Nocardioides cavernae* YIM A1136^T	99.36	Actinobacteria, Actinobacteria, Propionibacteriales, Nocardioidaceae		■	
CDVBN101	869 1/10	XIX	*Nocardioides cavernae* YIM A1136^T	–	Actinobacteria, Actinobacteria, Propionibacteriales, Nocardioidaceae			
CDVBN102 *	869 1/10	XX	*Micromonospora coxensis* DSM 45161^T	99.86	Actinobacteria; Actinobacteria; Micromonosporales; Micromonosporaceae			▢
PDABN18 *	869 1/10	XXI	*Flavobacterium pectinovorum* DSM6368^T	99.09	Bacteroidetes, Bacteroidetes, Flavobacteria, Flavobacteriales, Flavobacteriaceae		■	
PDABN27 *	869	XXII	*Staphylococcus cohnii* subsp. *cohnii* ATCC 29974^T	100	Firmicutes, Bacilli, Bacillales, Staphylococcaceae	▢	▢	
CDVBN19 *	869	XXIII	*Staphylococcus cohnii* subsp. *cohnii* ATCC 29974^T	99.93	Firmicutes, Bacilli, Bacillales, Staphylococcaceae	▢		▢
CDVBN54	869 1/10	XXIV	*Bacillus aryabhattai* JCM 13839^T	–	Firmicutes, Bacilli, Bacillales, Bacillaceae	▢	■	
CDVBN55	869 1/10	XXIV	*Bacillus aryabhattai* JCM 13839^T	–	Firmicutes, Bacilli, Bacillales, Bacillaceae	▢	■	
CDVBN58	869 1/10	XXIV	*Bacillus aryabhattai* JCM 13839^T	–	Firmicutes, Bacilli, Bacillales, Bacillaceae		■	
CDVBN68 *	YMA	XXIV	*Bacillus aryabhattai* JCM 13839^T	99.86	Firmicutes, Bacilli, Bacillales, Bacillaceae	■	■	

Table 1. *Cont.*

Strain	Bacterial Growth Medium	879F *	Most Closely Related Type Strain Based on the 16S rRNA Gene	% Similarity with the Most Closely Related Type Strain (16S rRNA)	Taxonomy	Siderophores	Cellulose	P Solub
CDVBN9 *	869 1/10	XXV	*Bacillus megaterium* NBRC 15308^T	100	Firmicutes, Bacilli, Bacillales, Bacillaceae			
CDVBN91 *	869 1/10	XXVI	*Bacillus niacini* IFO 15566^T	99.38	Firmicutes, Bacilli, Bacillales, Bacillaceae			
PDABN29 *	869 1/10	XXVII	*Bacillus safensis* FO-36B^T	99.93	Firmicutes, Bacilli, Bacillales, Bacillaceae			
PDABN11	TSA	XXVIII	*Bacillus siamensis* PD-A10^T	-	Firmicutes, Bacilli, Bacillales, Bacillaceae			
PDABN19B *	TSA	XXVIII	*Bacillus siamensis* PD-A10^T	99.86	Firmicutes, Bacilli, Bacillales, Bacillaceae			
CDVBN6 *	869	III	*Bacillus simplex* LMG 25856^T	99.93	Firmicutes, Bacilli, Bacillales, Bacillaceae			
CDVBN18 *	869	XXIX	*Pseudomonas baetica* A390^T	99.79	Proteobacteria, Gammaproteobacteria, Pseudomonadales, Pseudomonadaceae			
CDVBN66 *	YMA	XXX	*Pseudomonas baetica* A390^T	99.79	Proteobacteria, Gammaproteobacteria, Pseudomonadales, Pseudomonadaceae			
CDVBN28 *	YMA	XXXI	*Pseudomonas baetica* A390^T	99.79	Proteobacteria, Gammaproteobacteria, Pseudomonadales, Pseudomonadaceae			
CDVBN2	YMA	XXXII	*Pseudomonas baetica* A390^T	-	Proteobacteria, Gammaproteobacteria, Pseudomonadales, Pseudomonadaceae			
CDVBN4 *	YMA	XXXII	*Pseudomonas baetica* A390^T	99.79	Proteobacteria, Gammaproteobacteria, Pseudomonadales, Pseudomonadaceae			

Table 1. *Cont.*

Strain	Bacterial Growth Medium	879F *	Most Closely Related Type Strain Based on the 16S rRNA Gene	% Similarity with the Most Closely Related Type Strain (16S rRNA)	Taxonomy	Siderophores	Cellulose	P Solub
CDVBN8 *	869	XXXIII	*Pseudomonas baetica* A390^T	99.79	Proteobacteria, Gammaproteobacteria, Pseudomonadales, Pseudomonadaceae			
CDVBN41 *	869 1/10	XXXIV	*Pseudomonas baetica* A390^T	99.79	Proteobacteria, Gammaproteobacteria, Pseudomonadales, Pseudomonadaceae			
CDVBN45	869 1/10	XXXIV	*Pseudomonas baetica* A390^T	-	Proteobacteria, Gammaproteobacteria, Pseudomonadales, Pseudomonadaceae			
CDVBN38 *	869 1/10	XXXV	*Pseudomonas baetica* A390^T	99.79	Proteobacteria, Gammaproteobacteria, Pseudomonadales, Pseudomonadaceae			
CDVBN39	869 1/10	XXXV	*Pseudomonas baetica* A390^T	-	Proteobacteria, Gammaproteobacteria, Pseudomonadales, Pseudomonadaceae			
CDVBN37	869 1/10	XXXV	*Pseudomonas baetica* A390^T	-	Proteobacteria, Gammaproteobacteria, Pseudomonadales, Pseudomonadaceae			
CDVBN22 *	YMA	XXXVI	*Pseudomonas baetica* A390^T	99.79	Proteobacteria, Gammaproteobacteria, Pseudomonadales, Pseudomonadaceae			

Table 1. *Cont.*

Strain	Bacterial Growth Medium	879F *	Most Closely Related Type Strain Based on the 16S rRNA Gene	% Similarity with the Most Closely Related Type Strain (16S rRNA)	Taxonomy	Siderophores	Cellulose	P Solub
CDVBN23 *	YMA	XXXVII	*Pseudomonas baetica* A390^T	99.79	Proteobacteria, Gammaproteobacteria, Pseudomonadales, Pseudomonadaceae			
CDVBN70	YMA	XXXVIII	*Pseudomonas baetica* A390^T	–	Proteobacteria, Gammaproteobacteria, Pseudomonadales, Pseudomonadaceae			
CDVBN71 *	YMA	XXXVIII	*Pseudomonas baetica* A390^T	99.79	Proteobacteria, Gammaproteobacteria, Pseudomonadales, Pseudomonadaceae			
CDVBN13 *	TSA	XXXIX	*Pseudomonas brassicacearum* subsp. *brassicacearum* DBK11^T	99.72	Proteobacteria, Gammaproteobacteria, Pseudomonadales, Pseudomonadaceae			
CDVBN14	TSA	XXXIX	*Pseudomonas brassicacearum* subsp. *brassicacearum* DBK11^T	–	Proteobacteria, Gammaproteobacteria, Pseudomonadales, Pseudomonadaceae			
CDVBN62 *	YMA	XL	*Pseudomonas brassicacearum* subsp. *brassicacearum* DBK11^T	99.79	Proteobacteria, Gammaproteobacteria, Pseudomonadales, Pseudomonadaceae			
CDVBN47 *	869 1/10	XLI	*Pseudomonas brassicacearum* subsp. *brassicacearum* DBK11^T	99.79	Proteobacteria, Gammaproteobacteria, Pseudomonadales, Pseudomonadaceae			
CDVBN25 *	YMA	XLII	*Pseudomonas brassicacearum* subsp. *brassicacearum* DBK11^T	99.79	Proteobacteria, Gammaproteobacteria, Pseudomonadales, Pseudomonadaceae			

Table 1. *Cont.*

Strain	Bacterial Growth Medium	879F*	Most Closely Related Type Strain Based on the 16S rRNA Gene	% Similarity with the Most Closely Related Type Strain (16S rRNA)	Taxonomy	Siderophores	Cellulose	P Solub
CDVBN27	YMA	XLII	*Pseudomonas brassicacearum brassicacearum* DBK11^T	-	Proteobacteria, Gammaproteobacteria, Pseudomonadales, Pseudomonadaceae			
CDVBN52 *	869 1/10	XLIII	*Pseudomonas brassicacearum* subsp. *neoaurantiaca* ATCC 49054^T	99.79	Proteobacteria, Gammaproteobacteria, Pseudomonadales, Pseudomonadaceae			
CDVBN64 *	YMA	XLIV	*Pseudomonas brassicacearum* subsp. *neoaurantiaca* ATCC 49054^T	99.79	Proteobacteria, Gammaproteobacteria, Pseudomonadales, Pseudomonadaceae			
CDVBN26 *	YMA	XLV	*Pseudomonas brassicacearum* subsp. *neoaurantiaca* ATCC 49054^T	99.79	Proteobacteria, Gammaproteobacteria, Pseudomonadales, Pseudomonadaceae			
CDVBN108	TSA	XLVI	*Pseudomonas brassicacearum* subsp. *neoaurantiaca* ATCC 49054^T	-	Proteobacteria, Gammaproteobacteria, Pseudomonadales, Pseudomonadaceae			
CDVBN21 *	TSA	XLVI	*Pseudomonas brassicacearum* subsp. *neoaurantiaca* ATCC 49054^T	99.86	Proteobacteria, Gammaproteobacteria, Pseudomonadales, Pseudomonadaceae			
CDVBN10 *	869	XLVII	*Pseudomonas brassicacearum* subsp. *neoaurantiaca* ATCC 49054^T	99.86	Proteobacteria, Gammaproteobacteria, Pseudomonadales, Pseudomonadaceae			
CDVBN17	869	XLVII	*Pseudomonas brassicacearum* subsp. *neoaurantiaca* ATCC 49054^T	-	Proteobacteria, Gammaproteobacteria, Pseudomonadales, Pseudomonadaceae			

Table 1. *Cont.*

Strain	Bacterial Growth Medium	879F *	Most Closely Related Type Strain Based on the 16S rRNA Gene	% Similarity with the Most Closely Related Type Strain (16S rRNA)	Taxonomy	Siderophores	Cellulose	P Solub
CDVBN24	YMA	XLVII	*Pseudomonas brassicacearum* subsp. *neoaurantiaca* ATCC 49054^T	-	Proteobacteria, Gammaproteobacteria, Pseudomonadales, Pseudomonadaceae			
CDVBN11	TSA	XLVII	*Pseudomonas brassicacearum* subsp. *neoaurantiaca* ATCC 49054^T	-	Proteobacteria, Gammaproteobacteria, Pseudomonadales, Pseudomonadaceae			
CDVBN15	TSA	XLVII	*Pseudomonas brassicacearum* subsp. *neoaurantiaca* ATCC 49054^T	-	Proteobacteria, Gammaproteobacteria, Pseudomonadales, Pseudomonadaceae			
CDVBN1 *	YMA	XLVIII	*Pseudomonas brassicacearum* subsp. *neoaurantiaca* ATCC 49054^T	99.86	Proteobacteria, Gammaproteobacteria, Pseudomonadales, Pseudomonadaceae			
CDVBN69 *	YMA	XLIX	*Pseudomonas brassicacearum* subsp. *neoaurantiaca* ATCC 49054^T	99.86	Proteobacteria, Gammaproteobacteria, Pseudomonadales, Pseudomonadaceae			
CDVBN65 *	YMA	L	*Pseudomonas orientalis* CFML96-170^T	99.86	Proteobacteria, Gammaproteobacteria, Pseudomonadales, Pseudomonadaceae			
CDVBN3 *	YMA	LI	*Pseudomonas orientalis* CFML96-170^T	99.65	Proteobacteria, Gammaproteobacteria, Pseudomonadales, Pseudomonadaceae			
CDVBN20 *	869	LII	*Pseudomonas orientalis* CFML96-170^T	99.79	Proteobacteria, Gammaproteobacteria, Pseudomonadales, Pseudomonadaceae			

Table 1. Cont.

Strain	Bacterial Growth Medium	879F*	Most Closely Related Type Strain Based on the 16S rRNA Gene	% Similarity with the Most Closely Related Type Strain (16S rRNA)	Taxonomy	Siderophores	Cellulose	P Solub
PDABN1 *	TSA	LIII	Pseudomonas poae DSM 14936^T	100	Proteobacteria, Gammaproteobacteria, Pseudomonadales, Pseudomonadaceae			
PDABN14 *	YMA	LIV	Pseudomonas poae DSM 14936^T	100	Proteobacteria, Gammaproteobacteria, Pseudomonadales, Pseudomonadaceae			
PDABN5 *	869	LV	Pseudomonas thivervalensis DSM 13194^T	99.86	Proteobacteria, Gammaproteobacteria, Pseudomonadales, Pseudomonadaceae			
PDABN12	YMA	LV	Pseudomonas thivervalensis DSM 13194^T	-	Proteobacteria, Gammaproteobacteria, Pseudomonadales, Pseudomonadaceae			
PDABN3 *	869 1/10	LVI	Pseudomonas thivervalensis DSM 13194^T	99.86	Proteobacteria, Gammaproteobacteria, Pseudomonadales, Pseudomonadaceae			
PDABN4	869	LVI	Pseudomonas thivervalensis DSM 13194^T	-	Proteobacteria, Gammaproteobacteria, Pseudomonadales, Pseudomonadaceae			
PDABN6	YMA	LVI	Pseudomonas thivervalensis DSM 13194^T	-	Proteobacteria, Gammaproteobacteria, Pseudomonadales, Pseudomonadaceae			
PDABN7	YMA	LVI	Pseudomonas thivervalensis DSM 13194^T	-	Proteobacteria, Gammaproteobacteria, Pseudomonadales, Pseudomonadaceae			

Table 1. *Cont.*

Strain	Bacterial Growth Medium	879F *	Most Closely Related Type Strain Based on the 16S rRNA Gene	% Similarity with the Most Closely Related Type Strain (16S rRNA)	Taxonomy	Siderophores	Cellulose	P Solub
PDABN8	YMA	LVI	*Pseudomonas thivervalensis* DSM 13194^T	-	Proteobacteria, Gammaproteobacteria, Pseudomonadales, Pseudomonadaceae	▨	■	■
PDABN13	YMA	LVI	*Pseudomonas thivervalensis* DSM 13194^T	-	Proteobacteria, Gammaproteobacteria, Pseudomonadales, Pseudomonadaceae	▨	■	■
PDABN15	YMA	LVI	*Pseudomonas thivervalensis* DSM 13194^T	-	Proteobacteria, Gammaproteobacteria, Pseudomonadales, Pseudomonadaceae	▨	■	■
PDABN2	YMA	LVI	*Pseudomonas thivervalensis* DSM 13194^T	-	Proteobacteria, Gammaproteobacteria, Pseudomonadales, Pseudomonadaceae	▨	■	■
CDVBN16 *	869 1/10	LVII	*Pseudomonas thivervalensis* DSM 13194^T	99.86	Proteobacteria, Gammaproteobacteria, Pseudomonadales, Pseudomonadaceae	▨	▨	■
PDABN26 *	YMA	LVIII	*Bosea lathyri* DSM 26656^T	99.22	Proteobacteria, Alphaproteobacteria, Rhizobiales, Bradyrhizobiaceae	▨		
CDVBN78 *	869 1/10	LIX	*Decvosia psychrophila* Cr7-05^T	99.09	Proteobacteria, Alphaproteobacteria, Rhizobiales, Hyphomicrobiaceae		▨	
CDVBN77 *	869 1/10	LX	*Microvirga aerophila* KACC 12743^T	97.64	Proteobacteria, Alphaproteobacteria, Rhizobiales, Methylobacteriaceae	■	■	▨

Table 1. *Cont.*

Strain	Bacterial Growth Medium	879F*	Most Closely Related Type Strain Based on the 16S rRNA Gene	% Similarity with the Most Closely Related Type Strain (16S rRNA)	Taxonomy	Siderophores	Cellulose	P Solub
PDABN20 *	YMA	LXI	*Neorhizobium alkalisoli* CCBAU 01393ᵀ	99.76	Proteobacteria, Alphaproteobacteria, Rhizobiales, Rhizobiaceae	+	+	+
PDABN21	YMA	LXI	*Neorhizobium alkalisoli* CCBAU 01393ᵀ	-	Proteobacteria, Alphaproteobacteria, Rhizobiales, Rhizobiaceae		+	
PDABN21B *	869 1/10	LXII	*Agrobacterium nepotum* 39/7ᵀ	100	Proteobacteria, Alphaproteobacteria, Rhizobiales, Rhizobiaceae	+		
PDABN22B *	869 1/10	LXIII	*Agrobacterium nepotum* 39/7ᵀ	100	Proteobacteria, Alphaproteobacteria, Rhizobiales, Rhizobiaceae			
PDABN19A *	869	LXIV	*Shinella kummerowiae* CCBAU 25048ᵀ	98.53	Proteobacteria, Alphaproteobacteria, Rhizobiales, Rhizobiaceae	+	+	
PDABN23 *	869 1/10	LXV	*Shinella kummerowiae* CCBAU 25048ᵀ	98.76	Proteobacteria, Alphaproteobacteria, Rhizobiales, Rhizobiaceae	+	+	+
PDABN24B	YMA	LXV	*Shinella kummerowiae* CCBAU 25048ᵀ	-	Proteobacteria, Alphaproteobacteria, Rhizobiales, Rhizobiaceae	+	+	+
PDABN32 *	TSA	LXVI	*Shinella kummerowiae* CCBAU 25048ᵀ	99.76	Proteobacteria, Alphaproteobacteria, Rhizobiales, Rhizobiaceae	+	+	
PDABN23A *	TSA	LXVII	*Shinella kummerowiae* CCBAU 25048ᵀ	99.76	Proteobacteria, Alphaproteobacteria, Rhizobiales, Rhizobiaceae		+	
CDVBN83 *	YMA	LXVIII	*Sphingomonas faeni* DSM 14747ᵀ	99.78	Proteobacteria, Alphaproteobacteria, Sphingomonadales, Sphingomonadaceae	+		+

Table 1. *Cont.*

Strain	Bacterial Growth Medium	879F *	Most Closely Related Type Strain Based on the 16S rRNA Gene	% Similarity with the Most Closely Related Type Strain (16S rRNA)	Taxonomy	Siderophores	Cellulose	P Solub
CDVBN46B *	869 1/10	LXIX	*Massilia suwonensis* 5414S‑25[T]	99.01	Proteobacteria, Betaproteobacteria, Burkholderiales, Oxalobacteraceae			
CDVBN40 *	869 1/10	LXX	*Massilia yuzhufengensis* ZD1‑4[T]	98.59	Proteobacteria, Betaproteobacteria, Burkholderiales, Oxalobacteraceae			
PDABN9 *	YMA	LXXI	*Acidovorax radicis* N35[T]	99.65	Proteobacteria, Betaproteobacteria, Burkholderiales, Comamonadaceae			
CDVBN31 *	TSA	LXXII	*Variovorax paradoxus* NBRC 15149[T]	99.52	Proteobacteria, Betaproteobacteria, Burkholderiales, Comamonadaceae			
CDVBN59 *	869 1/10	LXXIII	*Herbaspirillum lusitanum* LMG 21710[T]	100	Proteobacteria, Betaproteobacteria, Burkholderiales, Oxalobacteraceae			
CDVBN63	YMA	LXXIII	*Herbaspirillum lusitanum* LMG 21710[T]	-	Proteobacteria, Betaproteobacteria, Burkholderiales, Oxalobacteraceae			
CDVBN67	YMA	LXXIII	*Herbaspirillum lusitanum* LMG 21710[T]	-	Proteobacteria, Betaproteobacteria, Burkholderiales, Oxalobacteraceae			
CDVBN32 *	TSA	LXXIV	*Herbaspirillum lusitanum* LMG 21710[T]	99.45	Proteobacteria, Betaproteobacteria, Burkholderiales, Oxalobacteraceae			

Table 1. *Cont.*

Strain	Bacterial Growth Medium	879F *	Most Closely Related Type Strain Based on the 16S rRNA Gene	% Similarity with the Most Closely Related Type Strain (16S rRNA)	Taxonomy	Siderophores	Cellulose	P Solub
PDABN25 *	YMA	LXXV	*Shigella flexneri* ATCC 29903^T	99.58	Proteobacteria, Gammaproteobacteria, Enterobacterales, Enterobacteriaceae			
CDVBN81 *	TSA	LXXVI	*Acinetobacter johnsonii* ATCC 17909^T	99.51	Proteobacteria, Gammaproteobacteria, Pseudomonadales, Moraxellaceae			

Representative strains from each of the 879F groups are marked with asterisks. Grey-highlighted names represent best performing strains regarding the PGP traits. CDV: Castellanos de Villiquera (Salamanca); PDA: Peleas de Arriba (Zamora). Color scale: Grey color means no growth. White color means negative result (growth but no activity). Different shades of blue mean a range from weak (light blue) to strong (dark blue).

Table 2. Results of plant growth-promoting (PGP) tests (IAA-like compounds, solubilization of bi- and tricalcium phosphate, nitrogen fixation, siderophore and cellulose production) performed with strains selected in the plant promotion assay. All the tests were performed in triplicate.

Strain	IAA-like Molecules ($\mu g \cdot mL^{-1}$)	P Solubilization ($Ca_3(PO_4)_2$)	N Fixation	Siderophores	Cellulose	P Solubilization ($CaHPO_4$)
CDVBN4	24.53	+	-	+++	++	+++
CDVBN6	5.34	-	+	+++	++	+++
CDVBN10 *	8.18	+	-	+++	++	+++
CDVBN20 *	75.19	w	-	+++	+	+++
CDVBN21	13.72	+	-	+++	++	+++
CDVBN65	5.14	+	-	+++	-	+++
CDVBN68	10.07	+	-	+++	++	+++
CDVBN69	8.45	+	-	+++	+	+++
CDVBN70	0.00	+	-	+++	++	++

* Selected for further assays. + to +++, positive (range of halo size); -, negative; w, weak.

The results of root and plant height at 5 and 15 days post inoculation (dpi) are shown in Figure 1. The best six bacterial strains according to these results from the 5 dpi samples were re-tested in planta, allowing the seedlings to grow to 15 dpi. All six strains but one significantly increased shoot length compared to the uninoculated control (Figure 1).

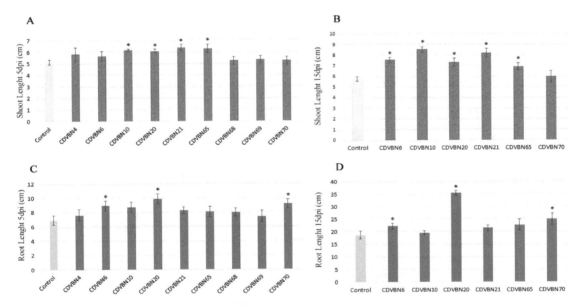

Figure 1. Growth promotion in rapeseed seedlings 5 and 15 days post inoculation (dpi): (**A**) plant height 5 dpi; (**B**) plant height 15 dpi; (**C**) root length 5 dpi; (**D**) root length 15 dpi. Bars indicate the standard error. Histogram bars marked with an asterisk indicate a value significantly different from the negative control ($p = 0.05$) according to Fisher's Protected LSD (Least Significant Differences).

Then, we selected *Pseudomonas brassicacearum* CDVBN10 and *P. orientalis* CDVBN20 to obtain their genome sequence and deepen the in silico study of their PGP capabilities. The reasons for the selection of these two strains are the following: (i) they presented good plant growth-promoting traits according to the in vitro assays, (ii) they presented a capability to promote plant growth at 5 and 15 dpi, and (iii) they belong to the genus *Pseudomonas*, the most abundant genus in plants from both locations, which might be related to a positive role of bacteria of this genus within their host plant (see discussion section).

3.4. Taxonomic Affiliation of the Best Performing Strains

General characteristics of strains CDVBN10 and CDVBN20 genomes are detailed in Table 3, as well as data from Subsystems Categories retrieved from the SeedViewer are shown in Table 4.

Table 3. General genome properties of the PGP strains CDVBN10 and CDVBN20.

Attributes	CDVBN10	CDVBN20
Genome size (bp)	6,180,897	5,666,760
GC Content (%)	60.8	60.6
N50 value	128,213	49,053
L50 value	15	34
Number of contigs (with PEGs)	85	271
Number of subsystems	403	393
Number of coding sequences	5773	5199
Number of RNAs	61	37

Table 4. Number of genes associated with specific functional categories in strains CDVBN10 and CDVBN20.

Number of Genes Related to:	CDVBN10	CDVBN20
Cofactors, vitamins, prosthetic groups, pigments	219	232
Cell wall and capsule	49	49
Virulence, disease and defense	58	60
Potassium metabolism	11	9
Miscellaneous	37	39
Phages, prophages, transposable elements, plasmids	8	3
Membrane transport	194	151
Iron acquisition and metabolism	19	52
RNA metabolism	50	52
Nucleosides and nucleotides	96	101
Protein metabolism	230	212
Motility and Chemotaxis	68	74
Regulation and cell signalling	55	61
Secondary metabolism	4	4
DNA metabolism	101	95
Fatty acids, lipids and isoprenoids	155	138
Nitrogen metabolism	55	19
Dormancy and sporulation	4	1
Respiration	133	111
Stress response	106	102
Metabolism of aromatic compounds	94	71
Amino acids and derivatives	548	481
Sulfur metabolism	24	14
Phosphorus metabolism	35	49
Carbohydrates	316	261

According to the 16S rRNA gene sequence, the most closely related type strains to CDVBN10 are *P. brassicacearum* subsp. *neurantiaca* CIP109457[T] (99.79%), *Pseudomonas corrugata* DSM7228[T] (99.65%) and *P. brassicacearum* subsp. *brassicacearum* DBK11[T] (99.59%). The *gyrB* gene sequence of strain CDVBN10 presented similarities of 94.99%, 92.92%, and 94.71% with those strains, respectively. In the case of the sequence of the *rpoB* gene, the similarities between the strain CDVBN10 and its closest related species were respectively 97.59%, 95.28%, and 97.00%. Thus, we can conclude that the most closely related type strain of CDVBN10 is *P. brassicacearum* subsp. *neurantiaca* CIP109457[T].

The comparison of the 16S rRNA gene sequence of strain CDVBN20 with the type strains available in databases showed that its most closely related type strains are *P. orientalis* CFML97-170[T] (99.66%), *Pseudomonas antarctica* CMS35[T] (99.31%), and *Pseudomonas meridiana* CMS38[T] (99.25%). In the case of the *gyrB* gene sequence, strain CDVBN20 showed the following similarities with the closest related type strain: 92.48%, 90.73%, and 90.73%, respectively. In the case of the sequence of the *rpoB* gene, the type strains of the most closely related species were not available in the databases. Therefore, according to the 16S rRNA and *gyrB* gene sequences, the most closely related type strain is *P. orientalis* CFML97-170[T].

3.5. Genome in Silico Analysis of Plant Growth-Promoting and Putative Colonization Related Mechanisms

The in silico analyses of the PGP mechanisms of strains CDVBN10 and CDVBN20 showed the presence of genes implicated in several interesting PGP pathways. Both genomes contain genes encoding enzymes involved in the solubilization of inorganic P or in the release of P from other molecules, such as exopolyphosphatases (EC 3.6.1.11), polyphosphate kinases (EC 2.7.4.1), inorganic triphosphatases (EC 3.6.1.25), inorganic pyrophosphatases (EC 3.6.1.1), pyrroloquinoline quinones (PQQ), glucose dehydrogenase PQQ-dependent (EC 1.1.5.2) and gluconate 2-dehydrogenase (EC 1.1.99.3), as well as genes of the Pst system (*pstSCAB*), which is the most conserved member of the Pho regulon [49], and some other genes related to unspecific uptake of this element [24].

Moreover, we found that both bacteria have genes involved in the metabolism of several acids that could solubilize both K and P, such as the genes encoding citrate synthase (EC 2.3.3.1) and malate synthase G (EC 2.3.3.9) responsible for the synthesis of citric acid and malic acid, respectively, genes related with the metabolism of malonic acid (malonate decarboxylase, malonate utilization transcriptional regulator, malonate-semialdehyde dehydrogenase), of gluconic acid (gluconate 2-dehydrogenase (EC 1.1.99.3), of 2-ketogluconic acid (2-ketogluconate kinase (EC 2.7.1.13), 2-ketogluconate transporter) and of lactic acid (D-lactate dehydrogenase, L-lactate dehydrogenase, L-lactate permease). We also found several genes implicated in K transport belonging to the Kup and Kef systems [50].

Regarding iron provision, we found a great number of genes linked with Fe uptake, metabolism and Fe efflux systems, as well as the ones related to the production of pyoverdine, a common siderophore in fluorescent *Pseudomonas* [51]. Regarding IAA, one of the main phytohormones responsible of many plant functions and directly related to plant growth, we found that both genomes have genes encoding for some enzymes related to IAA synthesis, such as the indole-3-glycerol phosphate synthase (EC 4.1.1.48) or the tryptophan synthase (alpha and beta chain; EC 4.2.1.20), amongst others. Nevertheless, we could not find a complete or clear pathway for the biosynthesis of IAA. In addition, using BLASTp search, we found genes encoding 1-aminocyclopropane-1-carboxylic acid (ACC) deaminase activity in both bacteria, an enzyme which catalyzes the conversion of ACC into ammonia and α-ketobutyrate, avoiding high levels of ethylene synthesis during abiotic stress situations.

Finally, both genomes showed genes involved in lipopolysaccharide (LPS) biosynthesis, such as *ipx*, *waa*, *kdt*, *ept* and *gmh* genes, or genes related to the LPS-assembly, such as *lptD* and *lptE*. Moreover, genes encoding enzymes involved in the synthesis of exopolysaccharides, such as a cyclic β-1,2-glucan synthetase, are in both genomes and *exo* genes, only in the strain CDVBN10. Both genomes also contained genes encoding glycosyl transferases and glycosyl hydrolases, enzymes involved in polysaccharide biosynthesis and biodegradation, and genes encoding transcriptional factors from AraC family.

3.6. Pseudomonas brassicacearum CDVBN10 Displays Beneficial Effects in Rapeseed Plants Cultivated in the Field

According to in vitro, in silico, and in vivo laboratory experiments, *Pseudomonas brassicacearum* CDVBN10 and *P. orientalis* CDVBN20 were shown to be promising plant growth-promoting bacteria. However, and taking into account that *P. brassicacearum* species had been isolated as a root endophyte from several different plants and that the preliminary hypothesis of this study was that bacteria with a good capability to enter plant roots will be more efficient under field conditions, we chose the bacterium *P. brassicacearum* CDVBN10 to tests its capability to promote plant growth in field conditions (a neighbor-joining phylogenetic tree based on the 16S sequence of the strain CDVBN10 and the closest related species of the genus *Pseudomonas* is available in the Supplementary Figure S2). Data from field experiments (Figures 2 and 3) showed a significant increase in both seed weight and shoot biomass in those plants inoculated with *P. brassicacearum* CDVBN10 compared to uninoculated plants. The percentages of the increase in pod number, pod dry weight and shoot dry weight in inoculated plants over the control plants were 216.0%, 174.3%, and 197.8%, respectively. Regarding the nutritional content of the plants, inoculated rapeseed plants present a significantly higher content in N, C and K, whereas uninoculated plants presented higher Fe content than those inoculated with *P. brassicacearum* CDVBN10 (Table 5).

Figure 2. Example of plant growth-promoting effect of *P. brassicacearum* CDVBN10 on *Brassica napus* plant in field experiment; (**A**) control not inoculated, (**B**) plant inoculated with *P. brassicacearum* CDVBN10. Bar represents 12 cm.

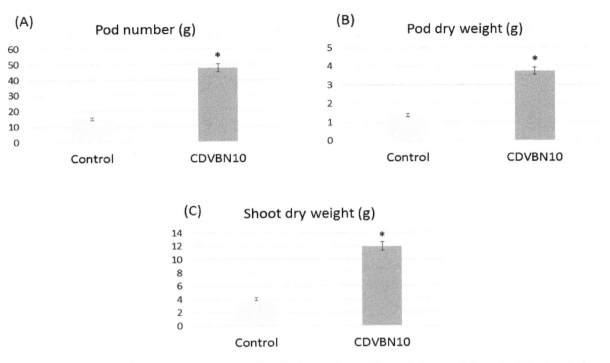

Figure 3. Results of field experiment. (**A**) Pod number, (**B**) pod dry weight (g), (**C**) shoot dry weight (g). Bars indicate the standard error. Histogram bars marked with an asterisk indicate a value significantly different from the negative control (p = 0.01) according to Fisher's protected least significant differences (LSD).

Table 5. Effects of *Pseudomonas brassicacearum* CDVBN10 inoculation on nutrient contents of rapeseed plants grown in the field experiment. Values marked with an asterisk indicate a value significantly different from the negative control (p = 0.05) according to Fisher's protected least significant differences (LSD).

Treatment	N (g/100g)	C (g/100g)	Fe (mg/kg)	K (g/100g)	P (g/100g)
Control	3.56 ± 0.05	53.69 ± 0.49	67.34 ± 2.52	1.04 ± 0.01	0.58 ± 0.02
CDVBN10	3.82 ± 0.07 *	54.89 ± 0.19 *	59.60 ± 1.50 *	0.99 ± 0.03	0.65 ± 0.03 *

3.7. CDVBN10 Inoculation Does Not Significantly Alter Bacterial Diversity in Rapeseed Roots Grown in the Field Trial

The SMRT PacBio sequencing produced a total of 376,370 reads for the eight samples (four uninoculated and four CDVBN10 inoculated). After the filtering, we obtained a total of 96,105 valid reads ($\geq$ 800 and $\leq$ 1600 bp), The minimum number of reads per sample was 2381 and the maximum was 21,274. We performed a clustering based on a threshold of 97% similarity and assigned taxonomic rank to generate a total of 3419 OTUs (Table 6). Underrepresented OTUs ($n \leq 2$) were also removed, being a final amount of 2130 OTUs in total.

Setting a level of similarity of 97% as the threshold and removing singletons and doubletons, the average number of OTUs among the samples was 552.2 ($\pm$56.81) and 541.2 ($\pm$120.73) for CDVBN10-inoculated and uninoculated treatments, respectively. The rarefaction curves for each sample (Figure S1) together with the different alpha diversity indexes (Table 6) show that the most common OTUs are present in the sequencing data. Both alpha (Table 6; Figure 4) and beta diversity (Supplementary Table S1; Figure 5) analyses revealed that there are no statistically significant differences among and within all samples from both treatments and that there are not associations between taxa and treatments (Supplementary Table S2).

Table 6. Number of sequences, OTUs and alpha diversity indexes of bacterial communities present in the 8 samples, 4 from uninoculated and 4 from CDVBN10-inoculated treatments. No significative differences were found ($p > 0.05$).

Samples		Raw Reads	Reads after Processing *	Observed OTUs	PD whole Tree	Chao-1	Shannon	Simpson	Good´s Coverage
CDVBN10	A1	40297	15884	540	69.68	2465.58	5.56	0.75	0.95
	A2	49926	15190	714	71.17	2416.30	8.35	0.97	0.96
	A3	50086	7353	537	44.89	1369.97	7.46	0.97	0.95
	A4	30213	14870	373	53.98	2083.64	3.95	0.61	0.96
Uninoculated	B1	47285	2381	532	31.42	782.39	7.51	0.97	0.90
	B2	61430	4491	525	35.55	1041.56	7.86	0.99	0.93
	B3	44494	14480	501	64.73	2201.01	5.51	0.77	0.95
	B4	52639	21274	648	81.05	2083.64	7.48	0.94	0.96
Total		376370	96105						

* after filtering (< 800 nt > 1300 nt) and chimera removal.

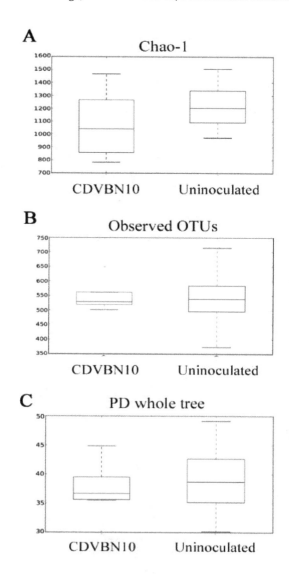

Figure 4. Comparison of alpha diversity between sampling sites; (**A**) boxplots represent Chao-1 index; (**B**) OTU richness/observed OTUs; (**C**) Phylogenetic Diversity (PD) whole tree index. T test was used to detect differences between treatments. No significant differences were found ($p > 0.05$).

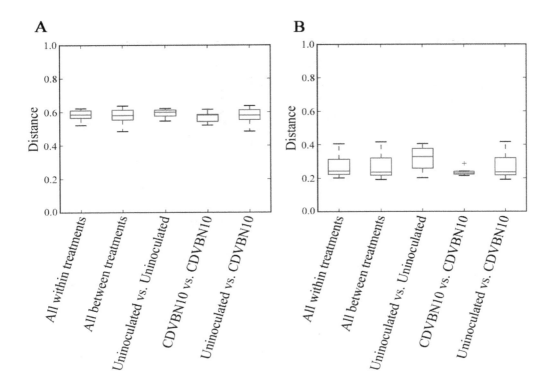

Figure 5. Comparison of beta diversity between sampling sites; (**A**) boxplots represent the unweighted Unifrac distances; (**B**) the weighted Unifrac distances. No significant differences were found among all the samples.

Eleven phyla were identified, with the phylum Proteobacteria, with four of the classes present (Alpha-, Beta-, Gamma- and Deltaproteobacteria), being the phylum with the highest relative abundance (27.8% in uninoculated treatment and 37.1% in CDVBN10-inoculated). The phyla Bacteroidetes (18.0% and 19.0%) and Verrucomicrobia (5.6% and 4.6%) were the second and the third in relative abundance, respectively (Figure 6A). There are more unassigned sequences in the uninoculated (42.6%) than in the CDVBN10-inoculated (34.7%) treatment. The class Betaproteobacteria is the most abundant within both treatments (18.9% and 20.9%), followed by the classes Flavobacteria (9.9% and 11.5%) and Gammaproteobacteria (4.2% and 5.6%) (Figure 6B). The orders Burkholderiales (15.6% and 18.5%), families Commamonadecae (8.8% and 10.6%) and Oxalobacteraceae (6.8% and 7.8%), genera *Polaromonas* (2.1% and 2.6%) and *Janthinobacterium* (2.9% and 3.4%); and Flavobacteriales (9.9% and 11.5%), the family Flavobacteriaceae (9.6% and 11.2%), and the genus *Flavobacterium* (9.6% and 11.2%) are those with the highest relative abundance in both treatments (Figure 6C–E). Other important taxa, such as the order Rhizobiales (2.1%) or the family Pseudomonadaceae (0.7%), showed similar relative abundances in both treatments. Indeed, the genus *Pseudomonas*, which is supposed to be enriched in the CDVBN10-inoculated treatment, showed the same relative abundance (0.7%) in both treatments (Figure 6E).

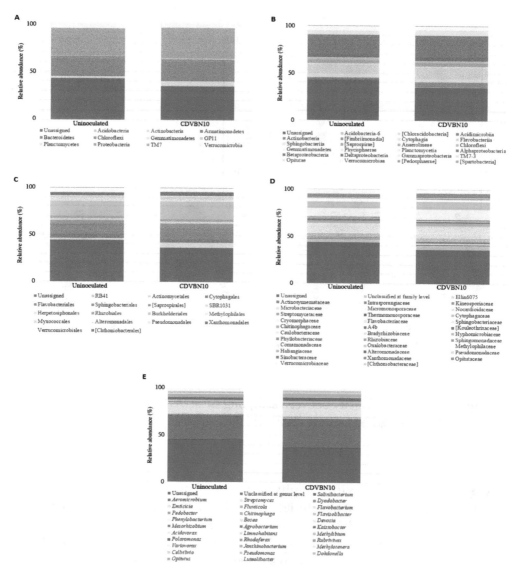

Figure 6. Relative abundance (%) of bacterial taxa found inside roots of rapeseed plants collected in the uninoculated and CDVBN10 inoculated treatments at different taxonomic levels: (**A**) phylum, (**B**) class, (**C**) order, (**D**) family and (**E**) genus. Taxa with relative abundances higher than 0.1% are represented in the charts.

4. Discussion

The results of the present study show a broad biodiversity of bacterial endophytic strains of *B. napus* roots in two soils from Northwest Spain: the 879F-RAPD fingerprinting, which had been proven to be a useful technique to generate different profiles at the intraspecific level in both Gram-positive and negative bacteria [24,26,52], showed the presence of several different profiles among the isolated strains, and the 16S rRNA sequence analysis showed a wide diversity of bacterial species and genera. The dominant genus was *Pseudomonas*, followed by *Pseudoarthrobacter* and *Bacillus*. The genera *Pseudomonas* and *Bacillus* appeared in samples from both localities, while all the other genera were location-specific. Strains from the genera *Pseudomonas*, *Bacillus*, *Rhizobium*, *Staphylococcus*, *Acidovorax*, *Micrococcus*, *Arthrobacter*, *Variovorax*, *Microbacterium*, *Sphingomonas*, *Acinetobacter*, *Devosia* and *Flavobacterium* had already been identified as rapeseed endophytes [19,53–59], while *Micromonospora*, *Massilia*, *Bosea*, *Shinella* and *Agromyces* had been found in soil or rhizosphere associated to *B. napus* roots [57–61]. However, to the best of our knowledge, this is the first report of the association of bacteria from genera *Neorhizobium*, *Microvirga*, *Herbaspirillum*, *Dermacoccus*, *Nocardioides*, *Isoptericola*, *Pseudoarthrobacter*, *Clavibacter* and *Shigella* to *B. napus* plants, although genera such as *Neorhizobium*,

Microvirga and *Herbaspirillum* are well-known PGP bacteria associated to different plants [62–64]. Considering that the plant endosphere is a much more restricted niche than the rhizosphere, these results show a great biodiversity within the isolated strains, probably due to the use of different isolation media.

Regarding the in vitro PGP potential, P is an essential plant nutrient and P deficiency is one of the most important limitations to plant development and crop production, it being estimated that more than 5 billion hectares of land are deficient in P [65]. On the other hand, iron (Fe) is essential for plants, forming part of chlorophyll. Siderophores are molecules that bacteria secrete to solubilize iron, forming a complex ferri-siderophore that can move by diffusion and be returned to the cell or captured by plants [66]. Finally, the production of polysaccharides is an advantage for the strain in order to colonize the plant roots. Amongst those polysaccharides, cellulose is involved in bacterial root colonization and biofilm production—preliminary steps prior to plant growth promotion—and thus, cellulose biosynthesis is important for biofertilizers efficiency [67]. Because of all the mentioned advantages of these PGP bacterial traits, the positive results found for our bacterial isolates suggest the presence of an advantageous endophytic microbiota in rapeseed roots. All isolates except *Nocardioides cavernae* CDVBN101, *Micromonospora coxensis* CDVBN102 and *Bosea lathyri* PDABN26 showed positive results for at least one of the in vitro assayed PGP traits. The best bacteria belonged to the species *Pseudomonas thivervalensis*, *P. poae*, *P. baetica*, *P. brassicacearum*, *Bacillus aryabhattai* and *Bacillus simplex*. Strains belonging to these species have been previously described as PGP of different plants [68–75].

Thus, we tested the capability of representative bacterial strains from those species to promote rapeseed seedling development. The results from these assays suggest that the strains CDVBN10 and CDVBN20, both belonging to the genus *Pseudomonas*, were the best rapeseed PGPs. The genome analysis of strains CDVBN10 and CDVBN20 showed an interesting genetic PGP potential, as both strains showed positive results in all the PGP traits tested (excepting growth in N-free media). According to the results obtained in the *in vitro* tests performed in this study and the analyses of other genomes of *Pseudomonas* strains [37,76–78], we found a great number of genes linked to Fe uptake, metabolism and efflux systems. In addition, in consonance with the in vitro tests and the results found for other *Pseudomonas* strains [79], both genomes contain gene sequences encoding enzymes that are involved in the solubilization P and K as well as the transport of these elements [20,24,50]. In addition, both bacterial genomes contain genes related to IAA biosynthesis. The lack of detection of a complete IAA biosynthetic pathway may be due to the biases of annotating draft genomes. On the other hand, as with other *Pseudomonas* strains [80], these two bacterial genomes encode the enzyme ACC deaminase; the synthesis of this enzyme would probably confer the plant a better resistance to abiotic stress conditions [81]. The synergy between both IAA synthesis and ACC deaminase activity could lead to a better performance of this plant–bacteria symbiosis [82]. Both bacteria also contain genes related to the biosynthesis of polysaccharides such as a cyclic β-1,2-glucan synthetase [83] in both genomes and *exo* genes [84], only in the strain CDVBN10 and genes encoding glycosyl transferases and glycosyl hydrolases, enzymes involved in polysaccharide biosynthesis and biodegradation [85]; polysaccharides have been proved to play a role in biofilm formation and the colonization of root surfaces [35,86,87]. Both genomes also have genes encoding transcriptional factors from AraC family, which are known as regulators of many processes including the ones involved in the interchange of signals among bacteria [88] and have been revealed as relevant for rhizosphere competition in rhizobia [89].

Strain CDVBN20 belongs to the species *Pseudomonas orientalis*, a bacterium not frequently associated with plant microbiomes, this being, to the best of our knowledge, the first time it has been described as a bacterial species associates to *B. napus*. However, the strain CDVBN10 belongs to the species *Pseudomonas brassicacearum*, which was originally described as a bacterial colonizer of *B. napus* rhizosphere [90]. Moreover, different strains of this species have been isolated as root endophytes from different plants, such as *Salvia miltiorrhiza* Bunge. [69], *Artemisia* sp. [91], *Lavandula dentata* L. [92] and nodules of the legume *Sphaerophysa salsula* (Pall.) DC. [93]. Some studies also reported how this species promotes the growth of *Pisum sativum* L. [94], *Solanum nigrum* L. [95] and *Medicago lupulina* L. [96] plants.

Moreover, the genome sequence analyses of other bacterial strains belonging to this species seem to indicate that this bacterium is a good plant growth promoter and a potential biocontrol agent [97,98]. Considering the results of this study and previous references of the species, we conclude that the strain *P. brassicacearum* CDVBN10 has a good potential as rapeseed biofertilizer and we decided to test its performance under field conditions. The results of our trial, performed with no addition of chemical fertilizers, show a significant increase not only in total plant biomass, but also in seed yields compared to the non-inoculated control plants, confirming that this bacterium has an interesting potential to be employed as a biofertilizer for *Brassica napus* crops, as it has been already for other *Pseudomonas* species inoculated in field trials [99–101], this being, to the best of our knowledge, the first report of a PGP bacterium with potential to specifically promote rapeseed/canola crops which showed an important yield increase in field trials

Interestingly, despite the significant differences in plant development and yields, the analysis of the biodiversity based on amplicon sequencing showed that there are no significant differences in the root bacterial communities of plants inoculated with the strain *Pseudomonas brassicacerarum* CDVBN10 nor in the associated functions of this community. In this sense, our results agree with those of Qiao et al. [102], which showed that the inoculation of a PGPB *Bacillus* strain does not alter the root bacterial microbiome on tomato plants. However, this effect might be strain-specific or context-specific, as suggested by Gadhave et al. [103]; these authors performed several inoculations with different PGPB strains belonging to the genus *Bacillus* and found that there is an infraspecific variation and competition issues within sprouting broccoli roots. The modulation of root microbiomes by addition of biofertilizers based on beneficial strains and other factors is not well-understood and further studies must be performed to elucidate these effects [104].

According to ecological theories [13,105], most bacteria living as root endophytes probably play important roles for the plant development and survival. Thus, in our opinion, the results obtained in this study are very positive: rapeseed plants from the plots inoculated with the strain CDVBN10 showed a clear benefit from the inoculation and their endophytic root microbiome was not altered by the inoculation, so there was not competition of potentially benefiting members of the plant microbiome.

There is an unexpected result in the PacBio data; we were not able to detect any OTU belonging to the phylum Firmicutes. This is a rare event, taking into account that members of this phylum were found within the root bacterial microbiome of *Brassica* plants [106]. However, Lay et al. [61] did not find any Firmicutes in canola roots. Some of the amplicon sequences appeared as unclassified at different taxonomic levels, which might be the reason for lacking some taxa in the amplicon sequencing analyses. These results highlight the importance of combining culturomics and metagenomics for biodiversity studies, because whereas isolated strains can be better identified, amplicon sequences allow us to decipher those members of the community which cannot grow in synthetic conditions or are inhibited by other members of the community in the selected growth conditions of the study.

As the bacterial communities associated to plants, both rhizospheric and endophytic, are strongly influenced by many factors [107–109], further studies on different soils and climate conditions should be performed in order to demonstrate the success of this strain as a biofertilizer for rapeseed crops and the lack of alteration of the root microbiome after its addition; furthermore, the best formulation of the strain to be commercialized as a biofertilizer should also be evaluated.

Supplementary Materials: Table S1: Beta diversity results. Statistics corresponding to distance boxplots of Figure 5 (from the main text) according to unweighted and weighted Unifrac distances. Supplementary Table S2: Statistic significance of the relatedness of each OTUs with each treatment group (control samples or CDVBN10 inoculated samples). Supplementary Figure S1: Rarefaction curve for observed bacterial OTUs clustering at 97% 16S rRNA sequence similarity. Curves represent number of observed OTUs from the uninoculated (A1-4) and CDVBN10 inoculated (B1-4) treatments. Supplementary Figure S2: Neighbour-joining phylogenetic tree based on the 16S rRNA gene sequences of strain *P. brassicacearum* CDVBN10 and its closest related type strains. Scale bar = 5 nucleotide (nt) substitutions per 1000 nt.

Author Contributions: Conceptualization, P.G.-F., R.R. and P.F.M.; methodology, A.J.-G. and Z.S.-S.; software, M.K. and Z.S.-S; validation, M.K., E.M. and Z.S.-S.; formal analysis, A.J.-G., Z.S.-S., E.M. and P.G.-F.; investigation, A.J.-G., Z.S.-S., M.K. and E.M.; resources, P.G.-F. and R.R.; data curation, M.K., E.M. and Z.S.-S.; writing—original draft preparation, P.G.-F. and E.M.; writing—review and editing, P.G.-F., E.V. and E.M.; visualization, P.G.-F. and E.M.; supervision, P.G.-F. and P.F.M.; project administration, P.G.-F.; funding acquisition, P.G.-F. All authors have read and agreed to the published version of the manuscript.

Acknowledgments: Authors thank the Strategic Research Programs for Units of Excellence from Junta de Castilla y León (CLU-2018-04) for funding equipment and facilities. The authors also thank Jose Antonio García Fernández for granting the land in which the field trial was performed.

References

1. Wu, W.; Ma, B. Integrated nutrient management (INM) for sustaining crop productivity and reducing environmental impact: A review. *Sci. Total Environ.* **2015**, *512*, 415–427. [CrossRef]
2. Adesemoye, A.O.; Torbert, H.A.; Kloepper, J.W. Plant Growth-Promoting Rhizobacteria Allow Reduced Application Rates of Chemical Fertilizers. *Microb. Ecol.* **2009**, *58*, 921–929. [CrossRef] [PubMed]
3. Bhardwaj, D.; Ansari, M.W.; Sahoo, R.K.; Tuteja, N. Biofertilizers function as key player in sustainable agriculture by improving soil fertility, plant tolerance and crop productivity. *Microb. Cell Fact.* **2014**, *13*, 66. [CrossRef] [PubMed]
4. Mahanty, T.; Bhattacharjee, S.; Goswami, M.; Bhattacharyya, P.; Das, B.; Ghosh, A. Biofertilizers: A potential approach for sustainable agriculture development. *Environ. Sci. Pollut. Res.* **2017**, *24*, 3315–3335. [CrossRef] [PubMed]
5. Olanrewaju, O.S.; Glick, B.R.; Babalola, O.O. Mechanisms of action of plant growth promoting bacteria. *World J. Microbiol. Biotechnol.* **2017**, *33*, 197. [CrossRef] [PubMed]
6. Menéndez, E.; Garcia-Fraile, P. Plant probiotic bacteria: Solutions to feed the world. *AIMS Microbiol.* **2017**, *3*, 502–524. [CrossRef] [PubMed]
7. Gaiero, J.R.; McCall, C.A.; Thompson, K.A.; Day, N.J.; Best, A.S.; Dunfield, K.E. Inside the root microbiome: Bacterial root endophytes and plant growth promotion. *Am. J. Bot.* **2013**, *100*, 1738–1750. [CrossRef]
8. Santoyo, G.; Moreno-Hagelsieb, G.; del Carmen Orozco-Mosqueda, M.; Glick, B.R. Plant growth-promoting bacterial endophytes. *Microbiol. Res.* **2016**, *183*, 92–99. [CrossRef]
9. Gopal, M.; Gupta, A. Microbiome selection could spur next-generation plant breeding strategies. *Front. Microbiol.* **2016**, *7*, 1971. [CrossRef]
10. Velázquez, E.; García-Fraile, P.; Ramírez-Bahena, M.H.; Rivas, R.; Martínez-Molina, E. Bacteria Involved in Nitrogen-Fixing Legume Symbiosis: Current Taxonomic Perspective. In *Microbes for Legume Improvement*; Springer: Vienna, Austria, 2010; pp. 1–5.
11. Brundrett, M.C.; Tedersoo, L. Evolutionary history of mycorrhizal symbioses and global host plant diversity. *New Phytol.* **2018**, *220*, 1108–1115. [CrossRef]
12. Clear, M.R.; Hom, E.F. The evolution of symbiotic plant-microbe signaling. *Ann. Plant. Rev. Online* **2019**, *2*, 1–52. [CrossRef]
13. Vorholt, J.A.; Vogel, C.; Carlström, C.I.; Müller, D.B. Establishing causality: Opportunities of synthetic communities for plant microbiome research. *Cell Host Microb.* **2017**, *22*, 142–155. [CrossRef] [PubMed]
14. Etesami, H.; Alikhani, H.A. Rhizosphere and endorhiza of oilseed rape (*Brassica napus* L.) plant harbor bacteria with multifaceted beneficial effects. *Biol. Cont.* **2016**, *94*, 11–24. [CrossRef]
15. Card, S.D.; Hume, D.E.; Roodi, D.; McGill, C.R.; Millner, J.P.; Johnson, R.D. Beneficial endophytic microorganisms of *Brassica*–A review. *Biol. Control.* **2015**, *90*, 102–112. [CrossRef]
16. Rathore, R.; Germaine, K.J.; Forristal, P.D.; Spink, J.; Dowling, D. Meta-Omics Approach to Unravel the Endophytic Bacterial Communities of *Brassica napus* and Other Agronomically Important Crops. In *Endophytes for a Growing World*; Hodkinson, T.R., Ed.; Cambridge University Press: Cambridge, UK, 2019; p. 232.

17. Farina, R.; Beneduzi, A.; Ambrosini, A.; de Campos, S.B.; Lisboa, B.B.; Wendisch, V. Diversity of plant growth-promoting rhizobacteria communities associated with the stages of canola growth. *Appl. Soil Ecol.* **2012**, *55*, 44–52. [CrossRef]

18. Bertrand, H.; Nalin, R.; Bally, R.; Cleyet-Marel, J.C. Isolation and identification of the most efficient plant growth-promoting bacteria associated with canola (*Brassica napus*). *Biol. Fertil. Soils* **2001**, *33*, 152–156. [CrossRef]

19. Sheng, X.F.; Xia, J.J.; Jiang, C.Y.; He, L.Y.; Qian, M. Characterization of heavy metal-resistant endophytic bacteria from rape (*Brassica napus*) roots and their potential in promoting the growth and lead accumulation of rape. *Environ. Pollut.* **2008**, *156*, 1164–1170. [CrossRef]

20. Etesami, H.; Emami, S.; Alikhani, H.A. Potassium solubilizing bacteria (KSB): Mechanisms, promotion of plant growth and future prospects A review. *J. Soil Sci. Plant. Nutr.* **2017**, *17*, 897–911. [CrossRef]

21. Puri, A.; Padda, K.P.; Chanway, C.P. Evidence of nitrogen fixation and growth promotion in canola (*Brassica napus* L.) by an endophytic diazotroph *Paenibacillus polymyxa* P2b-2R. *Biol. Fertil. Soils* **2016**, *52*, 119–125. [CrossRef]

22. Lally, R.D.; Galbally, P.; Moreira, A.S.; Spink, J.; Ryan, D.; Germaine, K.J. Application of endophytic *Pseudomonas fluorescens* and a bacterial consortium to *Brassica napus* can increase plant height and biomass under greenhouse and field conditions. *Front. Plant. Sci.* **2017**, *8*, 2193. [CrossRef]

23. Petrova, S.N.; Andronov, E.E.; Belimov, A.A.; Beregovaya, Y.V.; Denshchikov, V.A.; Minakov, D.L. Prokaryotic Community Structure in the Rapeseed (*Brassica napus* L.) Rhizosphere Depending on Addition of 1-Aminocyclopropane-1-Carboxylate-Utilizing Bacteria. *Microbiology* **2020**, *89*, 115–121. [CrossRef]

24. Bruto, M.; Prigent-Combaret, C.; Muller, D.; Moënne-Loccoz, Y. Analysis of genes contributing to plant-beneficial functions in plant growth-promoting rhizobacteria and related Proteobacteria. *Sci. Rep.* **2014**, *4*, 6261. [CrossRef] [PubMed]

25. López-Mondéjar, R.; Kostovčík, M.; Lladó, S.; Carro, L.; García-Fraile, P. Exploring the Plant Microbiome through Multi-Omics Approaches. In *Probiotics in Agroecosystem*; Springer: Singapore, 2017; pp. 233–268.

26. Igual, J.M.; Valverde, A.; Rivas, R.; Mateos, P.F.; Rodríguez-Barrueco, C.; Martínez-Molina, E. Genomic Fingerprinting of *Frankia* Strains by PCR-Based Techniques. Assessment of a Primer Based on the Sequence of 16S rRNA Gene of *Escherichia coli*. In *Frankia Symbiosis*; Normand, P., Pawlowski, K., Dawson, J.I., Eds.; Springer: Dordrecht, The Netherland, 2003; pp. 115–123.

27. Rivas, R.; Garcia-Fraile, P.; Peix, A.; Mateos, P.F.; Martínez-Molina, E.; Velazquez, E. *Alcanivorax balearicus* sp. nov., isolated from Lake Martel. *Int. J. Syst. Evol. Microbiol.* **2007**, *57*, 1331–1335. [CrossRef] [PubMed]

28. Poveda, J.; Jiménez-Gómez, A.; Saati-Santamaría, Z.; Usategui-Martín, R.; Rivas, R.; García-Fraile, P. Mealworm frass as a potential biofertilizer and abiotic stress tolerance-inductor in plants. *Appl. Soil Ecol.* **2019**, *142*, 110–122. [CrossRef]

29. Altschul, S.F.; Madden, T.L.; Schäffer, A.A.; Zhang, J.; Zhang, Z.; Miller, W. Gapped BLAST and PSI-BLAST: A new generation of protein database search programs. *Nucleic Acids Res.* **1997**, *25*, 3389–3402. [CrossRef]

30. Chun, J.; Lee, J.H.; Jung, Y.; Kim, M.; Kim, S.; Kim, B.K. EzTaxon: A web-based tool for the Identification of prokaryotes based on 16S ribosomal RNA gene sequences. *Int. J. Syst. Evol. Microbiol.* **2007**, *57*, 2259–2261. [CrossRef]

31. Jiménez-Gómez, A.; Saati-Santamaría, Z.; Igual, J.M.; Rivas, R.; Mateos, P.F.; García-Fraile, P. Genome Insights into the Novel Species *Microvirga brassicacearum*, a Rapeseed Endophyte with Biotechnological Potential. *Microorganisms* **2019**, *7*, 354. [CrossRef]

32. Schwyn, B.; Neilands, J.B. Universal chemical assay for the detection and determination of siderophores. *Anal. Biochem.* **1987**, *160*, 47–56. [CrossRef]

33. Alexander, D.B.; Zuberer, D.A. Use of Chrome Azurol S reagents to evaluate siderophore production by rhizosphere bacteria. *Biol. Fertil. Soils* **1991**, *12*, 39–45. [CrossRef]

34. Pikovskaya, R.I. Mobilization of phosphorus in soil connection with the vital activity of some microbial species. *Microbiologiya* **1948**, *17*, 362–370.

35. Robledo, M.; Rivera, L.; Jiménez-Zurdo, J.I.; Rivas, R.; Dazzo, F.; Velázquez, E.; Mateos, P.F. Role of *Rhizobium* endoglucanase CelC2 in cellulose biosynthesis and biofilm formation on plant roots and abiotic surfaces. *Microb. Cell Fact.* **2012**, *11*, 125. [CrossRef] [PubMed]

36. Khalid, A.; Arshad, M.; Zahir, Z.A. Screening plant growth-promoting rhizobacteria for improving growth and yield of wheat. *J. Appl. Microbiol.* **2004**, *96*, 473–480. [CrossRef] [PubMed]

37. Saati-Santamaría, Z.; López-Mondéjar, R.; Jiménez-Gómez, A.; Díez-Méndez, A.; Větrovský, T.; Igual, J.M.; Garcia-Fraile, P. Discovery of phloeophagus beetles as a source of *Pseudomonas* strains that produce potentially new bioactive substances and description of *Pseudomonas bohemica* sp. nov. *Front. Microbiol.* **2018**, *9*. [CrossRef] [PubMed]

38. Zerbino, D.R.; Birney, E. Velvet: Algorithms for de novo short read assembly using de Bruijn graphs. *Genome Res.* **2008**, *18*, 821–829. [CrossRef] [PubMed]

39. Aziz, R.K.; Bartels, D.; Best, A.A.; DeJongh, M.; Disz, T.; Edwards, R.A. The RAST Server: Rapid annotations using subsystems technology. *BMC Genomics* **2008**, *9*, 75. [CrossRef]

40. Dellavalle, N.B. Determination of Specific Conductance in Supernatant 1:2 soil:water solution. In *Handbook on Reference Methods for Soil Analysis*; Council, Inc.: Athens, GA, USA, 1992; pp. 44–50.

41. Větrovský, T.; Baldrian, P.; Morais, D. SEED 2: A user-friendly platform for amplicon high-throughput sequencing data analyses. *Bioinformatics* **2018**, *34*, 2292–2294. [CrossRef] [PubMed]

42. Caporaso, J.G.; Kuczynski, J.; Stombaugh, J.; Bittinger, K.; Bushman, F.D.; Costello, E.K. QIIME allows analysis of high-throughput community sequencing data. *Nat. Methods* **2010**, *7*, 335. [CrossRef]

43. DeSantis, T.Z.; Hugenholtz, P.; Larsen, N.; Rojas, M.; Brodie, E.L.; Keller, K. Greengenes, a chimera-checked 16S rRNA gene database and workbench compatible with ARB. *Appl. Environ. Microbiol.* **2006**, *72*, 5069–5072. [CrossRef]

44. Edgar, R.C.; Haas, B.J.; Clemente, J.C.; Quince, C.; Knight, R. UCHIME improves sensitivity and speed of chimera detection. *Bioinformatics* **2011**, *27*, 2194–2200. [CrossRef]

45. Kruskal, W.H.; Wallis, W.A. Use of ranks in one-criterion variance analysis. *J. Am. Stat. Assoc.* **1952**, *47*, 583–621. [CrossRef]

46. Benjamini, Y.; Hochberg, Y. Controlling the false discovery rate: A practical and powerful approach to multiple testing. *J. R. Stat. Soc. Ser. B* **1995**, *57*, 289–300. [CrossRef]

47. Bonferroni, C.E. *Teoria Statistica Delle Classi e Calcolo Delle Probabilità*; Libreria Internazionale Seeber: Florence, Italy, 1936.

48. Landau, S.; Rabe-Hesketh, S. Software review: StatView for windows, version 5.0. *Stat. Methods Med. Res.* **1999**, *8*, 337–341. [CrossRef] [PubMed]

49. Santos-Beneit, F. The Pho regulon: A huge regulatory network in bacteria. *Front. Microbiol.* **2015**, *6*, 402. [CrossRef] [PubMed]

50. Epstein, W. The roles and regulation of potassium in bacteria. *Prog. Nucleic Acid Res.* **2003**, *75*, 293–320.

51. Meyer, J.M. Pyoverdine Siderophores as Taxonomic and Phylogenic Markers. *Pseudomonas* **2010**, 201–233. [CrossRef]

52. Rivas, R.; Peix, A.; Mateos, P.F.; Trujillo, M.E.; Martínez-Molina, E.; Velázquez, E. Biodiversity of populations of phosphate solubilizing rhizobia that nodulates chickpea in different Spanish soils. *Plant. Soil* **2006**, *287*, 23–33. [CrossRef]

53. Germida, J.J.; Siciliano, S.D.; Renato de Freitas, J.; Seib, A.M. Diversity of root-associated bacteria associated with field-grown canola (*Brassica napus* L.) and wheat (*Triticum aestivum* L.). *FEMS Microbiol. Ecol.* **1998**, *26*, 43–50. [CrossRef]

54. Dunfield, K.E.; Germida, J.J. Diversity of bacterial communities in the rhizosphere and root interior of field-grown genetically modified *Brassica napus*. *FEMS Microbiol. Ecol.* **2001**, *38*, 1–9. [CrossRef]

55. Granér, G.; Persson, P.; Meijer, J.; Alström, S. A study on microbial diversity in different cultivars of *Brassica napus* in relation to its wilt pathogen, *Verticillium longisporum*. *FEMS Microbiol. Lett.* **2003**, *224*, 269–276. [CrossRef]

56. Zhang, Y.F.; He, L.Y.; Chen, Z.J.; Wang, Q.Y.; Qian, M.; Sheng, X.F. Characterization of ACC deaminase-producing endophytic bacteria isolated from copper-tolerant plants and their potential in promoting the growth and copper accumulation of *Brassica napus*. *Chemosphere* **2011**, *83*, 57–62. [CrossRef]

57. Croes, S.; Weyens, N.; Janssen, J.; Vercampt, H.; Colpaert, J.V.; Carleer, R. Bacterial communities associated with *Brassica napus* L. grown on trace element-contaminated and non-contaminated fields: A genotypic and phenotypic comparison. *Microb. Biotechnol.* **2013**, *6*, 371–384. [CrossRef] [PubMed]

58. Montalbán, B.; Croes, S.; Weyens, N.; Lobo, M.C.; Pérez-Sanz, A.; Vangronsveld, J. Characterization of bacterial communities associated with *Brassica napus* L. growing on a Zn-contaminated soil and their effects on root growth. *Int. J. Phytoremediat.* **2016**, *18*, 985–993. [CrossRef]

59.   Gkarmiri, K.; Mahmood, S.; Ekblad, A.; Alström, S.; Högberg, N.; Finlay, R. Identifying the active microbiome associated with roots and rhizosphere soil of oilseed rape. *Appl. Environ. Microbiol.* **2017**, *83*, e01938-17. [CrossRef] [PubMed]

60.   Larcher, M.; Rapior, S.; Cleyet-Marel, J.C. Bacteria from the rhizosphere and roots of *Brassica napus* influence its root growth promotion by *Phyllobacterium brassicacearum*. *Acta Bot. Gallica* **2008**, *155*, 355–366. [CrossRef]

61.   Lay, C.Y.; Bell, T.H.; Hamel, C.; Harker, K.N.; Mohr, R.; Greer, C.W. Canola Root–Associated Microbiomes in the Canadian Prairies. *Front. Microbiol.* **2018**, *9*, 1188. [CrossRef] [PubMed]

62.   Chen, L.; He, L.Y.; Wang, Q.; Sheng, X.F. Synergistic effects of plant growth-promoting *Neorhizobium huautlense* T1-17 and immobilizers on the growth and heavy metal accumulation of edible tissues of hot pepper. *J. Hazard. Mater.* **2016**, *312*, 123–131. [CrossRef]

63.   Msaddak, A.; Rejili, M.; Durán, D.; Rey, L.; Imperial, J.; Palacios, J.M. Members of *Microvirga* and *Bradyrhizobium* genera are native endosymbiotic bacteria nodulating *Lupinus luteus* in Northern Tunisian soils. *FEMS Microbiol. Ecol.* **2017**, *93*. [CrossRef] [PubMed]

64.   Dall'Asta, P.; Velho, A.C.; Pereira, T.P.; Stadnik, M.J.; Arisi, A.C.M. *Herbaspirillum seropedicae* promotes maize growth but fails to control the maize leaf anthracnose. *Physiol. Mol. Biol. Plant.* **2019**, *25*, 167–176. [CrossRef]

65.   Mouazen, A.M.; Kuang, B. On-line visible and near infrared spectroscopy for in-field phosphorus management. *Soil Tillage Res.* **2016**, *155*, 471–477. [CrossRef]

66.   Beneduzi, A.; Ambrosini, A.; Passaglia, L.M.P. Plant growth-promoting rhizobacteria (PGPR): Their potential as antagonists and biocontrol agents. *Genet. Mol. Biol.* **2012**, *35*, 1044–1051. [CrossRef]

67.   Jiménez-Gómez, A.; Flores-Félix, J.D.; García-Fraile, P.; Mateos, P.F.; Menéndez, E.; Velázquez, E.; Rivas, R. Probiotic activities of *Rhizobium laguerreae* on growth and quality of spinach. *Sci. Rep.* **2018**, *8*, 295. [CrossRef] [PubMed]

68.   Yang, P.X.; Li, M.A.; Ming-Hui, C.; Jia-Qin, X.I.; Feng, H.E.; Chang-Qun, D. Phosphate solubilizing ability and phylogenetic diversity of bacteria from P-rich soils around Dianchi Lake drainage area of China. *Pedosphere* **2012**, *22*, 707–716. [CrossRef]

69.   Li, X.J.; Tang, H.Y.; Duan, J.L.; Gao, J.M.; Xue, Q.H. Bioactive alkaloids produced by *Pseudomonas brassicacearum* subsp. *neoaurantiaca*, an endophytic bacterium from *Salvia miltiorrhiza*. *Nat. Prod. Res.* **2013**, *27*, 496–499. [CrossRef] [PubMed]

70.   Wang, X.; Mavrodi, D.V.; Ke, L.; Mavrodi, O.V.; Yang, M.; Thomashow, L.S. Biocontrol and plant growth-promoting activity of rhizobacteria from Chinese fields with contaminated soils. *Microb. Biotechnol.* **2015**, *8*, 404–418. [CrossRef] [PubMed]

71.   Matthijs, S.; Brandt, N.; Ongena, M.; Achouak, W.; Meyer, J.M.; Budzikiewicz, H. Pyoverdine and histicorrugatin-mediated iron acquisition in *Pseudomonas thivervalensis*. *Biometals* **2016**, *29*, 467–485. [CrossRef] [PubMed]

72.   Chakraborty, B.N.; Allay, S.; Chakraborty, A.P.; Chakraborty, U. PGPR in managing root rot disease and enhancing growth in mandarin (*Citrus reticulata* Blanco.) seedlings. *J. Hortic. Sci.* **2017**, *11*, 104–115.

73.   Ortiz-Ojeda, P.; Ogata-Gutiérrez, K.; Zúñiga-Dávila, D. Evaluation of plant growth promoting activity and heavy metal tolerance of psychrotrophic bacteria associated with maca (*Lepidium meyenii* Walp.) rhizosphere. *AIMS Microbiol.* **2017**, *3*, 279–292. [CrossRef]

74.   Park, Y.G.; Mun, B.G.; Kang, S.M.; Hussain, A.; Shahzad, R.; Seo, C.W. *Bacillus aryabhattai* SRB02 tolerates oxidative and nitrosative stress and promotes the growth of soybean by modulating the production of phytohormones. *PLoS ONE* **2017**, *12*, e0173203. [CrossRef]

75.   Naili, F.; Neifar, M.; Elhidri, D.; Cherif, H.; Bejaoui, B.; Aroua, M. Optimization of the effect of PGPR–based biofertlizer on wheat growth and yield. *Biom. Biostat. Int. J.* **2018**, *7*, 226–232. [CrossRef]

76.   Cézard, C.; Farvacques, N.; Sonnet, P. Chemistry and biology of pyoverdines, *Pseudomonas* primary siderophores. *Curr. Med. Chem.* **2015**, *22*, 165–186. [CrossRef]

77.   Trapet, P.; Avoscan, L.; Klinguer, A.; Pateyron, S.; Citerne, S.; Chervin, C. The *Pseudomonas fluorescens* siderophore pyoverdine weakens *Arabidopsis thaliana* defense in favor of growth in iron-deficient conditions. *Plant. Physiol.* **2016**, *171*, 675–693. [CrossRef] [PubMed]

78.   Kuzmanović, N.; Eltlbany, N.; Ding, G.; Baklawa, M.; Min, L.; Wei, L. Analysis of the genome sequence of plant beneficial strain *Pseudomonas* sp. RU47. *J. Biotechnol.* **2018**, *281*, 183–192. [CrossRef] [PubMed]

79.   Singh, S.K.; Singh, P.P.; Gupta, A.; Singh, A.K.; Keshri, J. Tolerance of Heavy Metal Toxicity Using PGPR

Strains of *Pseudomonas* Species. In *PGPR Amelioration in Sustainable Agriculture*; Singh, K., Ed.; Woodhead Publishing: Sawston, UK, 2019; pp. 239–252.

80. Nascimento, F.X.; Rossi, M.J.; Glick, B.R. Ethylene and 1-Aminocyclopropane-1-carboxylate (ACC) in plant–bacterial interactions. *Front. Plant. Sci.* **2018**, *9*, 114. [CrossRef] [PubMed]

81. Glick, B.R.; Todorovic, B.; Czarny, J.; Cheng, Z.; Duan, J.; McConkey, B. Promotion of plant growth by bacterial ACC deaminase. *Crit. Rev. Plant Sci.* **2007**, *26*, 227–242. [CrossRef]

82. Ciocchini, A.E.; Roset, M.S.; Briones, G.; de Iannino, N.I.; Ugalde, R.A. Identification of active site residues of the inverting glycosyltransferase Cgs required for the synthesis of cyclic β-1, 2-glucan, a Brucella abortus virulence factor. *Glycobiology* **2006**, *16*, 679–691. [CrossRef] [PubMed]

83. Glick, B.R. Bacteria with ACC deaminase can promote plant growth and help to feed the world. *Microbiol. Res.* **2014**, *169*, 30–39. [CrossRef] [PubMed]

84. Skorupska, A.; Janczarek, M.; Marczak, M.; Mazur, A.; Król, J. Rhizobial exopolysaccharides: Genetic control and symbiotic functions. *Microb. Cell Fact.* **2006**, *5*, 7. [CrossRef]

85. Henrissat, B.; Sulzenbacher, G.; Bourne, Y. Glycosyltransferases, glycoside hydrolases: Surprise. *Curr. Opin. Struct. Biol.* **2008**, *18*, 527–533. [CrossRef]

86. Rodríguez-Navarro, D.N.; Dardanelli, M.S.; Ruíz-Saínz, J.E. Attachment of bacteria to the roots of higher plants. *FEMS Microbiol. Lett.* **2007**, *272*, 127–136. [CrossRef]

87. Bogino, P.; Oliva, M.; Sorroche, F.; Giordano, W. The role of bacterial biofilms and surface components in plant-bacterial associations. *Int. J. Mol. Sci.* **2013**, *14*, 15838–15859. [CrossRef]

88. Yang, J.; Tauschek, M.; Robins-Browne, R.M. Control of bacterial virulence by AraC-like regulators that respond to chemical signals. *Trends Microbiol.* **2011**, *19*, 128–135. [CrossRef] [PubMed]

89. Garcia-Fraile, P.; Seaman, J.C.; Karunakaran, R.; Edwards, A.; Poole, P.S.; Downie, J.A. Arabinose and protocatechuate catabolism genes are important for growth of *Rhizobium leguminosarum* biovar viciae in the pea rhizosphere. *Plant Soil* **2015**, *390*, 251–264. [CrossRef]

90. Achouak, W.; Sutra, L.; Heulin, T.; Meyer, J.M.; Fromin, N.; Degraeve, S. *Pseudomonas brassicacearum* sp. nov. and *Pseudomonas thivervalensis* sp. nov., two root-associated bacteria isolated from *Brassica napus* and *Arabidopsis thaliana*. *Int. J. Syst. Evol. Microbiol.* **2000**, *50*, 9–18. [CrossRef] [PubMed]

91. Chung, B.S.; Aslam, Z.; Kim, S.W.; Kim, G.G.; Kang, H.S.; Ahn, J.W. A bacterial endophyte, *Pseudomonas brassicacearum* YC5480, isolated from the root of *Artemisia* sp. producing antifungal and phytotoxic compounds. *Plant Pathol. J.* **2008**, *24*, 461–468. [CrossRef]

92. Pereira, S.I.A.; Monteiro, C.; Vega, A.L.; Castro, P.M. Endophytic culturable bacteria colonizing *Lavandula dentata* L. plants: Isolation, characterization and evaluation of their plant growth-promoting activities. *Ecol. Eng.* **2016**, *87*, 91–97. [CrossRef]

93. Deng, Z.S.; Zhao, L.F.; Kong, Z.Y.; Yang, W.Q.; Lindström, K.; Wang, E.T. Diversity of endophytic bacteria within nodules of the *Sphaerophysa salsula* in different regions of Loess Plateau in China. *FEMS Microbiol. Ecol.* **2011**, *76*, 463–475. [CrossRef] [PubMed]

94. Safronova, V.I.; Stepanok, V.V.; Engqvist, G.L.; Alekseyev, Y.V.; Belimov, A.A. Root-associated bacteria containing 1-aminocyclopropane-1-carboxylate deaminase improve growth and nutrient uptake by pea genotypes cultivated in cadmium supplemented soil. *Biol. Fertil. Soil* **2006**, *42*, 267–272. [CrossRef]

95. Long, H.H.; Schmidt, D.D.; Baldwin, I.T. Native bacterial endophytes promote host growth in a species-specific manner; phytohormone manipulations do not result in common growth responses. *PLoS ONE* **2008**, *3*, e2702. [CrossRef]

96. Kong, Z.; Deng, Z.; Glick, B.R.; Wei, G.; Chou, M. A nodule endophytic plant growth-promoting *Pseudomonas* and its effects on growth, nodulation and metal uptake in *Medicago lupulina* under copper stress. *Ann. Microbiol.* **2017**, *67*, 49–58. [CrossRef]

97. Zachow, C.; Müller, H.; Monk, J.; Berg, G. Complete genome sequence of *Pseudomonas brassicacearum* strain L13-6-12, a biological control agent from the rhizosphere of potato. *Stand. Genomic Sci.* **2017**, *12*, 6. [CrossRef]

98. Zengerer, V.; Schmid, M.; Bieri, M.; Müller, D.C.; Remus-Emsermann, M.N.; Ahrens, C.H. *Pseudomonas orientalis* F9: A potent antagonist against phytopathogens with phytotoxic effect in the apple flower. *Front. Microbiol.* **2018**, *9*, 145. [CrossRef] [PubMed]

99. Madani, H.; Malboobi, M.A.; Bakhshkelarestaghi, K.; Stoklosa, A. Biological and Chemical Phosphorus Fertilizers Effect on Yield and P Accumulation in Rapeseed (*Brassica napus* L.). *Not. Bot. Hortic. Agrobot. Cluj-Napoca* **2012**, *40*, 210–214. [CrossRef]

100. Mohammadi, K.; Rokhzadi, A. An integrated fertilization system of canola (*Brassica napus* L.) production under different crop rotations. *Ind. Crops Prod.* **2012**, *37*, 264–269. [CrossRef]

101. Valetti, L.; Iriarte, L.; Fabra, A. Growth promotion of rapeseed (*Brassica napus*) associated with the inoculation of phosphate solubilizing bacteria. *Appl. Soil Ecol.* **2018**, *132*, 1–10. [CrossRef]

102. Qiao, J.; Yu, X.; Liang, X.; Liu, Y.; Borriss, R.; Liu, Y. Addition of plant-growth-promoting Bacillus subtilis PTS-394 on tomato rhizosphere has no durable impact on composition of root microbiome. *BMC Microbiol.* **2017**, *17*, 131. [CrossRef] [PubMed]

103. Gadhave, K.R.; Devlin, P.F.; Ebertz, A.; Ross, A.; Gange, A.C. Soil inoculation with Bacillus spp. modifies root endophytic bacterial diversity, evenness, and community composition in a context-specific manner. *Microb. Ecol.* **2018**, *76*, 741–750. [CrossRef] [PubMed]

104. Compant, S.; Samad, A.; Faist, H.; Sessitsch, A. A review on the plant microbiome: Ecology, functions and emerging trends in microbial application. *J. Adv. Res.* **2019**, *19*, 29–37. [CrossRef]

105. Van der Heijden, M.G.; Hartmann, M. Networking in the plant microbiome. *PLoS Biol.* **2016**, *14*, e1002378. [CrossRef]

106. Cordero, J.; de Freitas, J.R.; Germida, J.J. Bacterial microbiome associated with the rhizosphere and root interior of crops in Saskatchewan, Canada. *Can. J. Microbiol.* **2020**, *66*, 71–85. [CrossRef]

107. Berg, G.; Smalla, K. Plant species and soil type cooperatively shape the structure and function of microbial communities in the rhizosphere. *FEMS Microbiol. Ecol.* **2009**, *68*, 1–13. [CrossRef]

108. Hartmman, K.; Tringe, S.G. Interactions between plants and soil shaping the root microbiome under abiotic stress. *Biochem. J.* **2019**, *476*, 2705–2724. [CrossRef] [PubMed]

109. Naylor, D.; DeGraaf, S.; Purdom, E.; Coleman-Derr, D. Drought and host selection influence bacterial community dynamics in the grass root microbiome. *ISME J.* **2017**, *11*, 2691–2704. [CrossRef] [PubMed]

Evaluating Biochar-Microbe Synergies for Improved Growth, Yield of Maize and Post-Harvest Soil Characteristics in a Semi-Arid Climate

Maqshoof Ahmad [1,*]🆔, Xiukang Wang [2,*], Thomas H. Hilger [3]🆔, Muhammad Luqman [1], Farheen Nazli [4], Azhar Hussain [1], Zahir Ahmad Zahir [5], Muhammad Latif [6], Qudsia Saeed [7], Hina Ahmed Malik [8] and Adnan Mustafa [9]🆔

[1]   Department of Soil Science, The Islamia University of Bahawalpur, Bahawalpur 63100, Pakistan; luqmansidhu229@gmail.com (M.L.); azharhaseen@gmail.com (A.H.)
[2]   College of Life Sciences, Yan'an University, Yan'an 716000, China
[3]   Hans-Ruthenberg Institute, University of Hohenheim, 70593 Stuttgart, Germany; thomas.hilger@uni-hohenheim.de
[4]   Pesticide Quality Control Laboratory, Punjab Agriculture Department, Government of Punjab, Bahawalpur 63100, Pakistan; farheenmaqshoof@gmail.com
[5]   Institute of Soil and Environmental Sciences, University of Agriculture, Faisalabad 38000, Pakistan; zazahir@yahoo.com
[6]   Department of Agronomy, The Islamia University of Bahawalpur, Bahawalpur 63100, Pakistan; mlatifiub@gmail.com
[7]   College of Natural Resources and Environment, Northwest Agriculture and Forestry University, Xianyang 712100, China; syedaqudsia.saeed@yahoo.com
[8]   Institute of Food and Agriculture Sciences, University of Florida, Gainesville, FL 110690, USA; Hmalik@ufl.edu
[9]   National Engineering Laboratory for Improving Quality of Arable Land, Institute of Agricultural Resources and Regional Planning, Chinese Academy of Agricultural Sciences, Beijing 100081, China; adnanmustafa780@gmail.com
*   Correspondence: maqshoof_ahmad@yahoo.com (M.A.); wangxiukang@yau.edu.cn (X.W.)

Abstract: Arid and semi-arid regions are characterized by high temperature and low rainfall, leading to degraded agricultural soils of alkaline calcareous nature with low organic matter contents. Less availability of indigenous nutrients and efficacy of applied fertilizers are the major issues of crop production in these soils. Biochar application, in combination with plant growth promoting rhizobacteria with the ability to solubilize nutrients, can be an effective strategy for improving soil health and nutrient availability to crops under these conditions. Experiments were planned to evaluate the impact of biochar obtained from different sources in combination with acid-producing, nutrient-solubilizing *Bacillus* sp. ZM20 on soil biological properties and growth of maize (*Zea mays* L.) crops under natural conditions. Various biochar treatments, viz. wheat (*Triticum aestivum* L.) straw biochar, Egyptian acacia (*Vachellia nilotica* L.) biochar, and farm-yard manure biochar with and without *Bacillus* sp. ZM20, were used along with control. Soil used for pot and field trials was sandy loam in texture with poor water holding capacity and deficient in nutrients. Results of the pot trial showed that fresh and dry biomass, 1000 grain weight, and grain yield was significantly improved by application of biochar of different sources with and without *Bacillus* sp. ZM20. Application of biochar along with *Bacillus* sp. ZM20 also improved soil biological properties, i.e., soil organic matter, microbial biomass carbon, ammonium, and nitrate nitrogen. It was also observed that a combined application of biochar with *Bacillus* sp. ZM20 was more effective than a separate application of biochar. The results of wheat straw biochar along with *Bacillus* sp. ZM20 were better as compared to farm-yard manure biochar and Egyptian acacia biochar. Maximum increase (25.77%) in grain yield was observed in the treatment where wheat straw biochar (0.2%) was applied in combination with

Bacillus sp. ZM20. In conclusion, combined application of wheat straw biochar (0.2%) inoculated with *Bacillus* sp. ZM20 was the most effective treatment in improving the biological soil properties, plant growth, yield, and quality of maize crop as compared to all other treatments.

Keywords: aridity; *Bacillus* sp.; biochar; nutrient availability; organic matter; soil health

---

1. Introduction

The current world population is about 7.6 billion, which is increasing at an exponential rate and will be about 9.8 billion in 2050 and is further expected to rise to 11.2 billion in 2100, as reported by the United Nations [1]. About half of the added population will be concentrated in less developed countries. Due to this reason, there will be a marked decrease in agricultural lands, as most of the productive lands will be used for constructing new housing societies and infrastructure [2]. To feed the world population, utilization of less productive soils, and bringing such soils into the agricultural system by fighting desertification, salinization, and soil pollution is the major challenge for the scientific community [3]. Moreover, increasing per-hectare yield of the major crops along with exploring the unutilized arable lands can be helpful to meet the challenge of food requirements.

Maize, being the staple food of most of the world population, is an important cereal crop [4]. Its total production is even more than rice and wheat crops [5]. Maize has gained its popularity to meet the world food requirements due to higher yield per unit area as compared to other staple crops [6]. Although the per-acre yield of maize is adequate, it is an exhaustive crop that needs more nutrients and that is why it depletes more nutrients from the soil [7]. It has high demand for phosphatic- and zinc-containing fertilizers as compared to other major crops; therefore, nutrient deficiency is experienced more in the maize crop [8].

Biochar can be effective to rehabilitate degraded lands by improving the soil physical properties, nutrient-holding capacity, and soil carbon contents, leading to improvement in soil productivity [9]. It is a carbon-rich compound that is produced through a process known as pyrolysis and has beneficial implications as a potential soil amendment [10]. Use of biochar has gained popularity as a carbon negative material which resists environmental change as it draws carbon from the atmosphere into the soil and persists for hundreds to thousands of years [11]. Recent interest has been developed to use biochar as a soil amendment for improving soil quality through mitigation of soil salinization, soil acidity, and metal contamination, along with improvement in soil productivity [12–15]. Biochar application to soil positively affects the properties of soil, including soil structure, water retention capacity, fertility, and carbon sequestration of degraded soil [16,17]. It also improves soil microbial activity due to presence of micropores in biochar which allow the sorption of dissolved organic matter, thus, helping speed up the soil rehabilitation process [18]. However, the success highly depends upon the types and rates of biochar application, the nature of feedstock, and soil and climate variations. In this regard, utilizing biochar with other soil amendments such as plant growth-promoting rhizobacteria (PGPR) has proved to be a better approach to conserving the environment, resulting in increased efficacy and cost-effectiveness [3,9].

The use of microorganisms with the aim of improving nutrient availability for plants is an important practice and is considered necessary for agriculture these days [19]. The PGPR are the bacteria that inhabit either the rhizosphere, the soil in the immediate vicinity of plant roots, or inside the plant tissues, helping the plants with better growth through some direct and indirect mechanisms [20,21]. There are certain PGPR species which can solubilize insoluble mineral compounds in soil through the production of organic acids along with some other growth-promoting mechanisms [22,23]. Among these, phosphate solubilizing rhizobacteria [24], zinc solubilizing rhizobacteria [25], and potassium solubilizing rhizobacteria [26] are well documented. These nutrient-solubilizing bacterial species also have multiple plant growth-promoting traits such as siderophores production, chitin decomposition, hydrogen cyanide production, ammonia production, etc. [24]. They effectively colonize plant roots, thus helping the improvement of plant growth and nutrient acquisition [27,28]. These bacteria can also induce tolerance against different biotic and abiotic stresses in plants through several indirect mechanisms [27,29]. Moreover, bacterial inoculation improves soil health by fixing atmospheric nitrogen [30,31], production of plant hormones, siderophores and exopolysaccharides [32], and phytoremediation of heavy metals and other organic pollutants [33,34].

The integrated use of biochar and PGPR can reduce the use of chemical fertilizers for crop production in addition to improving soil health through increased soil organic matter contents, enhanced soil aggregation, better microbial activity, and increased soil fertility [35,36]. The improvement in soil health and maize growth has also been reported by the combined use of biochar and PGPR under water-stressed conditions [37]. Work on the use of biochar for increasing soil fertility and remediating the polluted soil has been carried out, but the use of biochar as soil amendment for improving the soil health, growth, and yield of maize in the degraded soils of arid and semi-arid regions has been least explored. It has been hypothesized that the use of biochar and PGPR can help improve barren desert soils to productive farmlands, and release the pressure off the ever-decreasing cultivated areas. Keeping with this view, current study was conducted to investigate the potential of biochar obtained from different sources along with acid-producing, nutrient-solubilizing *Bacillus* sp. for improving soil biological properties, growth, and yield of maize crop in desert regions.

2. Materials and Methods

2.1. Biochar Preparation and Characterization

Biochar was prepared from Egyptian acacia (*Vachellia nilotica* L.) stem, wheat straw, and dairy manure pyrolyzed at 450 °C following the method of Naeem et al. [38]. The dried branches of Egyptian acacia were chopped in small pieces of 2–3 inches and further dried at 105 °C for one hour. The oven-dried biomass was pyrolyzed at 450 °C. Finally, the prepared biochar was crushed into smaller particles for even distribution in soil. Before charring, the dairy manure was air-dried and sieved (≤2 mm), then pyrolyzed in a muffle furnace at 450 °C. Similarly, the air-dried, chopped wheat straw was also pyrolyzed at 450 °C. The weight of biomass used for each type of biochar was recorded prior and after pyrolysis. After cooling, the biochar was passed through a 250-µm sieve and stored in a refrigerator (at 4 °C) before use.

The biochar produced from different sources was analyzed for chemical characteristics (Table 1). Prepared biochar was analyzed for its turnover rate made from the pyrolysis of feedstock. The biochar production rate was calculated by using the total weight of raw material used to prepare that biochar. Biochar yield was estimated by following the method of Al-Wabel et al. [39]. The pH and electrical conductivity (EC) of the biochar was measured using a 1:20 solid/solution ratio after shaking for ninety minutes in deionized water in a mechanical shaker. The carbon contents of biochar were assessed using the loss-on-ignition approach [40,41]. Total nitrogen (N) contents were measured using Kjeldahl distillation equipment [42].

Table 1. Rate and physicochemical properties of biochar; values are the mean of three replications ± SE.

Parameters	Egyptian Acacia Biochar	Farmyard Manure Biochar	Wheat Straw Biochar
Turnover rate (%)	28.45 ± 1.27	23.31 ± 0.94	38.76 ± 1.78
pH	9.3 ± 0.25	8.67 ± 0.12	8.43 ± 0.21
EC (dS cm^{-1})	1.85 ± 0.03	2.1 ± 0.02	1.9 ± 0.03
Bulk density (g cm^{-3})	0.36 ± 0.01	0.43 ± 0.01	0.46 ± 0.03
Nitrogen (%)	0.31 ± 0.02	0.28 ± 0.01	0.47 ± 0.02
Carbon (%)	68.45 ± 3.24	44.21 ± 2.16	52.67 ± 1.89

2.2. Collection of Rhizobacterial Strains

Pre-selected and characterized rhizobacterial strain *Bacillus* sp. *ZM20*, accession number KX086260, with strong ability to produce organic acids [23] under zinc deficient conditions was collected from the Soil Microbiology and Biotechnology Laboratory, Department of Soil Science, the Islamia University of Bahawalpur.

2.3. Soil Sampling and Analysis

A bulk soil sample (0–15 cm) was taken from the experimental field and the soil that was used for pot trial. The soil samples were air-dried and sieved through 2-mm sieve followed by analysis for basic soil characteristics (Table 2) as per standard protocols. The pH, EC, and organic matter were measured according to the method of Nelson and Sommers [43]. The available N was analyzed by the Kjeldhal method [42], while for available phosphorus (P), Olsen's method [44] was used. The extractable potassium (K) was measured using a flame photometer (Model; Model BWB-XP, BWB Technologies, UK). The saturation percentage referring to the field capacity of soil was estimated by oven-drying the soil sample at 105 °C to a constant weight, followed by calculations according to the method as described by Sarfraz et al. [45]. All the chemicals were of analytical grade (Sigma-Aldrich, Unichem, Merck) supplied by Wahid Scientific Store, Lahore, Pakistan.

Table 2. Characteristics of the soils; values are the mean of three replications ± SE.

Parameter	Pot Trial	Field Trial
EC_e (dS m^{-1})	1.6 ± 0.01	1.8 ± 0.01
pH	8.1 ± 0.04	7.9 ± 0.02
Organic matter (%)	0.39 ± 0.02	0.47 ± 0.02
Available N (%)	0.024 ± 0.001	0.059 ± 0.003
Available P (mg kg^{-1})	3.7 ± 0.01	4.5 ± 0.03
Extractable K (mg kg^{-1})	53 ± 1.68	77 ± 3.21
Saturation percentage (%)	33 ± 0.76	36 ± 0.71
Water-holding capacity (Inches ft^{-1})	1.27 ± 0.05	1.29 ± 0.03
Textural class	Sandy loam	Sandy loam

2.4. Pot Trial

A pot trial was conducted in the wire house to evaluate the impact of biochar obtained from different sources in combination with acid-producing, nutrient-solubilizing *Bacillus* sp. *ZM20* on soil biological properties, and the growth and yield of maize crops (Pioneer-30Y80) under natural environmental conditions in the February–March sowing season. The experiment was conducted in a wire house with natural growth conditions, protecting the experimental units from animals and birds with wire only. Various biochar treatments viz. Egyptian acacia biochar (0.1%), Egyptian acacia biochar (0.2%), farmyard manure (FYM) biochar (0.1%), FYM biochar (0.2%), wheat straw biochar (0.1%), and wheat straw biochar (0.2%) with and without *Bacillus* sp. *ZM20* were used along with controls. Soil (8 kg per pot^{-1}) used to fill the pots (height 12″, diameter 12″) was sandy loam in texture with poor water-holding capacity (1.27 inches ft^{-1}) and deficient in nutrients (Table 2), as analyzed by following the standard protocols as defined by Ryan et al. [46]. The pots were arranged in the wire house following a completely randomized design (CRD) in factorial arrangement with three replications. Maize seeds were inoculated with a slurry of *Bacillus* sp. *ZM20* prepared by mixing the inoculum, sugar solution, and peat in the ratio (04:01:05). The inoculated seeds were used to sow in one set of treatments while in the other set un-inoculated maize seeds were sown. The recommended doses of P and K at the rate of 90 kg ha^{-1} and 60 kg ha^{-1} while half of the recommended dose of N (120 kg ha^{-1}) were applied as basal doses in the form of diammonium phosphate, sulfate of potash, and urea. The remaining dose of N was applied in two splits. Good quality tap water meeting the irrigation quality criteria [47] was used to irrigate pots, and all other agronomic practices were carried out according to requirements. Growth and yield parameters were recorded at the time of harvesting and grain samples were collected to analyze for N, P, and K.

2.5. Field Trial

A field trial was conducted in the February–March (Spring 2019) sowing season to verify the results of the pot trial and further recommendation to farming community. The same treatment plan was followed as observed in the pot trial. The field trial was conducted in the field area of the Department of Soil Science, The Islamia University of Bahawalpur, Pakistan. The soil of the experimental field was sandy loam in texture with poor water-holding capacity and deficient in nutrients (Table 2). The randomized complete block design was used for the field trial with a factorial arrangement and three replications. The size of the plots was 22' × 16' with a row-to-row distance of 2.5'. Maize seeds were inoculated before sowing by following the same procedure as described above in the pot trial. The recommended doses of P and K at the rate of 90 and 60 kg ha^{-1} while half of the recommended dose of N (120 kg ha^{-1}) were applied as basal doses in the form of diammonium phosphate, sulfate of potash, and urea. The remaining dose of N was applied in two splits. Canal water was used for irrigation purposes and all other agronomic practices were carried out according to requirements. Growth and yield parameters were recorded at the time of harvesting and grain samples were collected to analyze for N, P, and K.

2.6. Nutrient Analyses in Grains

Grains were digested according to the protocol as described by Wolf [48]. The P in grain samples was analyzed using a UV-visible spectrophotometer (Agilent Carry 60, USA), while K in grains was determined on a flame photometer (Model; Model BWB-XP, BWB Technologies, Newbury, UK) by following the standard methods [46]. For the analysis of N in grains, an automatic digestion unit (DK 6), semi-automatic distillation unit (UDK 126) of Kjeldahl apparatus (VELP Sci., Italy) was used, followed by standard titration as described in the Kjeldahl method [49]. All the chemicals were of analytical grade (Sigma-Aldrich, Unichem, Merck) supplied by Wahid Scientific Store, Lahore, Pakistan.

2.7. Post-Harvest Soil Sample Collection and Analysis

The post-harvest soils samples were collected from the pot (harvested in July) and field trials (harvested in July), and analyzed for organic matter, microbial biomass carbon (MBC), ammonium N, and nitrate N under pot and field conditions. The composite soil sampling method was used, and the samples were air-dried and sieved through a 2-mm sieve before analysis. The prepared soil samples were stored in a refrigerator at 4 °C and analyzed within seven days. The organic matter contents were measured according to the method of Nelson and Sommers [43]. For the analysis of microbial biomass carbon (MBC), chloroform fumigation and extraction methods were used [50,51]. For the analyses of ammoniacal N and nitrate N in soil, the methods of Kamphake et al. [52] and Sims and Jackson [53], respectively, were used. All the chemicals were of analytical grade (Sigma-Aldrich, Unichem, Merck) supplied by Wahid Scientific Store, Lahore, Pakistan. Replicated measurements were always performed to ensure the accuracy of the data.

2.8. Statistical Analysis

All data reported here are means of three replicates which were analyzed using one-way analysis of variance (ANOVA) in Statistix 8.1. The mean values were compared through a least significant difference (LSD) test as described by Steel et al. [54].

3. Results

3.1. Pot Trial

Integrated use of biochar and *Bacillus* sp. ZM20 improved soil properties in the pot trial. Results (Figure 1A) showed that FYM (Farm yard manure) biochar treatments increased the organic matter in the pot trial. The application of biochar without inoculation increased the soil organic matter contents, but the results of Egyptian acacia biochar at both levels were non-significant when compared with the control under un-inoculated and inoculated sets of treatments. Under inoculated treatments, the maximum organic matter (0.449%) was observed in the treatment where wheat straw biochar (0.2%) was applied in combination with *Bacillus* sp. ZM20; this treatment was, however, non-significant with FYM biochar application (0.2%), and wheat straw biochar application (0.1%) under inoculated treatments. Combined inoculation of biochar and *Bacillus* sp. ZM20 showed better results than separate application of biochar in all treatments.

The application of biochar from different sources significantly improved the MBC in the soil (Figure 1B). Maximum improvement in MBC under an un-inoculated set of treatments was observed by the application of wheat straw biochar (0.2%), which was statistically at par with the application of wheat straw biochar (0.1%) and FYM biochar (0.2%). These treatments, however, were significantly better than all other treatments under un-inoculated conditions. The application of biochar from all sources in the presence of *Bacillus* sp. ZM20 was significantly better than separate use, except for Egyptian acacia biochar (0.1%), where the increase was non-significant with that of the respective un-inoculated treatment. Maximum MBC (342 mg kg^{-1}) was observed in the treatment where wheat straw biochar was applied (0.2%), and it was statistically similar with that of the wheat straw biochar (0.1%) treatment.

The sole and combined application of biochar and inoculated with *Bacillus* sp. ZM20 to improve the ammonium N and nitrate N in the pot trial was observed (Figure 1C,D). The application of biochar separately, and in combination with *Bacillus* sp. ZM20, significantly enhanced the ammonium N and nitrate N, except for Egyptian acacia biochar treatments, which gave non-significant improvement in both cases as compared to the control. A maximum increase in ammonium N and nitrate N was recorded due to the combined application of wheat straw biochar (0.2%) and *Bacillus* sp. ZM20 as compared to the inoculated control. Overall, inoculated treatments showed better results regarding ammonium N and nitrate N than un-inoculated treatments.

The results of the impact of the integrated use of biochar and *Bacillus* sp. ZM20 on plant height (Figure 2A) revealed that the separate as well as combined use of biochar with *Bacillus* sp. ZM20 significantly improved plant height in the pot trial, except for Egyptian acacia biochar, at both levels, which was statistically non-significant compared to the control. In the inoculated treatment, the combined use of wheat straw biochar (0.2%) and *Bacillus* sp. ZM20 was carried out and showed the maximum plant height. In the case of root length, the results of separate applications of wheat straw biochar at both levels and FYM biochar (0.2%) were significantly better than those of the control; however, other treatments gave non-significant improvement in root length when compared with the control. The combined use of biochar and *Bacillus* sp. ZM20 was better than the separate use of biochar in improving the root length, but the results were non-significant with un-inoculated treatments in all cases. Maximum improvement in root length as compared to control was observed with the combined use of wheat straw biochar (0.2%) and *Bacillus* sp. ZM20; however, it was statistically non-significant with the treatments of wheat straw biochar (0.1%) and FYM biochar (0.2%) in combination with *Bacillus* sp. ZM20 (Figure 2B). Application of biochar significantly improved the shoot fresh and dry biomass separately and in combination with *Bacillus* sp. ZM20 as compared to respective controls. Maximum improvement in shoot fresh biomass and shoot dry biomass were observed with the application of wheat straw biochar in combination with *Bacillus* sp. ZM20 (Figure 2C,D). The improvement due to the inoculation of *Bacillus* sp. ZM20 in both the parameters over the respective un-inoculated treatment was non-significant in all the cases.

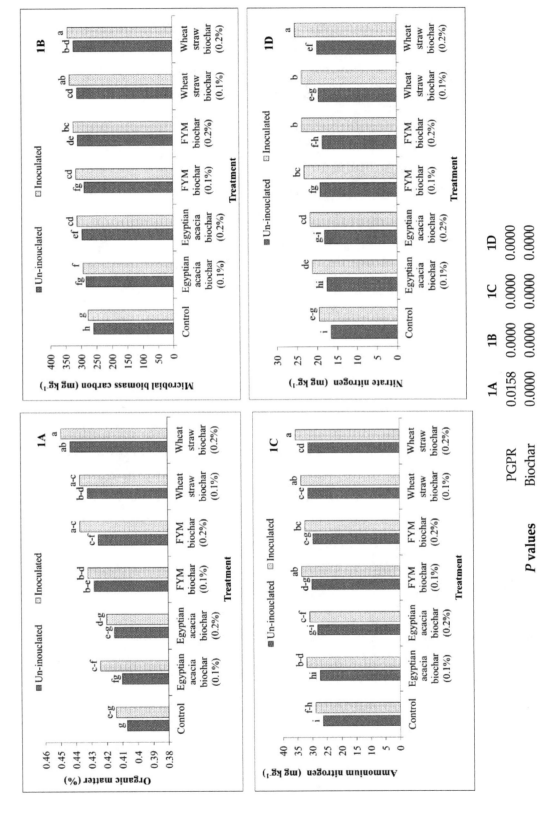

Figure 1. Effects of biochar with and without *Bacillus* sp. ZM20 on organic matter (**A**): Least significant difference (LSD) 0.0156, microbial biomass carbon (**B**): LSD 14.290, ammonium N (**C**): LSD 2.096, and nitrate N (**D**): LSD 1.6386 under pot trial; (*n* = 3); bars sharing same letters are statistically not different from each other at *p* ≤ 0.05.

	1A	1B	1C	1D
PGPR	0.0158	0.0000	0.0000	0.0000
P values Biochar	0.0000	0.0000	0.0000	0.0000
PGPR + Biochar	0.9401	0.6795	0.6394	0.3057

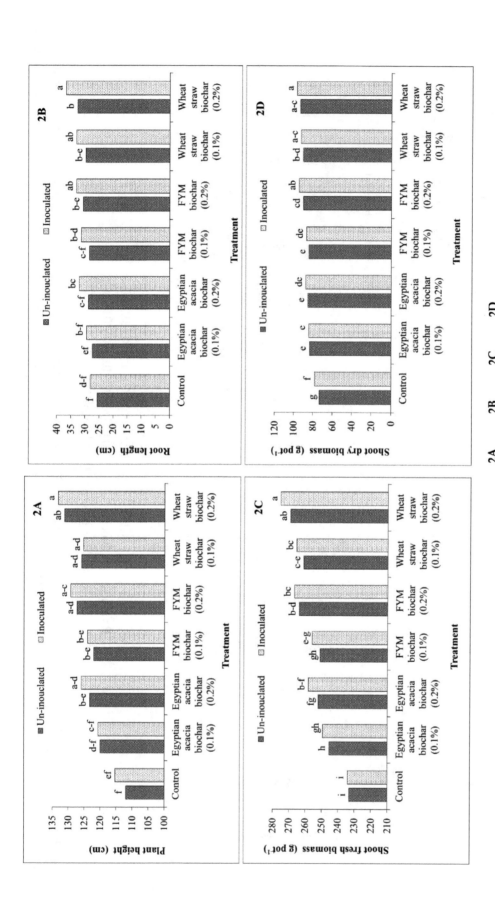

Figure 2. Effects of biochar with and without *Bacillus* sp. ZM20 on plant height (**A**): LSD 3.6996, shoot fresh biomass (**C**): LSD 6.8333 and shoot dry biomass (**D**): 4.6543 of maize under pot trial; (*n* = 3); bars sharing same letters are statistically not different from each other at *p* ≤ 0.05.

Results of the effects of separate as well as combined applications of biochar inoculated with *Bacillus* sp. ZM20 significantly improved the root fresh biomass in the pot trial (Table 3). A maximum increase (29.63%) in root fresh biomass was observed due to the combined use of wheat straw biochar (0.2%) with *Bacillus* sp. ZM20. The results of the separate use of biochar (all treatments) without inoculum were, however, non-significant with the control in most of the cases, except wheat straw biochar (0.2%), which gave significantly better results than the control. The results of the improvement in root dry biomass were non-significant with the control due to the biochar application with and without in most of the cases, except for FYM biochar (0.2%) and wheat straw biochar (0.2%) in both sets of treatments, i.e., inoculated and un-inoculated. A maximum increase (23.36%) in root dry biomass was observed due to the combined use of wheat straw biochar (0.2%) with *Bacillus* sp. ZM20 (Table 3).

Table 3. Effects of biochar with and without *Bacillus* sp. ZM20 on root fresh biomass, root dry biomass, 100 grain weight, and grain yield of maize in pot trial.

Treatment	Un-Inoculated	Inoculated	Un-Inoculated	Inoculated
	Root Fresh Biomass (g pot^{-1})		Root Dry Biomass (g pot^{-1})	
Control	28.00 ^{f}	36.00 $^{c-e}$	14.33 ^{d}	15.67 ^{c,d}
Egyptian acacia biochar (0.1%)	29.33 ^{f}	38.33 $^{b-d}$	15.33 ^{c,d}	17.33 $^{a-d}$
Egyptian acacia biochar (0.2%)	31.67 ^{e,f}	42.00 ^{a,b}	16.33 $^{a-d}$	17.67 $^{a-c}$
FYM biochar (0.1%)	30.67 ^{e,f}	41.33 $^{a-c}$	16.00 $^{b-d}$	17.67 $^{a-c}$
FYM biochar (0.2%)	33.00 $^{d-f}$	43.67 ^{a,b}	17.67 $^{a-c}$	19.00 ^{a,b}
Wheat straw biochar (0.1%)	31.00 ^{e,f}	43.00 ^{a,b}	17.33 $^{a-d}$	18.33 $^{a-c}$
Wheat straw biochar (0.2%)	35.67 ^{d,e}	46.67 ^{a}	18.00 $^{a-c}$	19.33 ^{a}
LSD (*p* ≤ 0.05)	5.5636		3.1110	
p value PGPR	0.0000		0.0192	
Biochar	0.0020		0.0314	
PGPR + Biochar	0.9586		0.9996	
	100 Grain Weight (g)		Grain Yield (g pot^{-1})	
Control	18.67 ^{c}	19.67 ^{b,c}	103.3 ^{f}	106.3 ^{e,f}
Egyptian acacia biochar (0.1%)	20.33 ^{b,c}	21.33 $^{a-c}$	114.3 ^{d,e}	116.0 ^{d}
Egyptian acacia biochar (0.2%)	21.00 $^{a-c}$	21.67 $^{a-c}$	116.3 ^{d}	121.0 $^{b-d}$
FYM biochar (0.1%)	21.00 $^{a-c}$	21.67 $^{a-c}$	116.0 ^{d}	121.7 $^{b-d}$
FYM biochar (0.2%)	21.67 $^{a-c}$	23.67 ^{a,b}	122.0 $^{b-d}$	127.0 $^{a-c}$
Wheat straw biochar (0.1%)	21.33 $^{a-c}$	23.00 ^{a,b}	120.0 ^{c,d}	126.0 $^{a-c}$
Wheat straw biochar (0.2%)	22.33 $^{a-c}$	24.67 ^{a}	129.0 ^{a,b}	133.7 ^{a}
LSD (*p* ≤ 0.05)	4.3030		8.9777	
p value PGPR	0.1039		0.0134	
Biochar	0.1421		0.0000	
PGPR + Biochar	0.9956		0.9755	

Values sharing same letter(s) within a parameter are statistically non-significant with each other at 5% level of probability; values are the mean of three replications ± SE. Lower case words show difference in treatment means.

Results regarding the effects of separate and combined applications of biochar with *Bacillus* sp. ZM20 on 100-grain weight are presented in (Table 3). In most of the cases, regarding the use of all types of biochar, individually as well as in combination with *Bacillus* sp. ZM20, the results were nonsignificant with the control, except for the wheat straw biochar (0.2%) application in combination with *Bacillus* sp. ZM20. Statistical analyses showed that all treatments of the separate and combined uses of biochar showed significantly better results than the control in improving the grain yield of maize in the pot trial. A maximum increase (25.77%) in grain yield was observed in the treatment where wheat straw biochar (0.2%) was applied in combination with *Bacillus* sp. ZM20 (Table 3). Data (Table 4) showed that the use of all types of biochar, individually as well as in combination with *Bacillus* sp. ZM20, gave non-significant results in improving the stover yield in all the cases when compared to the control.

Table 4. Effects of biochar with and without *Bacillus* sp. ZM20 on stover yield, and nitrogen, phosphorus and potassium concentration in grains of maize in pot trial.

Treatment	Un-Inoculated	Inoculated	Un-Inoculated	Inoculated
	Stover Yield (g pot^{-1})		Nitrogen Conc. in Grains (%)	
Control	20.33 ^{c}	23.00 $^{a-c}$	2.17 ^{g}	2.19 ^{f,g}
Egyptian acacia biochar (0.1%)	22.00 ^{b,c}	24.33 $^{a-c}$	2.21 ^{f,g}	2.21 $^{e-g}$
Egyptian acacia biochar (0.2%)	23.00 $^{a-c}$	26.33 ^{a,b}	2.24 $^{c-g}$	2.25 $^{c-f}$
FYM biochar (0.1%)	26.00 $^{a-c}$	25.67 $^{a-c}$	2.22 $^{d-g}$	2.25 $^{c-f}$
FYM biochar (0.2%)	25.00 $^{a-c}$	27.00 ^{a,b}	2.28 $^{a-e}$	2.30 $^{a-c}$
Wheat straw biochar (0.1%)	24.33 $^{a-c}$	26.33 ^{a,b}	2.28 $^{b-e}$	2.29 $^{a-d}$
Wheat straw biochar (0.2%)	25.67 $^{a-c}$	28.00 ^{a}	2.32 ^{a,b}	2.35 ^{a}
LSD (*p* ≤ 0.05)	5.9763		0.0705	
p value PGPR	0.0737		0.1857	
Biochar	0.2071		0.0000	
PGPR + Biochar	0.9862		0.9971	
	Phosphorus Conc. in Grains (%)		Potassium Conc. in Grains (%)	
Control	0.373 ^{e}	0.390 $^{c-e}$	2.57 ^{h}	2.60 ^{g,h}
Egyptian acacia biochar (0.1%)	0.380 ^{d,e}	0.403 $^{b-d}$	2.61 ^{g,h}	2.63 ^{f,g}
Egyptian acacia biochar (0.2%)	0.390 $^{c-e}$	0.410 $^{a-c}$	2.64 $^{e-g}$	2.69 $^{c-e}$
FYM biochar (0.1%)	0.387 $^{c-e}$	0.407 $^{a-c}$	2.62 ^{g,h}	2.67 $^{d-f}$
FYM biochar (0.2%)	0.397 $^{b-e}$	0.417 ^{a,b}	2.70 $^{b-d}$	2.74 ^{a,b}
Wheat straw biochar (0.1%)	0.400 $^{b-d}$	0.410 $^{a-c}$	2.67 $^{d-f}$	2.72 ^{b,c}
Wheat straw biochar (0.2%)	0.407 $^{a-c}$	0.430 ^{a}	2.74 ^{a,b}	2.77 ^{a}
LSD (*p* ≤ 0.05)	0.0250		0.0499	
p value PGPR	0.0000		0.0003	
Biochar	0.0000		0.0000	
PGPR + Biochar	0.0000		0.9546	

Values sharing same letter(s) within a parameter are statistically non-significant with each other at 5% level of probability; values are the mean of three replications ± SE. Lower case words show difference in treatment means.

Results (Table 4) showed that N concentration in grains of maize was significantly improved due to the separate as well as combined application of all types of biochar and *Bacillus* sp. ZM20. The results of Egyptian acacia biochar (both levels) and FYM biochar (0.1%) gave non-significant results in both sets of treatments. A maximum increase (7.30%) in N concentration in maize grains was observed due to the combined use of wheat straw biochar (0.2%) with *Bacillus* sp. ZM20. The results of the impact of biochar (all treatments), with and without inoculum, on P concentration in maize grains were non-significant with the control in most of the cases, except for the wheat straw biochar (at both levels) application in un-inoculated set of treatments, and the FYM biochar (0.1%) and wheat straw biochar (0.2%) application in combination with *Bacillus* sp. ZM20. Similar results were observed in case of K concentration in maize grains, where the maximum improvement (6.5%) over control was observed due to combined use of wheat straw biochar (0.2%) and *Bacillus* sp. ZM20 (Table 4).

3.2. Field Trial

Results (Figure 3) showed that all treatments significantly increased the organic matter and MBC under field conditions, except the application of Egyptian acacia biochar (0.1%), which gave non-significant improvement in organic matter when compared with the control under the un-inoculated treatment (Figure 3A). Under inoculated treatments, maximum improvement (5.78%) in organic matter contents over the control was observed in treatment where wheat straw biochar (0.2%) was applied in combination with *Bacillus* sp. ZM20; this treatment was, however, non-significant with the use of wheat straw biochar (0.1%) under inoculated treatments. The combined inoculation of biochar and *Bacillus* sp. ZM20 showed better results than the separate application of biochar in all treatments.

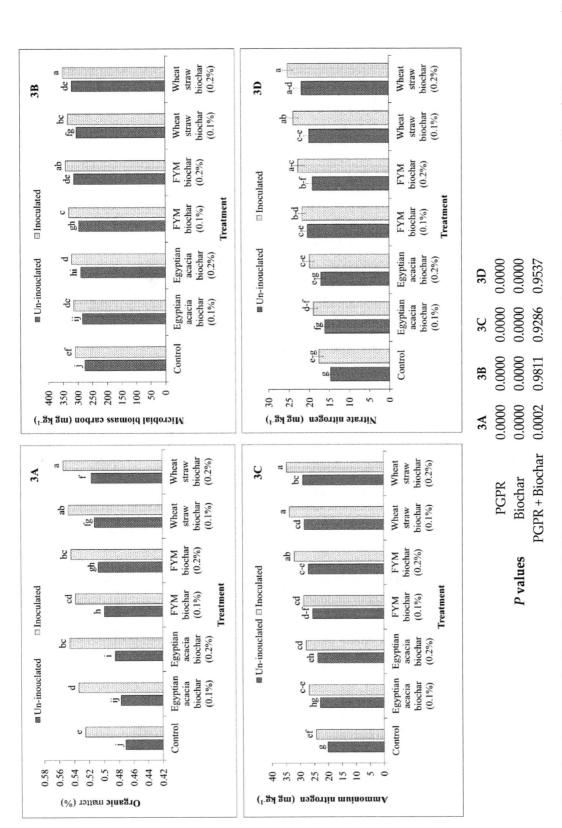

Figure 3. Effects of biochar with and without *Bacillus* sp. ZM20 on organic matter (**A**), microbial biomass carbon (**B**), ammonium nitrogen (**C**), and nitrate nitrogen (**D**) under field trial; (*n* = 3); bars sharing same letters are statistically not different from each other at *p* ≤ 0.05.

The application of biochar from different sources also significantly improved the MBC under field conditions in a semi-arid climate (Figure 3B). A maximum improvement (23.39%) in MBC under an un-inoculated set of treatments was observed by the application of wheat straw biochar (0.2%), which was statistically at par with application of FYM biochar (0.2%). These treatments, however, were significantly better than all other treatments under un-inoculated conditions. The application of biochar from all sources in the presence of *Bacillus* sp. ZM20 was significantly better than the separate use, except for Egyptian acacia biochar (0.1%), where the increase was non-significant with that of respective un-inoculated treatment. A maximum MBC (22.89%) was observed in the treatment where the wheat straw biochar (0.2%) was applied in combination with *Bacillus* sp. ZM20, and it was statistically similar with that of the FYM biochar (0.2%) treatment.

The application of biochar separately and in combination with *Bacillus* sp. ZM20 significantly enhanced the ammonium N and nitrate N, except for the Egyptian acacia biochar treatments (both levels), which gave non-significant improvement in both cases as compared to the control (Figure 3C,D). A maximum increase in ammonium N (22.61%) and nitrate N (29.59%) as compared to the inoculated control was recorded due to the combined application of wheat straw biochar (0.2%) and *Bacillus* sp. ZM20. Overall, inoculated treatments showed better results regarding ammonium N and nitrate N than un-inoculated treatments.

The results of the impact of integrated use of biochar and *Bacillus* sp. ZM20 on plant height (Table 5) under field conditions revealed that the separate as well as combined use of biochar with *Bacillus* sp. ZM20 improved plant height, but that this improvement was statistically non-significant with the control in most of the cases. In the inoculated treatment, the combined use of wheat straw biochar (0.2%) and *Bacillus* sp. ZM20 was carried out and showed the maximum improvement (12.7%) in plant height. The application of biochar separately and in combination with *Bacillus* sp. ZM20 significantly improved the shoot fresh and dry biomass as compared to respective controls. Maximum improvement in shoot fresh biomass (16.6%) and shoot dry biomass (20.75%) was observed with the application of wheat straw biochar (0.2%) in combination with *Bacillus* sp. ZM20 (Table 5). Results regarding the effects of separate and combined applications of biochar with *Bacillus* sp. ZM20 on 1000-grain weight and grain yield showed a significant improvement in most of the cases, except for Egyptian acacia biochar (0.1%), which gave non-significant improvement when compared with the control. A maximum increase (21.9%) in grain yield was observed in the treatment where wheat straw biochar (0.2%) was applied in combination with *Bacillus* sp. ZM20 (Table 6).

Table 5. Effects of biochar with and without *Bacillus* sp. ZM20 on plant height, shoot fresh biomass, shoot dry biomass, and 1000-grain weight of maize in field trial.

Treatment	Un-Inoculated	Inoculated	Un-Inoculated	Inoculated
	Plant Height (cm)		Shoot Fresh Biomass (g pot^{-1})	
Control	136.3 ^{f}	139.3 ^{e,f}	243.6 ^{g}	245.3 ^{g}
Egyptian acacia biochar (0.1%)	141.0 $^{d-f}$	143.3 $^{c-f}$	255.0 ^{f}	260.3 ^{e,f}
Egyptian acacia biochar (0.2%)	144.0 ^{e,f}	150.3 $^{a-c}$	262.3 ^{e}	266.0 ^{d,e}
FYM biochar (0.1%)	143.7 $^{c-f}$	146.7 $^{b-e}$	261.7 ^{e,f}	265.7 ^{d,e}
FYM biochar (0.2%)	147.0 $^{b-e}$	154.3 ^{a,b}	273.3 ^{b,c}	278.0 ^{b,c}
Wheat straw biochar (0.1%)	146.3 $^{b-e}$	149.7 $^{a-d}$	271.0 ^{c,d}	276.3 ^{b,c}
Wheat straw biochar (0.2%)	153.0 ^{a,b}	157.0 ^{a}	279.0 ^{a,b}	286.0 ^{a}
LSD (*p* ≤ 0.05)	8.9702		7.2668	
p value — PGPR	0.0174		0.0022	
p value — Biochar	0.0003		0.0000	
p value — PGPR + Biochar	0.9774		0.9678	

Table 5. *Cont.*

Treatment	Un-Inoculated	Inoculated	Un-Inoculated	Inoculated
	Plant Height (cm)		Shoot Fresh Biomass (g pot^{-1})	
	Shoot Dry Biomass (g pot^{-1})		1000-Grain Weight (g)	
Control	79.67 ^{g}	80.33 ^{f,g}	222.33 ^{g}	232.67 ^{e,f}
Egyptian acacia biochar (0.1%)	84.33 $^{d-f}$	83.67 $^{e-g}$	229.00 ^{f,g}	237.67 ^{d,e}
Egyptian acacia biochar (0.2%)	85.00 ^{d,e}	87.67 $^{c-e}$	236.00 $^{b-f}$	243.33 $^{b-d}$
FYM biochar (0.1%)	84.33 $^{d-f}$	87.00 $^{c-e}$	234.00 ^{e,f}	242.67 $^{b-d}$
FYM biochar (0.2%)	88.67 ^{c,d}	95.00 ^{a,b}	243.00 $^{b-d}$	248.33 ^{a,b}
Wheat straw biochar (0.1%)	88.67 ^{c,d}	93.67 ^{a,b}	239.33 $^{c-e}$	246.33 $^{a-c}$
Wheat straw biochar (0.2%)	91.33 ^{b,c}	97.00 ^{a}	247.67 ^{a,b}	251.33 ^{a}
LSD ($p \leq 0.05$)	4.5478		7.6490	
p value — PGPR	0.0008		0.0000	
Biochar	0.0000		0.0000	
PGPR + Biochar	0.2527		0.8961	

Values sharing same letter(s) with in a parameter are statistically non-significant with each other at 5% level of probability; values are the mean of three replications ± SE.

Table 6. Effects of biochar with and without *Bacillus* sp. ZM20 on grain yield, and nitrogen, phosphorus, and potassium concentrations in grain of maize in field trial.

Treatment	Un-Inoculated	Inoculated	Un-Inoculated	Inoculated
	Grain Yield (t ha^{-1})		Nitrogen Conc. in Grains (%)	
Control	7.40 ^{f}	7.90 ^{d,e}	2.20 ^{i}	2.24 $^{g-i}$
Egyptian acacia biochar (0.1%)	7.57 ^{e,f}	8.23 ^{c,d}	2.23 ^{h,i}	2.31 ^{e,f}
Egyptian acacia biochar (0.2%)	8.07 ^{d}	8.60 ^{c}	2.28 ^{f,g}	2.34 $^{b-e}$
FYM biochar (0.1%)	7.93 ^{d,e}	8.57 ^{c}	2.27 $^{f-h}$	2.34 $^{b-e}$
FYM biochar (0.2%)	8.33 ^{c,d}	9.10 ^{b}	2.33 $^{c-e}$	2.38 ^{a,b}
Wheat straw biochar (0.1%)	8.27 ^{c,d}	9.10 ^{b}	2.31 $^{d-f}$	2.35 $^{b-d}$
Wheat straw biochar (0.2%)	8.63 ^{c}	9.63 ^{a}	2.37 $^{a-c}$	2.41 ^{a}
LSD ($p \leq 0.05$)	0.4485		0.0434	
p value — PGPR	0.0000		0.0000	
Biochar	0.0000		0.0000	
PGPR + Biochar	0.6905		0.7951	
	Phosphorus Conc. in Grain (%)		Potassium Conc. in Grain (%)	
Control	0.393 ^{g}	0.403 $^{e-g}$	2.64 ^{i}	2.69 $^{g-i}$
Egyptian acacia biochar (0.1%)	0.400 ^{fg}	0.417 $^{c-f}$	2.68 ^{h,i}	2.74 $^{d-g}$
Egyptian acacia biochar (0.2%)	0.410 $^{d-g}$	0.423 $^{b-d}$	2.71 $^{f-h}$	2.80 $^{b-d}$
FYM biochar (0.1%)	0.407 $^{d-g}$	0.420 $^{b-e}$	2.71 $^{f-h}$	2.78 $^{c-e}$
FYM biochar (0.2%)	0.420 $^{b-e}$	0.437 ^{a,b}	2.76 $^{c-f}$	2.85 ^{a,b}
Wheat straw biochar (0.1%)	0.413 $^{c-f}$	0.430 $^{a-c}$	2.74 $^{e-h}$	2.82 ^{b,c}
Wheat straw biochar (0.2%)	0.430 $^{a-c}$	0.447 ^{a}	2.82 ^{b,c}	2.89 ^{a}
LSD ($p \leq 0.05$)	0.0197		0.0622	
p value — PGPR	0.0004		0.0000	
Biochar	0.0001		0.0000	
PGPR + Biochar	0.9981		0.9359	

Values sharing same letter(s) with in a parameter are statistically non-significant with each other at 5% level of probability; values are the mean of three replications ± SE. Lower case words show difference in treatment means.

The results in Table 6 show that the N concentration in grains of maize was significantly improved due to the separate as well as combined application of all types of biochar and *Bacillus* sp. ZM20, except for Egyptian acacia biochar (0.1%), which gave non-significant improvement in the

N concentration in grains under an un-inoculated set of treatments. A maximum increase (7.6%) in the N concentration in maize grains was observed due to the combined use of wheat straw biochar (0.2%) with *Bacillus* sp. ZM20. The results of the impact of biochar (all treatments), with and without inoculum, on P and K concentrations in maize grains were significantly better than the control in most of the cases. A maximum improvement in both the parameters over the control was observed due to combined use of wheat straw biochar (0.2%) and *Bacillus* sp. ZM20 (Table 6).

4. Discussion

The present study was conducted in arid and semi-arid regions on sandy loam soil characterized by low rainfall and high temperature, associated with low organic matter content. Due to low organic matter, biochar in combination with bacterial inoculation can have the ability to improve the soil health and crop yield under such a scenario. The application of biochar can be effective at rehabilitating degraded lands by improving the soil structure, nutrient- and water-holding capacity, and soil carbon contents, leading to improvement in soil productivity [9,55]. A carbon-rich compound called charcoal is produced through a process known as pyrolysis and has beneficial implications such as soil amendment for improving soil health and crop yield [10,56]. The physicochemical properties of biochar are crucial in determining its functionality and impact on plant growth and soil health [57]. It was observed that biochar contains a high carbon-to-nitrogen ratio (Table 1), which makes it stable against decomposition. The carbon contents of Egyptian acacia biochar were higher compared to the other two sources, but wheat straw biochar had a higher turnover rate as compared to the other sources. In previous studies, scientists have also reported that biochar is rich in carbon contents along with other nutrients like C, N, and S [58,59] which have shown promising results in improving crop growth and yield characteristics similar to the findings of the current study.

In this study, the application of biochar improved the soil biological properties (soil organic matter contents, MBC), along with improvement in ammonium and nitrate N contents in soil (Figures 1 and 3). The increase in the levels of biochar increased the content of organic matter and MBC in studied soil. The presence of high carbon and other nutrients might have helped in the improvement of soil fertility as reported by Oni et al. [17], suggesting that biochar application positively affects the soil structure, water retention capacity, fertility, and soil carbon sequestration, leading to improvement in crop growth and productivity. Similarly, biochar application increased the ratio of below-ground biomass to above-ground biomass due to an increase in water-holding capacity, as reported previously [60], and a reduction in soil strength [61]. The integration of biochar and PGPR is a win-win strategy as biochar provides a niche for microbes due to its microporous structure, which in turn increases microbial activity and hence the sorption of dissolved organic matter [18]. The increase in carbon and organic matter contents in the present study due to the addition of different biochar types is in good agreement with Shenbagavalli and Mahimairaja [62]. The integrated use of biochar and PGPR can improve soil health through increasing soil organic matter contents, enhancing soil aggregation, promoting better microbial activity, and increasing soil fertility [35,36]. In our study, the integrated use of biochar inoculated with *Bacillus* sp. ZM20 was significantly better in improving soil organic matter and MBC, which might have supported crop growth. Our results are in good agreement with previous reports by Ullah et al. [37], in which they reported the increased growth, physiology, and production of crops under the combined application of biochar and PGPR. This increase in growth and yield of wheat in present study might be attributed to enhanced supply of nutrients that are scarcely available in the soil including nitrogen, phosphorus, zinc, and iron. This may also be due to the positive effects of applied PGPR which are well recognized candidates equipped with plenty of mechanisms, i.e., the production of siderophores that helps in iron acquisition, synthesis of plant growth regulators, and exopolysaccharides [15,32,36].

Biochar application as a soil amendment increases the growth parameters of plants (plant root and shoot growth), and their nutrient uptake by improving the water status of plants and water-use efficiency [63,64], thus leading to improved yield of crop plants. In the present study, the application of biochar from different sources improved the maize root and shoot growth and nutrient uptake, along with the yield and yield contributing factors (Tables 3 and 5). A maximum increase (25.77%) in grain yield was observed in the treatment where wheat straw biochar (0.2%) was applied in combination with *Bacillus* sp. ZM20. This might be due to the enhanced water-holding capacity of the soil [65] that resulted in enhanced nutrient availability [65], thus improving the growth of crop plants under the applied biochar [38] and PGPR [28]. As stated by Hussain et al. [31], the combined use of PGPR and biochar at the rate of 0.5 tons/ha have shown enhanced water-holding capacity of the soil, and hence the growth and yield of maize (*Zea mays* L.). Recently, Shen et al. [66] reported that biochar application improved plant growth; however, willow woodchip biochar was significantly better than pine-based biochar in improving plant growth and nutrient uptake of *Lotus pedunculatus*. The improved growth and yield characteristics of maize under the applied biochar are in good agreement with previous studies [14,67,68]. The enhanced soil characteristics and crop growth responses in the present study under the application of biochar and PGPR might be attributed to the differences in soil characteristics and the alkaline nature of biochar in the soil studied here.

The PGPR inhabits either the rhizosphere, the soil in the immediate vicinity of plant roots, or inside the plant tissues, and helps the plants exhibit better growth through some direct and indirect mechanisms [20,21]. The phosphorus-solubilizing bacteria (PSB), zinc-solubilizing bacteria (ZSB), and potassium-solubilizing bacteria (KSB) can increase plant nutrient availability along multiple plant growth promoting traits, such as siderophores production, chitin decomposition, hydrogen cyanide production, and ammonia production [25,26,69]. In the current study, the combined use of biochar and *Bacillus* sp. ZM20 improved maize growth, the uptake of N, P, and K, and the yield, which might be due to solubilization of nutrients through acid production, along with other growth-promoting characteristics such as siderophore production, exopolysaccharides production, and HCN production exhibited by this strain, as reported in previous studies [14,25]. The application of biochar improves the quality of soil and makes it conducive for better microbial activity [70]. Previous studies have reported that the integrated use of biochar and *Pseudomonas fluorescens* enhanced the growth of cucumber by improving plant–water relations under water deficit conditions. It has been reported that PGPR effectively colonize plant rhizosphere, thus helping in improving the growth, yield, and nutrient acquisition [29]. One possible reason behind increased uptake of N, P, and K in the present study (Tables 4 and 6) might be due to the promoting effects of PGPR and the applied biochar, which resulted in enhanced nutrient use efficiency, as has been reported previously [19,28,36]. Moreover, the presence of biochar in addition to PGPR might have helped to increase the sorption capacity of soil, resulting in higher mineral (NPK) concentration in wheat grains (Tables 4 and 6). These results are substantiated with those reported previously [37].

5. Conclusions

Low organic matter and depleted nutrients are the major issues of agricultural soils in arid and semi-arid regions. In the present study, the application of biochar from different sources significantly improved soil biological properties, growth, yield, and quality of maize grains. The integrated use of biochar and *Bacillus* sp. ZM20 was more effective as compared to the separate application. Biochar application along with *Bacillus* sp. ZM20 also improved soil biological properties, i.e., soil organic matter, MBC. Moreover, the biochar source and rate also influenced the soil properties and plant growth with different degrees of efficacy. The use of wheat straw biochar along with inoculated with the *Bacillus* sp. ZM20 bacterial strain was better as compared to farm-yard manure biochar and Egyptian acacia biochar. It is concluded that the combined application of wheat straw biochar (0.2%) and *Bacillus* sp. ZM20 was the most effective treatment in improving the soil properties, plant growth, yield, and quality of maize crops as compared to all other treatments in the pot and field trials.

Author Contributions: Conceptualization, M.A. and A.M.; methodology, M.A., A.H. and Z.A.Z.; software, H.A.M. and X.W.; validation, Z.A., Q.S., X.W. and A.M.; formal analysis, A.H., Q.S. and F.N.; investigation, M.L. (Muhammad Latif), A.M., H.A.M. and M.A.; resources, A.M.; data curation, M.L. (Muhammad Luqman), M.A. and F.N.; writing—original draft preparation, M.A.; writing—review and editing, A.M., X.W., T.H.H. and Z.A.Z.; visualization, Z.A.Z., A.H., T.H.H. and A.M.; supervision, Z.A.Z. All authors have read and agreed to the published version of the manuscript.

Acknowledgments: The authors acknowledge the Soil Microbiology and Biotechnology Laboratory, Department of Soil Science, and the Islamia University of Bahawalpur, Pakistan for providing research facilities. This work was supported by the Higher Education Commission of Pakistan (grant number SRGP-785).

References

1. United Nations. The World Population Prospects: The 2017 Revision, Published by the UN Department of Economic and Social Affairs. 2017. Available online: https://www.un.org/development/desa/en/news/population/world-population-prospects-2017.html (accessed on 2 April 2020).

2. Peerzado, M.B.; Magsi, H.; Sheikh, M.J. Land use conflicts and urban sprawl: Conversion of agriculture lands into urbanization in Hyderabad, Pakistan. *J. Saudi Soc. Agric. Sci.* **2019**, *18*, 423–428. [CrossRef]

3. Shrivastava, P.; Kumar, R. Soil salinity: A serious environmental issue and plant growth promoting bacteria as one of the tools for its alleviation. *Saudi J. Biol. Sci.* **2015**, *22*, 123–131. [CrossRef] [PubMed]

4. Nuss, E.T.; Tanumihardjo, S.A. Maize: A paramount staple crop in the context of global nutrition. *Compr. Rev. Food Sci. Food Saf.* **2010**, *9*, 417–436. [CrossRef]

5. Wang, Y.; Gao, F.; Gao, G.; Zhao, J.; Wang, X.; Zhang, R. Production and cultivated area variation in cereal, rice, wheat and maize in China (1998–2016). *Agronomy* **2019**, *9*, 222. [CrossRef]

6. Tandzi, L.N.; Mutengwa, C.S. Estimation of maize (*Zea mays* L.) yield per harvest area: Appropriate methods. *Agronomy* **2020**, *10*, 29. [CrossRef]

7. Das, A.; Patel, D.; Munda, G.C.; Ghosh, P.K. Effect of organic and inorganic sources of nutrients on yield, nutrient uptake and soil fertility of maize (*Zea mays*)—Mustard (*Brassica campestris*) cropping system. *Indian J. Agric. Sci.* **2010**, *80*, 85–88.

8. Thilakarathna, M.S.; Raizada, M.N. A review of nutrient management studies involving finger millet in the semi-arid tropics of Asia and Africa. *Agronomy* **2015**, *5*, 262–290. [CrossRef]

9. Cybulak, M.; Sokołowska, Z.; Boguta, P. Impact of biochar on physicochemical properties of haplic luvisol soil under different land use: A plot experiment. *Agronomy* **2019**, *9*, 531. [CrossRef]

10. Glaser, B.; Wiedne, K.; Seeling, S.; Schmidt, H.P.; Gerber, H. Biochar organic fertilizers from natural resources as substitute for mineral fertilizers. *Agron. Sustain. Dev.* **2015**, *35*, 667–678. [CrossRef]

11. Duku, H.M.; Gu, S.; Hagan, E.B. Biochar production potentials in Ghana-a review. *Renew. Sustain. Energy Rev.* **2011**, *15*, 3539–3551. [CrossRef]

12. Ding, Y.; Liu, Y.; Liu, S.; Huang, X.; Li, Z.; Tan, X.; Zeng, G.; Zhou, L. Potential benefits of biochar in agricultural soils: A review. *Pedosphere* **2017**, *27*, 645–661. [CrossRef]

13. Palansooriya, K.N.; Ok, Y.S.; Award, Y.M.; Lee, S.S.; Sung, J.K.; Kautsospyros, A.; Moon, D.H. Impact of biochar application on upland agriculture: A review. *J. Environ. Manag.* **2019**, *234*, 52–64. [CrossRef] [PubMed]

14. Naveed, M.; Ramzan, N.; Mustafa, A.; Samad, A.; Niamat, B.; Yaseen, M.; Ahmad, Z.; Hasanuzzaman, M.; Sun, N.; Shi, W.; et al. Alleviation of salinity induced oxidative stress in *Chenopodium quinoa* by Fe biofortification and biochar-endophyte interaction. *Agronomy* **2020**, *10*, 168. [CrossRef]

15. Sabir, A.; Naveed, M.; Bashir, M.A.; Hussain, A.; Mustafa, A.; Zahir, Z.A.; Kamran, M.; Ditta, A.; Núñez-Delgado, A.; Saeed, Q.; et al. Cadmium mediated phytotoxic impacts in *Brassica napus*: Managing growth, physiological and oxidative disturbances through combined use of biochar and *Enterobacter* sp. MN17. *J. Environ. Manag.* **2020**, *265*, 110522. [CrossRef] [PubMed]

16. Awad, Y.M.; Lee, S.S.; Kim, K.H.; Ok, Y.S.; Kuzyakov, Y. Carbon and nitrogen mineralization and enzyme activities in soil aggregate-size classes: Effects of biochar, oyster shells, and polymers. *Chemosphere* **2018**, *198*, 40–48. [CrossRef] [PubMed]

17. Oni, B.A.; Oziegbe, O.; Olawole, O.O. Significance of biochar application to the environment and economy. *Ann. Agric. Sci.* **2019**, *64*, 222–236. [CrossRef]

18. Hameed, A.; Hussain, S.A.; Yang, J.; Ijaz, M.U.; Liu, Q.; Suleria, H.A.R.; Song, Y. Antioxidants potential of the filamentous fungi (*Mucor circinelloides*). *Nutrients* **2017**, *9*, 1101. [CrossRef]

19. Bargaz, A.; Karim, L.; Chtouki, M.; Zeroual, Y.; Driss, D. Soil microbial resources for improving fertilizers efficiency in an integrated plant nutrient management system. *Front. Microbiol.* **2018**, *9*, 1606. [CrossRef]

20. Santoyo, G.; Moreno-Hagelsieb, G.; Orozco-Mosqueda, M.C.; Glick, B.R. Plant growth-promoting bacterial endophytes. *Microbiol. Res.* **2016**, *183*, 92–99. [CrossRef]

21. Ahmad, M.; Nadeem, S.M.; Zahir, Z.A. Plant-microbiome interactions in agroecosystem: An application. In *Microbiome in Plant Health and Disease*; Kumar, V., Ed.; Springer Nature: Singapore, 2019; pp. 251–291.

22. Hussain, A.; Zahir, Z.A.; Asghar, H.N.; Ahmad, M.; Jamil, M.; Naveed, M.; Akhtar, M.F.Z. Zinc solubilizing bacteria for zinc biofortification in cereals: A step towards sustainable nutritional security. In *Role of Rhizospheric Microbes in Soil. Volume 2: Nutrient Management and Crop Improvement*; Meena, V.S., Ed.; Springer: New Delhi, India, 2018; pp. 203–227.

23. Mumtaz, M.Z.; Barry, K.M.; Baker, A.L.; Nichols, D.S.; Ahmad, M.; Zahir, Z.A.; Britz, M.L. Production of lactic and acetic acids by *Bacillus* sp. ZM20 and *Bacillus cereus* following exposure to zinc oxide: A possible mechanism for Zn solubilization. *Rhizosphere* **2019**, *12*, 100170. [CrossRef]

24. Ahmad, M.; Ahmad, I.; Hilger, T.H.; Nadeem, S.M.; Akhtar, M.F.; Jamil, M.; Hussain, A.; Zahir, Z.A. Preliminary study on phosphate solubilizing *Bacillus subtilis* strain Q3 and *Paenibacillus* sp. strain Q6 for improving cotton growth under alkaline conditions. *PeerJ* **2018**, *6*, e5122. [CrossRef] [PubMed]

25. Mumtaz, M.Z.; Ahmad, M.; Jamil, M.; Asad, S.A.; Hafeez, F. *Bacillus* strains as potential alternate for zinc biofortification of maize grains. *Int. J. Agric. Biol.* **2018**, *20*, 1779–1786.

26. Saha, M.; Maurya, B.R.; Meena, V.S.; Bahadur, I.; Kumar, A. Identification and characterization of potassium solubilizing bacteria (KSB) from Indo-Gangetic Plains of India. *Biocatal. Agric. Biotechnol.* **2016**, *7*, 202–209. [CrossRef]

27. Ahmad, M.; Naseer, I.; Hussain, A.; Mumtaz, M.Z.; Mustafa, A.; Hilger, T.H.; Zahir, Z.A.; Minggang, X. Appraising endophyte—Plant symbiosis for improved growth, nodulation, nitrogen fixation and abiotic stress tolerance: An experimental investigation with chickpea (*Cicer arietinum* L.). *Agronomy* **2019**, *9*, 621. [CrossRef]

28. Ali, M.A.; Naveed, M.; Mustafa, A.; Abbas, A. The good, the bad, and the ugly of rhizosphere microbiome. In *Probiotics and Plant Health*; Springer: Singapore, 2017; pp. 253–290.

29. Nazli, F.; Najm-ul-Seher; Khan, M.Y.; Jamil, M.; Nadeem, S.M.; Ahmad, M. Soil microbes and plant health. In *Plant Disease Management Strategies for Sustainable Agriculture through Traditional and Modern Approaches, Sustainability in Plant and Crop Protection*; IUl Haq, I., Ijaz, S., Eds.; Springer Nature: Basel, Switzerland, 2020; pp. 111–135.

30. Naseer, I.; Ahmad, M.; Nadeem, S.M.; Ahmad, I.; Najm-ul-Seher; Zahir, Z.A. Rhizobial inoculants for sustainable agriculture: Prospects and applications. In *Biofertilizers for Sustainable Agriculture and Environment, Soil Biology*; Giri, B., Ed.; Springer Nature: Basel, Switzerland, 2019; pp. 245–284.

31. Hussain, A.; Ahmad, M.; Mumtaz, M.Z.; Ali, S.; Sarfraz, R.; Naveed, M.; Jamil, M.; Damalas, C.A. Integrated application of organic amendments with *Alcaligenes* sp. AZ9 improves nutrient uptake and yield of maize (*Zea mays*). *J. Plant Growth Regul.* **2020**, *9*, 1–16. [CrossRef]

32. Khan, N.; Bano, A. Exopolysaccharide producing rhizobacteria and their impact on growth and drought tolerance of wheat grown under rainfed conditions. *PLoS ONE* **2019**, *14*, e0222302. [CrossRef] [PubMed]

33. Pramanik, K.; Mitra, S.; Sarkar, A.; Soren, T.; Maiti, T.K. Characterization of cadmium resistant *Klebsiella pneumoniae* MCC 3091 promoted rice seedling growth by alleviating phytotoxicity of cadmium. *Environ. Sci. Pollut. Res.* **2017**, *24*, 24419–24437. [CrossRef]

34. Saeed, Z.; Naveed, M.; Imran, M.; Bashir, M.A.; Sattar, A.; Mustafa, A.; Hussain, A.; Xu, M. Combined use of *Enterobacter* sp. MN17 and zeolite reverts the adverse effects of cadmium on growth, physiology and antioxidant activity of *Brassica napus*. *PLoS ONE* **2019**, *14*, e0213016. [CrossRef]

35. Ijaz, M.; Tahir, M.; Shahid, M.; Ul-Allah, S.; Sattar, A.; Sher, A.; Mahmood, K.; Hussain, M. Combined application of biochar and PGPR consortia for sustainable production of wheat under semiarid conditions with a reduced dose of synthetic fertilizer. *Braz. J. Microbiol.* **2019**, *50*, 449–458. [CrossRef]

36. Hussain, A.; Ahmad, M.; Mumtaz, M.Z.; Nazli, F.; Farooqi, M.A.; Khalid, I.; Iqbal, Z.; Arshad, H. Impact of integrated use of enriched compost, biochar, humic acid and *Alcaligenes* sp. AZ9 on maize productivity and soil biological attributes in natural field conditions. *Ital. J. Agron.* **2019**, *14*, 101–107. [CrossRef]

37. Ullah, N.; Ditta, A.; Khalid, A.; Mehmood, S.; Rizwan, M.S.; Ashraf, M.; Mubeen, F.; Imtiaz, M.; Iqbal, M.M. Integrated effect of algal biochar and plant growth promoting rhizobacteria on physiology and growth of maize under deficit irrigations. *J. Soil Sci. Plant Nutr.* **2019**, *20*, 346–356. [CrossRef]

38. Naeem, M.A.; Khalid, M.; Ahmad, Z.; Naveed, M. Low pyrolysis temperature biochar improve growth and nutrient availability of maize on typic Calciargid. *Commun. Soil Sci. Plant Anal.* **2015**, *47*, 41–51. [CrossRef]

39. Al-Wabel, M.I.; Al-Omran, A.; El-Naggar, A.H.; Nadeem, M.; Usman, A.R.A. Pyrolysis temperature induced changes in characteristics and chemical composition of biochar produced from conocarpus wastes. *Bioresour. Technol.* **2013**, *131*, 374–379. [CrossRef] [PubMed]

40. Nelson, D.W.; Sommers, L.E. Total carbon, organic carbon, and organic matter. In *Methods of Soil Analysis. Part 3. Chemical Methods. Soil Science of America and American Society of Agronomy*; Black, C.A., Ed.; ACSESS: Madison, WI, USA, 1996; pp. 961–1010.

41. Ryan, J.; Estefan, G.; Rashid, A. *Soil and Plant Analysis Laboratory Manual*, 2nd ed.; International Center for Agriculture in Dry Areas (ICARDA): Aleppo, Syria, 2001; p. 172.

42. Jackson, M.L. *Soil Chemical Analysis*; Prentice Hall Inc.: New York, NY, USA, 1962.

43. Nelson, D.W.; Sommers, L.E. Total carbon, organic carbon and organic matter. In *Methods of Soil Analysis. Part 2: Chemical and Microbiological Properties*; Agronomy Monographs; SSSA: Madison, WI, USA, 1982; pp. 570–571.

44. Watanabe, F.S.; Olsen, S.R. Test of an ascorbic acid method for determining phosphorus in water and NaHCO$_3$ extracts from soil. *Soil Sci. Soc. Am. Proc.* **1965**, *29*, 677–678. [CrossRef]

45. Sarfraz, M.; Ashraf, Y.; Ashraf, S. A Review: Prevalence and antimicrobial susceptibility profile of listeria species in milk products. *Matrix Sci. Media* **2017**, *1*, 3–9. [CrossRef]

46. Ryan, J. *Methods of Soil, Plant, and Water Analysis: A Manual for the West Asia and North Africa Region*; International Center for Agricultural Research in the Dry Areas (ICARDA): Beirut, Lebanon, 2017.

47. Ayers, R.S.; Westcot, D.W. *Water Quality for Agriculture, FAO Irrigation and Drainage Papers 29 (Rev.1)*; Food and Agriculture Organization of the United Nations: Rome, Italy, 1994; pp. 1–11.

48. Wolf, B. The comprehensive system of leaf analysis and its use for diagnosing crop nutrient status. *Commun. Soil Sci. Plant Anal.* **1982**, *13*, 1035–1059. [CrossRef]

49. Bullock, D.; Moore, K. Protein and Fat Determination in Corn. In *Seed Analysis. Modern Methods of Plant Analysis*; Linskens, H.F., Jackson, J.F., Eds.; Springer: Berlin/Heidelberg, Germany, 1992; Volume 14, pp. 181–197.

50. Jenkinson, D.S.; Ladd, J.N. *Microbial Biomass in Soil, Measurement and Turn Over*; Marcel Dekker: New York, NY, USA, 1981.

51. Bremner, E.; Kessel, V. Extractability of microbial ^{14}C and ^{15}N following addition of variable rates of labeled glucose and ammonium sulphate to soil. *Soil Biol. Biochem.* **1990**, *22*, 707–713. [CrossRef]

52. Kamphake, L.J.; Hannah, S.A.; Cohen, J.M. Automated analysis for nitrate by hydrazine reduction. *Water Res.* **1967**, *1*, 205–216. [CrossRef]

53. Sims, J.R.; Jackson, D.G. Rapid analysis of soil nitrate with chromotropic acid. *Soil Sci. Soc. Am. J.* **1971**, *35*, 603–606. [CrossRef]

54. Steel, R.G.D.; Torrie, J.H.; Dickey, D.A. Principles and Procedures of Statistics. In *A Biometrical Approach*, 3rd ed.; McGraw Hill Book Co.: New York, NY, USA, 2007.

55. Bashir, M.A.; Naveed, M.; Ahmad, Z.; Gao, B.; Mustafa, A.; Núñez-Delgado, A. Combined application of biochar and sulfur regulated growth, physiological, antioxidant responses and Cr removal capacity of maize (*Zea mays* L.) in tannery polluted soils. *J. Environ. Manag.* **2020**, *259*, 110051. [CrossRef]

56. Fidel, R.B.; Laird, D.A.; Thompson, M.L.; Lawrinenko, M. Characterization and quantification of biochar alkalinity. *Chemosphere* **2017**, *167*, 367–373. [CrossRef] [PubMed]

57. Zimmerman, R.A.; Gao, B.; Ahn, M.Y. Positive and negative carbon mineralization priming effects among a variety of biochar-amended soils. *Soil Biol. Biochem.* **2011**, *43*, 1169–1179. [CrossRef]

58. Novak, J.M.; Cantrell, K.B.; Watts, D.W. Compositional and thermal evaluation of lignocellulosic and poultry litter chars via high and low temperature pyrolysis. *Bioenergy Res.* **2013**, *6*, 114–130. [CrossRef]

59. Mierzwa-Hersztek, M.; Gondek, K.; Limkowicz-Pawlas, A.; Baran, A.; Bajda, T. Sewage sludge biochars management-ecotoxicity, mobility of heavy metals, and soil microbial biomass. *Environ. Toxicol. Chem.* **2017**, *37*, 1197–1207. [CrossRef] [PubMed]

60. Karhu, K.; Mattila, T.; Irina, B.; Regina, K. Biochar addition to agricultural soil increased CH$_4$ uptake and water holding capacity—Results from a short-term pilot field study. *Agric. Ecosyst. Environ.* **2011**, *140*, 309–313. [CrossRef]

61. Steinbeiss, S.; Gleixner, G.; Antonietti, M. Effect of biochar amendment on soil carbon balance and soil microbial activity. *Soil Biol. Biochem.* **2009**, *41*, 1301–1310. [CrossRef]

62. Shenbagavalli, S.; Mahimairaja, S. Characterization and effect of biochar on nitrogen and carbon dynamics in soil. *Int. J. Adv. Biol. Res.* **2012**, *2*, 249–255.

63. Haider, G.; Koyro, H.W.; Azam, F.; Steffens, D.; Müller, C.; Kam-mann, C. Biochar but not humic acid product amendment affected maize yields via improving plant-soil moisture relations. *Plant Soil* **2015**, *395*, 141–157. [CrossRef]

64. Bruun, E.W.; Petersen, C.T.; Hansen, E.; Holm, J.K.; Hauggaard-Nielsen, H. Biochar amendment to coarse sandy subsoil improves root growth and increases water retention. *Soil Use Manag.* **2014**, *30*, 109–118. [CrossRef]

65. Rogovska, N.; Laird, D.A.; Rathke, S.J.; Karlen, D.L. Biochar impact on midwestern mollisols and maize nutrient availability. *Geoderma* **2014**, *23*, 340–347. [CrossRef]

66. Shen, Q.; Hedley, M.; Arbestain, M.C.; Kirschbaum, M.U.F. Can biochar increase the bioavailability of phosphorus? *J. Soil Sci. Plant Nutr.* **2016**, *16*, 268–286. [CrossRef]

67. Kamran, M.; Malik, Z.; Parveen, A.; Zong, Y.; Abbasi, G.H.; Rafiq, M.T.; Shaaban, M.; Mustafa, A.; Bashir, S.; Rafay, M.; et al. Biochar alleviates Cd phytotoxicity by minimizing bioavailability and oxidative stress in pak choi (*Brassica chinensis* L.) cultivated in Cd-polluted soil. *J. Environ. Manag.* **2019**, *250*, 109500. [CrossRef]

68. Naveed, M.; Mustafa, A.; Azhar, A.Q.; Kamran, M.; Zahir, Z.A.; Núñez-Delgado, A. *Burkholderia phytofirmans* PsJN and tree twigs derived biochar together retrieved Pb-induced growth, physiological and biochemical disturbances by minimizing its uptake and translocation in mung bean (*Vigna radiata* L.). *J. Environ. Manag.* **2020**, *257*, 109974. [CrossRef]

69. Mustafa, A.; Naveed, M.; Saeed, Q.; Ashraf, M.N.; Hussain, A.; Abbas, T.; Kamran, M.; Minggang, X. Application potentials of plant growth promoting rhizobacteria and fungi as an alternative to conventional weed control methods. In *Crop Production*; IntechOpen: London, UK, 2019.

70. Yang, S.; Chen, X.; Jiang, Z.; Ding, J.; Sun, X.; Xu, J. Effects of biochar application on soil organic carbon composition and enzyme activity in paddy soil under water-saving irrigation. *Int. J. Environ. Res. Public Health* **2020**, *17*, 333. [CrossRef]

Large Scale Screening of Rhizospheric Allelopathic Bacteria and their Potential for the Biocontrol of Wheat-Associated Weeds

Tasawar Abbas [1], Zahir Ahmad Zahir [1], Muhammad Naveed [1,*], Sana Abbas [2], Mona S. Alwahibi [3], Mohamed Soliman Elshikh [3] and Adnan Mustafa [4]

[1]   Soil Microbiology and Biochemistry Laboratory, Institute of Soil and Environmental Sciences, University of Agriculture, Faisalabad 38040, Pakistan; tasawarabbasuaf@gmail.com (T.A.); zazahir@yahoo.com (Z.A.Z.)

[2]   Department of Chemistry, Government College Women University Faisalabad, Faisalabad 38040, Pakistan; sanabas.gc@gmail.com

[3]   Department of Botany and Microbiology, College of Science, King Saud University, Riyadh 11451, Saudi Arabia; malwahibi@ksu.edu.sa (M.S.A.); Melshikh@ksu.edu.sa (M.S.E.)

[4]   National Engineering Laboratory for Improving Quality of Arable Land, Institute of Agricultural Resources and Regional Planning, Chinese Academy of Agricultural Sciences, Beijing 100081, China; adnanmustafa780@gmail.com

*   Correspondence: muhammad.naveed@uaf.edu.pk.

Abstract: Conventional weed control practices have generated serious issues related to the environment and human health. Therefore, there is a demand for the development of alternative techniques for sustainable agriculture. The present study performed a large-scale screening of allelopathic bacteria from the rhizosphere of weeds and wheat to obtain biological weed control inoculants in the cultivation of wheat. Initially, around 400 strains of rhizobacteria were isolated from the rhizosphere of weeds as well as wheat that grows in areas of chronic weed invasions. A series of the screen was performed on these strains, including the release of phytotoxic metabolites, growth inhibition of sensitive *Escherichia coli*, growth inhibition of indicator plant of lettuce, agar bioassays on five weeds, and agar bioassay on wheat. Firstly, 22.6% (89 strains) of the total strains were cyanogenic, and among the cyanogenic strains, 21.3% (19 strains) were inhibitory to the growth of sensitive *E. coli*. Then, these 19 strains were tested using lettuce seedling bioassay to show that eight strains suppressed, nine strains promoted, and two strains remained ineffective on the growth. These 19 strains were further applied to weeds and wheat on agar bioassays. The results indicated that dry matter of broad-leaved dock, wild oat, little seed canary grass, and common lambs' quarter were reduced by eight strains (23.1–68.1%), seven strains (38.5–80.2%), eight strains (16.5–69.4%), and three strains (27.5–50.0%), respectively. Five strains suppressed the growth of wheat, nine strains increased its dry matter (12.8–47.9%), and five remained ineffective. Altogether, the strains that selectively inhibit weeds, while retaining normal growth of wheat, can offer good opportunities for the development of biological weed control in the cultivation of wheat.

Keywords: allelopathic bacteria; antimetabolites; biological control; phytotoxic metabolites; rhizobacteria; weed invasion

1. Introduction

Dramatic increases in food production have been observed in the latter half of the twentieth century owing to the use of agro-chemicals, mechanization, irrigation, high yielding varieties, and post-harvest technology. The production of wheat in Pakistan has increased to ~25 m ton from 4.55 m ton in 1965 [1,2]. The pest attacks continue to incur losses to crop production owing to the diversity of pests and their resistance to prevailing control practices. The use of pesticides has increased from 15 to 20-fold over the last fifty years [3]. Chemical herbicides have gained importance in crop production in the face of a shortage of labor and limited application of mechanical control [4]. The mechanical control is known to contribute to soil erosion and its degradation [5]. Herbicides have led to the emergence of resistant biotypes of weeds, making the herbicide compounds useless to control these weeds [6]. Hence, the discovery of new compounds with novel modes of action is needed to replace these herbicides with more effective compounds to control such weeds. The discovery of such compounds, having herbicidal properties, has reduced over time. Further, the control of one type of weeds with herbicides has provided space to the proliferation of other weed species, which were less problematic for crop production in the past [7]. They have caused losses of biodiversity in the environment. It has deprived the ecosystems of some of their vital functions. Herbicides have aggravated the loss of biodiversity by killing the susceptible species, restricting the growth of others and the degradation of natural resources [8]. Poisoning, growth retardation, sterility, and deaths of wildlife owing to herbicide exposure have been reported by [9]. The residues of herbicides, apart from polluting the natural resources and destroying life forms, may also accumulate in the edible portions of plants, which facilitate their entry to the food chain and bodies of humans. It causes poisoning and chronic diseases in human beings, leading to deaths [10]. Human health disorders caused by herbicides include disorders of the nervous system, malformation of the embryo, loss of fertility, loss of immunity, kidney disorders, and liver disorders [11].

Farmers pay only the costs of manufacturing and marketing of herbicides, which provides economic access to farmers to adopt chemical weed control. The additional costs incurred on the treatment of human illnesses, degradation of natural resources and environment, and loss of biodiversity also need to be paid by farmers, society, or governments. Hence, the scenario of economic, environmental, and biological costs of chemical weed control pushes the researchers towards finding out safer weed control techniques. The importance of biological control has dramatically increased in the present situation. It presents a safer, inexpensive, and easier solution to the above-discussed issues of other control practices. It relies on increasing the strength, population, and activities of the organisms, resulting in growth reduction of weeds [12].

The past efforts in this area were focused on pathogens causing diseases in weeds [13] and insects feeding on weeds [14]. The success of insect biocontrol agents is limited by the existence of multiple hosts of insects in nature, which may cause the emergence of new pests of crops [15]. The pathogens of weeds used for biocontrol wait for suitable environmental conditions to cause infections and diseases in weed plants [13]. It may usually lead to delayed disease development, even after the weeds have caused economic losses of crops. Plant allelochemicals have also been investigated for biological weed control [16]. Their efficacy for weed control is reduced owing to the soil reactions, biodegradation, and mobility. It reduces their bioavailability and phytotoxicity on weeds [17]. These limitations of conventional biological weed control have discouraged researchers of this field, and the popularity of chemical weed control has increased dramatically.

The low success rate in conventional biological weed control has driven scientists to explore the characteristics of the rhizosphere inhabiting bacteria of weeds and crops for the development of novel weed biocontrol techniques. However, researchers have made efforts to explore the type of rhizobacteria, which produce substances inhibitory to the growth of weeds and are the least explored candidates for biological weed control. They release their secondary metabolites (phytotoxic in nature) in the rhizosphere, which is followed by their absorption in weeds. It results in a growth reduction of these weeds. The nature of this interaction between plants and microorganisms may be termed as plant-microbe allelopathy, and the bacteria responsible for these interactions may be called as allelopathic bacteria (AB) [18]. The discovery of host specificity in such microbial interactions with plants by [19] has opened ways for their potential application in crops for weed control. It reflects the properties of non-inhibition or even promotion of growth of crops among these rhizobacteria [20]. Therefore, the present study was conducted to explore such bacteria from the rhizosphere of weeds and wheat growing in fields facing weed invasions chronically, characterize them for the biological weed control, and evaluate their effects on the growth of wheat and weeds species of wheat.

2. Materials and Methods

2.1. Isolation of Rhizobacteria

We collected a large pool of samples of wheat and five weeds along with earth ball across the District of Faisalabad, Punjab, Pakistan. The sampling field was selected based on chronic weed invasions over the last 5 years. The weed species sampled included field bindweed, little seed canary grass, common lambs' quarter, wild oat, and broad-leaved dock. The scientific names of these weeds are *Convolvulus arvensis*, *Phalaris minor*, *Chenopodium album*, *Avena fatua*, and *Rumex dentatus*, respectively. These samples were transferred to the laboratory in an icebox and stored at 4 °C. The rhizosphere soil of these samples was used for the isolation of rhizobacteria using the dilution plating technique. A hundred microliters of each of the serial dilutions (10^{-1}–10^{-8}) were spread on the sterilized King's B agar media in Petri plates aseptically. This media was prepared by adding 1.5-g K_2HPO_4, 10 mL glycerol, 20 gm proteose peptone, 1.5 gm $MgSO_4.7H_2O$, and 20-g agar and making up the volume of one liter with distilled water following King et al. [21]. The growth of rhizobacterial colonies was obtained after 48 h of incubation of these plates at 28 ± 1 °C. The fast-growing colonies were picked and transferred to other Petri plates containing sterilized King's B agar media. These colonies were, hence, purified after some streaking. In this way, 393 strains were purified and preserved at −20 °C in 40% glycerol.

2.2. Cyanide Production Assay on Strains of Rhizobacteria

The method given by Bakker and Schipper [22] was followed for the qualitative determination of the production of hydrogen cyanide (HCN) by the isolated strains of rhizobacteria. The pieces of filter paper to the sizes of Petri plates were made, autoclaved for sterilization, and soaked in a 1% solution of picric acid for 12 h. These soaked filter papers were dried aseptically. Glycine amended media was prepared by adding 0.35 gm K_2HPO_4, 2.5 mL glycerol, 5 gm proteose peptone, 0.35 gm $MgSO_4.7H_2O$, 5 gm glycine, and 20-g agar and making up the volume to one liter with distilled water. It gave out quarter strength media with glycine amendment. It was autoclaved and poured in Petri plates. The fresh culture of the strains was used to make a layer on the surface of the media and placing the picric acid-soaked paper on the inner side of the Petri plate lid. The paper was fastened with the help of a 10% solution of Na_2CO_3. The plates were closed and tightened with parafilm to avoid the leakage of gas. The plates were incubated at 28 °C and periodically observed for a change in the color of filter paper. The turning of color to brown indicated the production of HCN, while the intensity of brown color indicated the level of its production (Figure 1).

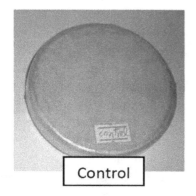 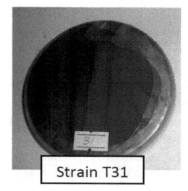 

Control   Strain T31   Strain T35

Figure 1. Pictorial view of cyanide production by rhizobacteria.

2.3. Antimetabolite Assay on E. coli

The bacterial production of toxic metabolites in extracellular spaces can be tested in a simple test based on the growth retardation of sensitive bacteria, *E. coli* [23]. All the strains (393) were tested for HCN production, while this assay was performed on only those strains that produced HCN to any level in step 1. These were 89 strains. Strain K12 of *E. coli* was cultured on LB agar media and placed in an incubator at 28 °C. After 2 days, the gentle rubbing of the surface and mixing with sterilized 0.01 M $MgSO_4$ solution formed the culture suspension of *E. coli*. The population of cells of bacteria in the suspension was maintained at 10^8 cells mL^{-1} through the measurement of optical density at 600 nm and the addition of 0.01 M $MgSO_4$ to get the value around 0.55–0.6. A layer of the harvested cell suspension was made on the Petri plates containing sterilized media (King's B). The culture of strains of cyanogenic rhizobacteria was spot inoculated at 3 points of equal distance on the plates pre-inoculated with *E. coli*. The plates were placed in an incubator at <40 °C. The production and release of toxic substances by the strains were evident from the zone of clearing around the spot of inoculation of strain. It indicated that the extracellular release of toxic compounds by the strains killed the growth of *E. coli* around its growth. The diameters of the zone of the clearing were recorded.

2.4. Antimetabolite Assay on Lettuce (Lectuca sativa L.) Seedlings

Nineteen strains restricted the growth of *E. coli* in the previous test. These strains were tested on the seedlings of lettuce as lettuce is considered sensitive to any type of phytotoxic substances and, hence, can be used as an indicator plant [24]. The fresh culture of the selected strains was prepared in Petri plates on KB media. This culture was suspended with the help of a sterilized buffer solution of $MgSO_4$ (0.01 M) by shaking gently. The suspension was collected in test tubes, and the cell population was maintained using optical density measurement at 600 nm with a value of 0.33. It established the population at 10^6 cells mL^{-1}.

The seeds of lettuce were disinfected on their surface in a parallel activity. The surface disinfection process comprised of seed dipping in ethanol for a moment, followed by the treatment with sodium hypochlorite (5%) for three minutes and complete rinsing of the seed with autoclaved water [25]. These seeds were allowed to germinate in the growth chamber.

Water agar was used as a medium for the growth of lettuce seedlings, where agar was added into the water at the rate of 1%. It was sterilized and poured in large-sized Petri plates, having a diameter of 15 cm. Seeds with good germination were picked up and transferred to the surface of these plates aseptically. Twenty germinating seeds of lettuce were placed on each plate.

Thirty microliters of the bacterial cell suspension were dispensed to each seed for inoculation. Three Petri plates were prepared for each strain in the same way. The control plates were treated with 30 µL buffer (0.01 M $MgSO_4$) per seed. The plates were placed at ambient temperature in the dark for 4 days. Then, the seedlings were removed from the plates and blotted. The measurements of masses and lengths of roots and shoots were done. The data were analyzed statistically to determine the significant differences [26].

2.5. Antimetabolite Assay on Weeds Using Presumed Allelopathic Bacteria

The strains of rhizobacteria obtained after the above-mentioned steps of the screening process were now called as presumed allelopathic bacteria. These strains were, now, used for testing on weeds. We selected four weeds of wheat for this assay i.e., wild oat, broad-leaved dock, common lambs' quarter, and little seed canary grass. These weeds cause maximum economic losses in the wheat crop in Pakistan. Nineteen strains were used to conduct this study in an experimental set up similar to the one used for bioassay on lettuce seedlings in Section 2.4. The culture of each strain was prepared in King's B broth. The culture was centrifuged to get the supernatant and form the bacterial pellets. These pellets were mixed in a sterilized buffer (0.01 M $MgSO_4$) to adjust the optical density value of 0.55 at 600 nm. It gave out the bacterial cell population at 10^8 cells mL^{-1}.

Water agar was prepared by adding 10 g of agar in 1 L distilled water and sterilizing in an autoclave at 121 °C and 15 PSI pressure for 20 min. The water agar was poured on large-sized Petri plates. It served as a medium for the growth of seedlings (Figure 2).

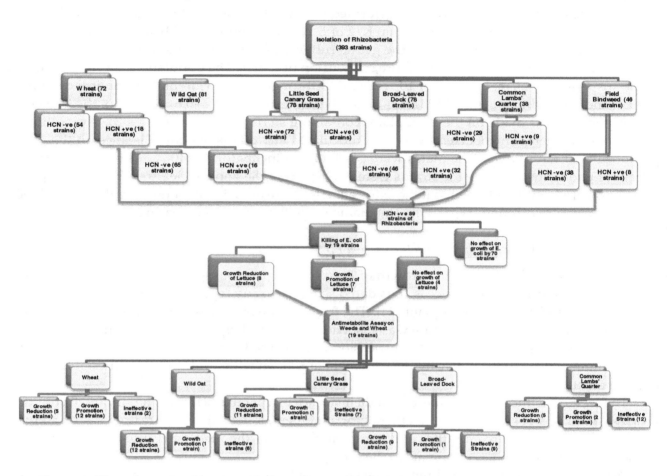

Figure 2. Flow chart of isolation and large-scale screening of allelopathic bacteria for the biocontrol of wheat-associated weeds.

The seeds of the selected weeds were surface disinfected by washing with ethanol (70%) momentarily, followed by washing with sodium hypochlorite (5%) and rinsing of seeds in plenty of sterilized water [25]. These seeds were placed in the growth chamber for germination.

Twenty germinated seeds were placed inside each prepared Petri plates aseptically. The culture suspension of each strain was applied at the rate of 30 μL per seed. For the control treatment, the sterilized buffer (0.01 M MgSO$_4$) was applied at the same rate. The plates were placed at ambient temperature in the dark. Each treatment in the experiment was replicated four times. After 7 days, the seedlings were uprooted from the water agar plates and blotted. These seedlings were measured for the lengths and weights of roots and shoots. The data were analyzed statistically to determine the significant differences following Steel et al. [26].

2.6. Antimetabolite Assay on Wheat Using Presumed Allelopathic Bacteria

The same nineteen strains were also tested for their effects on the growth of seedlings of wheat in a similar agar bioassay (Figure 3). The culture suspension of the strains was prepared following the same method as above. The large-sized Petri plates containing water agar were prepared as in previous bioassays. The surface of seeds of wheat was disinfected following Abd-Alla et al. [25]. Then, the seeds were placed for germination. The germinated seeds were placed on the already prepared Petri plates aseptically. The culture suspension of each strain was dispensed at the rate of 30 μL per seed. For the control treatment, the sterilized buffer (0.01 M MgSO$_4$) was dispensed to each seed at the rate of 30 μL. Each treatment was replicated four times. The seedlings were uprooted after five days and blotted. The data of lengths and weights of roots and shoots were taken and analyzed statistically to determine the significant differences following Steel et al. [26]. These analyses were carried out using *Statistix 8.1* software. All the data were first subjected to analysis of variance (ANOVA) test in this software, followed by multiple comparisons of means using the linear model. The least significant difference (LSD) test was then applied to determine the significant difference among treatments at $p < 0.05$.

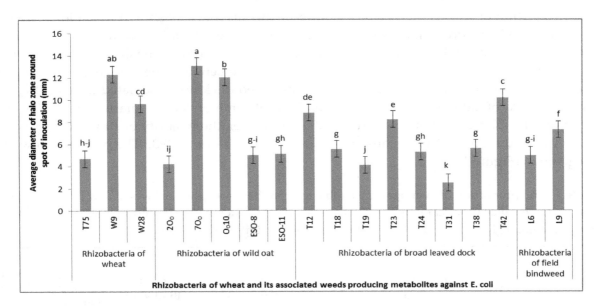

Figure 3. Cyanogenic rhizobacteria of weeds and wheat-producing metabolites against *E. coli* in the antimetabolite assay.

2.7. Cluster Analysis for the Screening of Biological Weed Control Agents

Cluster analysis was carried out for the grouping of strains applied in antimetabolite assays on weeds and wheat. The strains were categorized as non-selective biological weed control agents (the strains that reduced the growth of all the tested weeds and wheat), selective (the strains that reduced the growth of some of the tested plants and also wheat), selective (the strains that reduced the growth of some of the weeds but not wheat), and selective (the strains that reduced the growth of one more weed but promoted the growth of wheat). Five most efficient strains of allelopathic bacteria obtained from this study were identified through 16s rDNA sequencing as *Pseudomonas* strain T42 as *Pseudomonas putida*, strains L9 and 7O$_0$ as *P. fluorescens*, strain O$_0$10 as *P. aeruginosa*, and strain W9 as *P. alcaligenes*.

3. Results

The present study explored the rhizosphere of wheat and five weeds of wheat in search of allelopathic bacteria for the development of biological weed control agents. The selected weeds cause huge economic losses to the production of wheat in Pakistan annually [27]. These weed species were wild oat, common lambs' quarter, little seed canary grass, broad-leaved dock, and field bindweed. The job was carried out by the isolation of a large number of strains of rhizobacteria (393) from the weeds and wheat growing in areas of high weed invasion. Multiple bioassays were conducted on these strains to evaluate if they produced some phytotoxic substances, whether the release of such substances resulted in growth suppression of weeds, and if they were selective to inhibit the growth of weeds but not crop. The screening process of rhizobacteria to find out allelopathic bacteria from the rhizosphere of weeds and wheat is shown in the form of a flow chart (Figure 2).

3.1. Isolation of Rhizobacteria

We isolated 78 strains from the rhizosphere of wild oat, 81 from the broad-leaved dock, 78 from common lambs' quarter, 46 from field bindweed, 38 from little seed canary grass, and 72 from wheat. The total number of strains was 393. Multiple screening tests were conducted on these strains to characterize weed suppressive allelopathic bacteria.

3.2. Production of HCN by Rhizobacteria

The proportion of strains producing cyanide to various levels is shown in Table 1. We got 89 strains, which could produce cyanide to any level. Among these, 33 strains produced a low amount of cyanide, 25 medium, 20 high, and 11 very high, depending upon the intensity of change of color of picrate-treated filter paper inside the Petri plates and the time taken to change the color. The proportion of cyanogenic strains in the rhizosphere of the broad-leaved dock was calculated to be 41.0%, that of wild oat was 19.8%, of little seed canary grass was 7.7%, of common lambs' quarter was 23.7%, of field bindweed was 17.4%, and that of wheat was 25.0%. However, the majority of strains (77.6%) did not produce HCN in this study. These counted to 304 in number out of 393. The pictorial view of this assay is given in (Figure 1).

Table 1. The proportion of cyanogenic rhizobacteria in the rhizosphere of wheat and its associated weeds. The cyanide production by the strains was indicated after 48, 36, 24, and 12 h of incubation for low, medium, high, and very high cyanide production activity, respectively.

Category	Rhizosphere of						Total Strains
	Wheat	Broad-Leaved Dock	Wild Oat	Little Seed Canary Grass	Field Bindweed	Common Lambs' Quarter	
Non-cyanogenic strains	54	46	65	72	38	29	304
Low cyanide activity strains	8	3	8	5	3	6	33
Medium cyanide activity strains	6	12	3	0	2	2	25
High cyanide activity strains	2	14	0	1	2	1	20
Very high cyanide activity strains	2	3	5	0	1	0	11
Total strains	72	78	81	78	46	38	393

3.3. Antimetabolite Assay on E. coli

Clearing zones were produced around the inoculation spot of some strains, while the growth of most of the strains was mixed with the growth of E. coli, i.e., mutualistic strains. The clearing zones indicated the killing of E. coli, which occurred with nineteen strains. The diameter of these clearing or halo zones indicated the level of inhibition of growth of E. coli (Figure 3). Strain $7O_0$ produced the maximum diameter of the halo zone, which was followed by strains W9, O_010, T42, W28, T12, T23, and L9. The average diameter of zones produced by these strains was measured to be 1.3 ± 0.08, 1.23 ± 0.13, 1.21 ± 0.08, 1.01 ± 0.10, 0.96 ± 0.09, 0.88 ± 0.06, 0.82 ± 0.06, and 0.72 ± 0.07 cm, respectively. The remaining strains showed positive interaction with the growth of E. coli.

3.4. Antimetabolite Assay on Lettuce Seedlings

Results indicated that the strains imparted mixed effects on the growth of lettuce seedlings (Table 2). Five of the application strains significantly reduced the dry matter, root length, and shoot length of lettuce seedlings from 18.8 to 38.9%, 19.7 to 36.3%, and 17.3 to 24.3%, respectively. These strains were T18, T12, W9, W28, and O_010. The strains L6 and T31 caused a significant reduction in root length only. However, the strain T38 caused a significant reduction in the length of root and shoot. There were seven strains, which increased the dry matter, root length, and shoot length of lettuce seedlings from 15.7 to 41.5%, 16.7 to 61.4%, and 26.2 to 43.4%, respectively. These strains were T23, T42, T19, $2O_0$, T24, L9, and $7O_0$. The strains B11 and ESO-8 increased the shoot length only. The other strains remained ineffective on the growth of seedlings of lettuce.

Table 2. The effect of cyanogenic *E. coli* inhibiting rhizobacteria on lettuce seedlings in agar bioassay. Values sharing the same letter(s) in a column do not differ significantly from each other at $p < 0.05$. Values in a column indicate mean ± standard error.

Treatments	Root Length (cm)	Shoot Length (cm)	Dry Matter (mg)
Control	5.08 ± 0.14 [d,e]	4.01 ± 0.13 [f,g]	51.49 ± 0.006 [f,g]
T12	4.07 ± 0.16 [g,h]	3.19 ± 0.10 [i]	38.94 ± 0.006 [i,j]
T18	3.79 ± 0.16 [h,i]	3.32 ± 0.09 [i]	41.81 ± 0.006 [h,i,j]
T19	7.1 ± 0.13 [b]	5.39 ± 0.03 [b,c]	67.13 ± 0.006 [a,b]
T23	6.08 ± 0.12 [c]	5.22 ± 0.13 [c]	61.7 ± 0.010 [b,c,d]
T24	5.92 ± 0.27 [c]	5.07 ± 0.15 [c]	59.57 ± 0.007 [c,d,e]
T31	4.36 ± 0.20 [f,g]	3.68 ± 0.12 [g,h]	44.2 ± 0.007 [h,i]
T38	4.06 ± 0.16 [g,h]	3.37 ± 0.08 [h,i]	42.07 ± 0.007 [hij]
T42	6.35 ± 0.22 [c]	4.63 ± 0.15 [d]	58.42 ± 0.007 [d,e]
T75	4.8 ± 0.26 [e,f]	4.02 ± 0.15 [f]	50.7 ± 0.003 [f,g]
$2O_0$	6.19 ± 0.20 [c]	5.16 ± 0.14 [c]	65.26 ± 0.007 [b,c]
$7O_0$	8.19 ± 0.17 [a]	5.76 ± 0.11 [a]	72.19 ± 0.006 [a]
$O_0 10$	3.4 ± 0.14 [i,j]	3.18 ± 0.09 [i]	39.78 ± 0.006 [hij]
ESO-8	5.37 ± 0.20 [d]	4.43 ± 0.17 [d]	55.0 ± 0.003 [e,f]
ESO-11	5.03 ± 0.25 [d,e]	4.06 ± 0.11 [e,f]	52.0 ± 0.007 [f,g]
L6	4.48 ± 0.13 [f,g]	3.79 ± 0.11 [f,g]	45.79 ± 0.003 [g,h]
L9	8.01 ± 0.19 [a]	5.73 ± 0.05 [a,b]	72.88 ± 0.007 [a]
B11	5.26 ± 0.11 [d,e]	4.38 ± 0.18 [d,e]	54.53 ± 0.006 [e,f]
W9	3.34 ± 0.12 [i,j]	3.11 ± 0.08 [i]	35.99 ± 0.003 [j]
W28	3.231 ± 0.19 [j]	3.04 ± 0.14 [i]	38.08 ± 0.003 [i,j]
LSD	0.517	0.345	6.42

3.5. Antimetabolite Assay on Broad-Leaved Dock

The effects of the applied strains on the growth of the seedling of the broad-leaved dock were mixed, i.e., inhibiting, promoting, and neutral (Table 3). The dry matter, root length, and germination rate of the broad-leaved dock were significantly reduced by eight of the applied strains from 23.1 to 68.1%, 23.9 to 61.8%, and 26.7 to 64.4% than control, respectively. These strains were T42, $O_0 10$, L9, T38, $7O_0$, ESO-11, W9, and W28. The strain T19 caused a reduction in root length and germination rate only. The strain T31 caused a significant increase in root length and germination rate of the dock. The other strains remained ineffective on the growth of the seedlings of the dock.

3.6. Antimetabolite Assay on Wild Oat

Seven strains significantly reduced the dry matter, root length, and germination rate of wild oat from 38.5 to 80.2%, 19.4 to 60.2%, and 25.4 to 70.9%, respectively (Table 3). These strains were $2O_0$, ESO-8, $O_0 10$, T42, W28, W9, and $7O_0$. The strains T18, T12, ESO-11, and T75 significantly inhibited the germination rate from 14.5 to 25.4% but no other parameters. The strain T24 only reduced the root length of wild oat. The root length and germination rate of the weed were significantly increased by strain T19 up to 13.3 and 14.5%, respectively. The other strains remained ineffective on the growth of the seedlings of wild oat.

3.7. Antimetabolite Assay on Little Seed Canary Grass

Eight of the nineteen applied strains caused a significant reduction in dry matter, root length, and germination rate of little seed canary grass from 16.5 to 69.4%, 24.2 to 63.6%, and 20 to 52.7%, respectively (Table 4). These eight strains were T75, $7O_0$, T42, ESO-11, O_010, W9, L9, and W28. The strains T18 and T12 reduced only the root length (10.5–20%) and germination rate (18.2–25.4%). The strain $2O_0$ significantly reduced the dry matter (21.2%) and root length (10.8%) of the weed. However, the strain T19 significantly increased the dry matter (23.5%) and root length (10.4%) of the weed. Other strains remained ineffective on the growth of the seedlings of this weed. The pictorial view of the assay is available in (Figure 4).

Table 3. The effect of presumed allelopathic bacteria on the germination and seedling growth of broad-leaved dock and wild oat in agar bioassay. Values sharing the same letter(s) in a column do not differ significantly from each other at $p < 0.05$. Values in a column indicate mean ± standard error.

Treatment	Broad-Leaved Dock			Wild Oat		
	Germination Rate (%)	Root Length (cm)	Dry Matter (g)	Germination Rate (%)	Root Length (cm)	Dry Matter (g)
Control	75.0 ± 0.58 [b,c]	3.52 ± 0.13 [b]	0.307 ± 0.014 [a,b,c]	73.3 ± 0.33 [b,c]	6.0 ± 0.16 [b,c,d]	0.32 ± 0.03 [b,c,d]
T12	73.4 ± 0.88 [b,c]	3.48 ± 0.12 [b]	0.29 ± 0.035 [a,b,c,d]	62.7 ± 0.88 [d,e,f]	5.6 ± 0.17 [c,d,e]	0.29 ± 0.02 [b,c,d]
T18	80.0 ± 1.00 [a,b]	3.5 ± 0.10 [b]	0.303 ± 0.026 [a,b,c]	58.7 ± 0.67 [f,g]	5.52 ± 0.24 [d,e]	0.28 ± 0.02 [c,d]
T19	63.4 ± 0.67 [d,e]	2.98 ± 0.18 [c,d]	0.247 ± 0.013 [c,d,e]	84.0 ± 0.58 [a]	6.79 ± 0.13 [a]	0.41 ± 0.03 [a]
T23	68.3 ± 0.33 [c,d]	3.33 ± 0.11 [b,c]	0.277 ± 0.018 [b,c,d]	66.7 ± 0.33 [c,d,e]	6.16 ± 0.16 [b]	0.30 ± 0.02 [b,c,d]
T24	71.6 ± 0.33 [b,c,d]	3.46 ± 0.11 [b]	0.297 ± 0.023 [a,b,cd]	68.0 ± 0.58 [c,d]	5.28 ± 0.14 [e,f]	0.29 ± 0.02 [b,c,d]
T31	86.6 ± 0.33 [a]	4.1 ± 0.23 [a]	0.353 ± 0.014 [a]	73.3 ± 0.33 [b,c]	5.75 ± 0.15 [b,c,d,e]	0.32 ± 0.02 [b,c,d]
T38	55.0 ± 0.58 [e,f]	2.68 ± 0.28 [d,e]	0.233 ± 0.017 [d,e,f]	77.3 ± 0.33 [a,b]	5.96 ± 0.08 [b,c,d]	0.33 ± 0.02 [b,c]
T42	33.3 ± 0.67 [i,j]	1.73 ± 0.29 [h,i]	0.13 ± 0.020 [h,i]	32.0 ± 0.58 [k]	2.74 ± 0.18 [j]	0.11 ± 0.01 [h,i]
T75	70.0 ± 1.00 [c,d]	3.33 ± 0.06 [b,c]	0.28 ± 0.036 [b,c,d]	60.0 ± 1.00 [e,f,g]	5.88 ± 0.33 [b,c,d]	0.28 ± 0.02 [b,c,d]
$2O_0$	73.4 ± 0.33 [b,c]	3.61 ± 0.14 [b]	0.303 ± 0.022 [a,b,c]	54.7 ± 0.88 [g,h]	4.55 ± 0.18 [g]	0.20 ± 0.01 [e,f]
$7O_0$	41.7 ± 0.33 [h,i]	1.96 ± 0.06 [g,h]	0.17 ± 0.023 [f,g,h]	49.3 ± 0.67 [h,i]	3.77 ± 0.22 [h,i]	0.16 ± 0.003 [f,g,h]
O_010	51.6 ± 0.33 [f,g]	2.17 ± 0.10 [f,g]	0.203 ± 0.027 [e,f,g]	45.3 ± 0.67 [i,j]	3.6 ± 0.19 [i]	0.15 ± 0.03 [f,g,h]
ESO-8	73.4 ± 0.67 [b,c]	3.41 ± 0.21 [b]	0.29 ± 0.026 [a,c,d]	60.0 ± 0.58 [e,f,g]	4.83 ± 0.18 [f,g]	0.20 ± 0.01 [f,g]
ESO-11	48.4 ± 1.20 [f,g,h]	2.53 ± 0.06 [e,f]	0.203 ± 0.018 [e,f,g]	54.7 ± 0.88 [g,h]	5.56 ± 0.29 [d,e]	0.26 ± 0.02 [d,e]
L6	56.6 ± 0.67 [e,f]	2.77 ± 0.08 [d,e]	0.24 ± 0.020 [c,d,e]	82.7 ± 0.33 [a]	6.13 ± 0.17 [b,c]	0.35 0.03 [a,b]
L9	41.6 ± 0.67 [h,i]	1.88 ± 0.10 [g,h]	0.157 ± 0.022 [g,h,i]	21.3 ± 0.33 [l]	2.39 ± 0.23 [j]	0.63 ± 0.01 [i]
B11	76.6 ± 0.67 [b,c]	3.65 ± 0.11 [b]	0.317 ± 0.033 [a,b]	70.7 ± 0.33 [b,c]	6.0 ± 0.19 [b–d]	0.32 ± 0.02 [b,c,d]
W9	26.6 ± 0.67 [j]	1.34 ± 0.12 [i]	0.097 ± 0.018 [i]	41.3 ± 0.33 [j]	3.42 ± 0.13 [i]	0.13 ± 0.03 [g,h]
W28	43.4 ± 0.88 [g,h]	1.41 ± 0.08 [i]	0.14 ± 0.029 [g,h,i]	44.0 ± 1.16 [i,j]	4.3 ± 0.13 [g,h]	0.17 ± 0.02 [f,g,h]
LSD	9.8205	0.429	0.068	7.32	0.545	0.0642

Table 4. The effect of presumed allelopathic bacteria on the germination and seedling growth of little seed canary grass and common lambs' quarter. Values sharing the same letter(s) in a column do not differ significantly from each other at $p < 0.05$. Values in a column indicate mean ± standard error.

Treatments	Little Seed Canary Grass			Common Lambs' Quarter		
	Germination Rate (%)	Root Length (cm)	Dry Matter (g)	Germination Rate (%)	Root Length (cm)	Dry Matter (g)
Control	73.3 ± 1.20 [a,b]	4.59 ± 0.22 [b,c]	0.283 ± 0.026 [b,c]	63.3 ± 1.00 [c,d,e]	2.87 ± 0.19 [b,c,d,e]	0.27 ± 0.01 [a,b,c]
T12	54.7 ± 1.45 [d,e,f,g]	4.11 ± 0.06 [d,e]	0.287 ± 0.024 [b,c]	61.0 ± 0.67 [c,d,e]	2.57 ± 0.12 [d,e,f,g]	0.25 ± 0.02 [b,c,d]
T18	60.0 ± 1.53 [c,d,e,f]	3.68 ± 0.08 [e,f]	0.267 ± 0.026 [b,c,d]	61.5 ± 0.88 [c,d,e]	2.29 ± 0.11 [f,g,h]	0.24 ± 0.01 [b,c,d]
T19	81.3 ± 0.67 [a]	5.07 ± 0.11 [a]	0.35 ± 0.021 [a]	63.2 ± 1.00 [c,d,e]	2.50 ± 0.17 [e,f,g,h]	0.27 ± 0.01 [a,b,c]
T23	66.7 ± 0.33 [b,c]	4.64 ± 0.15 [a,b]	0.26 ± 0.015 [b,c,d]	58.7 ± 1.45 [c,d,e]	2.73 ± 0.11 [c,d,e,f]	0.25 ± 0.02 [b,c,d]
T24	68.0 ± 0 [b,c]	4.28 ± 0.16 [b,c,d]	0.237 ± 0.013 [c,d,e]	66.7 ± 1.00 [b,c]	3.14 ± 0.22 [a,b,c]	0.29 ± 0.01 [a,b]
T31	74.7 ± 0.33 [a,b]	4.32 ± 0.08 [b,c,d]	0.287 ± 0.014 [b,c]	57.7 ± 0.33 [d,e,f]	2.9 ± 0.22 [b,c,d,e]	0.28 ± 0.06 [a,b]
T38	74.7 ± 0.67 [a,b]	4.58 ± 0.11 [b,c]	0.277 ± 0.024 [b,c]	57.2 ± 0.33 [d,e,f]	2.61 ± 0.28 [d,e,f]	0.24 ± 0.03 [b,c,d]
T42	54.7 ± 0.88 [d,e,f,g]	2.89 ± 0.10 [g,h]	0.153 ± 0.017 [g,h,i]	65.7 ± 0.67 [b,c,d]	3.04 ± 0.19 [a,b,c,d]	0.24 ± 0.04 [b,c,d]
T75	53.3 ± 1.67 [e,f,g,h]	2.34 ± 0.09 [i]	0.175 ± 0.005 [f,g,h]	47.7 ± 0.67 [g]	2.04 ± 0.17 [h,i]	0.19 ± 0.02 [d,e,f]
$2O_0$	64.0 ± 1.00 [b,c,d,e]	4.1 ± 0.24 [d,e]	0.223 ± 0.007 [d,e,f]	75.6 ± 0.33 [a]	3.51 ± 0.09 [a]	0.32 ± 0.01 [a]
$7O_0$	42.7 ± 0.67 [h,i]	2.37 ± 0.18 [i]	0.100 ± 0.006 [j]	55.7 ± 0.33 [e,f,g]	2.90 ± 0.17 [b,c,d,e]	0.24 ± 0.01 [b,c,d,e]
O_010	34.7 ± 0.67 [i]	1.8 ± 0.19 [j]	0.087 ± 0.019 [j]	63.3 ± 0.58 [c,d,e]	2.93 ± 0.11 [b,c,d,e]	0.24 ± 0.02 [b,c,d]
ESO-8	65.3 ± 0.88 [b,c,d]	4.16 ± 0.21 [c,d]	0.237 ± 0.012 [c,d,e]	64.3 ± 0.67 [b,c,d]	2.93 ± 0.27 [b,c,d,e]	0.28 ± 0.02 [a,b,c]
ESO-11	48.0 ± 1.53 [g,h]	3.48 ± 0.08 [f]	0.237 ± 0.022 [c,d,e]	49.0 ± 0.67 [g]	1.69 ± 0.14 [i,j]	0.17 ± 0.02 [e,f]
L6	80.0 ± 0.58 [a]	4.57 ± 0.11 [b,c]	0.293 ± 0.013 [b]	72.3 ± 0.88 [a,b]	3.32 ± 0.25 [a,b]	0.30 ± 0.01 [a,b]
L9	49.3 ± 0.67 [f,g,h]	2.54 ± 0.20 [h,i]	0.123 ± 0.007 [i,j]	63.3 ± 1.53 [c,d,e]	2.69 ± 0.12 [c,d,e,f]	0.27 ± 0.04 [a,b,c]
B11	74.7 ± 0.67 [a,b]	4.52 ± 0.11 [b,c,d]	0.257 ± 0.012 [b,c,d]	49.0 ± 0.88 [g]	2.09 ± 0.11 [g,h,i]	0.21 ± 0.01 [c,d,e]
W9	58.7 ± 0.33 [c,d,e,f,g]	3.26 ± 0.27 [f,g]	0.197 ± 0.032 [e,f,g]	59.0 ± 0.67 [c,d,e]	2.98 ± 0.29 [b,c,d,e]	0.26 ± 0.03 [a,b,c,d]
W28	49.3 ± 1.77 [f,g,h]	1.67 ± 0.09 [j]	0.137 ± 0.003 [h,i,j]	50.0 ± 1.16 [f,g]	1.29 ± 0.13 [j]	0.13 ± 0.02 [f]
LSD	11.432	0.445	0.0507	8.127	0.524	0.073

Figure 4. The pictorial view of seedlings of little seed canary grass growing on water agar in agar bioassay.

3.8. Antimetabolite Assay on Common Lambs' Quarter

The present study reported a decrease in dry matter, root length, and germination rate of common lambs' quarter by three of the applied strains from 27.5 to 50.0%, 29.0 to 55.0%, and 21.0 to 24.6%, respectively (Table 4). These strains were W28, ESO-11, and T75. The strain B11 caused a reduction in root length (27.3%) and germination rate (22.8%) only. The strain T18 caused a reduction in root length only, which was 20.3% lesser than the control. However, a significant increase in root length (13.0%) and germination rate (19.3%) was observed with the inoculation of strain $2O_0$. The strain L6 increased the germination rate of the weed by 14%. The other strains remained ineffective on the growth of the seedlings of this weed.

3.9. Antimetabolite Assay on Wheat

There were three strains in the whole lot, which significantly reduced the dry matter, shoot length, root length, and germination rate of wheat from 23.4 to 34%, 21.0 to 38.5%, 27.2 to 52.8%, and 8.3 to 10.4%, respectively (Figure 5, Table 5). These three strains were ESO-11, W28, and T18. Two strains (T75 and T12) reduced the dry matter (23.4 and 26.6%), root length (24.8 and 50.1%), and shoot length (18.9 and 35.5%) of the crop. However, there were six strains, which significantly increased the dry matter, shoot length, root length, and germination rate of wheat from 24.5 to 47.9%, 14.6 to 29.7%, 19.4 to 37.7%, and 12.5 to 18.8%, respectively. These strains were T23, $7O_0$, $2O_0$, L9, T24, and T19. The strains L6, O_010, and B11 caused an increment in dry matter of the crop up to 13.8, 12.8, and 27.7% than control, respectively. The strains T38 and T31 caused a significant increase in shoot length of the crop up to 18.9 and 18.7% than control, respectively. The strain T42, however, increased the germination rate of the crop up to 8.3%. The other strains remained ineffective on the growth of the seedlings of wheat.

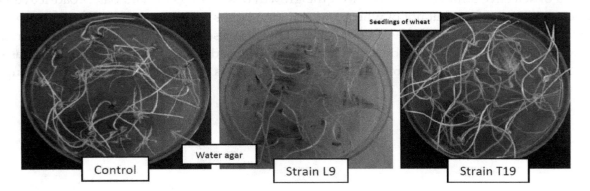

Figure 5. The pictorial view of seedlings of wheat growing on water agar in agar bioassay.

Table 5. The effect of presumed allelopathic bacteria on the germination and seedling growth of wheat. Values sharing the same letter(s) in a column do not differ significantly from each other at $p < 0.05$. Values in a column indicate mean ± standard error.

Treatments	Germination Rate (%)	Root Length (cm)	Shoot Length (cm)	Dry Matter (g)
Control	80.0 ± 0.58 [d,e]	6.60 ± 0.50 [c,d,e]	8.58 ± 0.22 [c,d]	0.313 ± 0.018 [f,g]
T12	75.0 ± 0.58 [e,f]	4.97 ± 0.08 [f]	6.96 ± 0.27 [e]	0.233 ± 0.003 [h]
T18	73.35 ± 0.33 [f]	4.81 ± 0.10 [f]	6.78 ± 0.16 [e]	0.240 ± 0.015 [h]
T19	93.4 ± 0.33 [a]	8.22 ± 0.15 [b]	9.83 ± 0.37 [b]	0.407 ± 0.012 [c]
T23	90.0 ± 0.58 [a,b]	8.13 ± 0.14 [b]	10.13 ± 0.34 [b]	0.390 ± 0.006 [c,d]
T24	90.0 ± 0 [a,b]	7.92 ± 0.13 [b]	10.05 ± 0.21 [b]	0.400 ± 0.015 [c]
T31	80.0 ± 0.58 [d,e]	6.47 ± 0.23 [d,e]	10.19 ± 0.20 [b]	0.300 ± 0.020 [g]
T38	80.0 ± 0.58 [d,e]	6.67 ± 0.27 [c,d,e]	10.21 ± 0.17 [b]	0.310 ± 0.015 [f,g]
T42	86.5 ± 0.33 [b,c]	7.15 ± 0.25 [c]	9.097 ± 0.38 [c]	0.343 ± 0.012 [e,f]
T75	75.0 ± 0 [e,f]	3.3 ± 0.17 [g]	5.53 ± 0.10 [f]	0.230 ± 0.012 [h]
$2O_0$	93.4 ± 0.33 [a]	7.89 ± 0.26 [b]	10.03 ± 0.18 [b]	0.413 ± 0.003 [b,c]
$7O_0$	93.4 ± 0.33 [a]	9.09 ± 0.21 [a]	11.13 ± 0.25 [a]	0.450 ± 0.010 [a,b]
O_010	85 ± 0 [b–d]	7.17 ± 0.15 [c]	8.69 ± 0.19 [c,d]	0.353 ± 0.003 [d,e]
ESO-8	81.5 ± 0.67 [c,d]	6.53 ± 0.17 [c–e]	8.52 ± 0.16 [c,d]	0.347 ± 0.014 [e,f]
ESO-11	73.4 ± 0.67 [f]	3.38 ± 0.33 [g]	5.56 ± 0.15 [f]	0.220 ± 0.006 [h]
L6	83.4 ± 0.33 [c,d]	6.41 ± 0.29 [e]	8.6 ± 0.14 [c,d]	0.357 ± 0.019 [d,e]
L9	95 ± 0 [a]	9.09 ± 0.12 [a]	11.13 ± 0.20 [a]	0.463 ± 0.020 [a]
B11	83.4 ± 0.67 [c,d]	6.32 ± 0.23 [e]	8.42 ± 0.17 [d]	0.400 ± 0.006 [c]
W9	81.7 ± 0.67 [c,d]	7.14 ± 0.31 [c,d]	8.62 ± 0.38 [c,d]	0.337 ± 0.014 [e,f,g]
W28	71.6 ± 0.67 [f]	3.12 ± 0.26 [g]	5.28 ± 0.17 [f]	0.207 ± 0.014 [h]
LSD	6.565	0.682	0.673	0.0376

3.10. Cluster Analysis

The cluster analysis was performed to categorize the tested strains of this study based on the objectives of this study (Table 6). All the strains were categorized into four groups: the first group of two strains (W28 and ESO-11) comprised of non-selective strains, which reduced the growth of seedlings of all the tested plants; the second group of three strains (T75, T18, and T12) comprised of selective strains, which reduced the growth of little seed canary grass, wild oat, common lambs' quarter, and wheat but not of the broad-leaved dock; the third group of three strains (W9, ESO-8, and T38) comprised of selective strains, which reduced the growth of seedlings of wild oat, broad-leaved dock, and little seed canary grass but not of wheat and common lambs' quarter; and the fourth group of nine strains (T24, $2O_0$, O_010, L9, B11, T19, T42, $7O_0$, and L6) comprised of selective strains, which reduced the growth of seedlings of little seed canary grass, broad-leaved dock, and wild oat but increased the growth of seedlings of wheat. The remaining two strains of this study (T23 and T31) did not suppress the growth of any weed or wheat.

Table 6. Cluster analysis for the selection of bioherbicidal agents based on the response of rhizobacteria in wheat and its associated weeds in agar bioassays. Candidate strains for biological weed control in wheat are indicated in bold.

Category of Strains	Strain	Effects on Weeds and Wheat		
		Inhibition	Promotion	No Effect
Non-selective	ESO-11	All the tested weeds and wheat	–	–
	W28			
Selective and inhibitory to wheat	T12	Wheat, wild oat, and little seed canary grass	–	Broad-leaved dock and common lambs' quarter
	T18	Wheat, wild oat, little seed canary grass, and common lambs' quarter	–	Broad-leaved dock
	T75			
Selective and non-inhibitory to wheat	**T38**	Broad-leaved dock	–	Wheat, wild oat, little seed canary grass, and common lambs' quarter
	ESO-8	Wild oat	–	Wheat, little seed canary grass, broad-leaved dock, and common lambs' quarter
	W9	Wild oat, little seed canary grass, and broad-leaved dock	–	Wheat and common lambs' quarter
Selective and promotory to wheat	T19	Broad-leaved dock	Wheat, wild oat, and little seed canary grass	Common lambs' quarter
	T24	Wild oat	Wheat	Little seed canary grass, broad-leaved dock, and common lambs' quarter
	T42	Wild oat, little seed canary grass, and broad-leaved dock	Wheat	Common lambs' quarter
	7O$_0$			
	O$_0$10			
	L9			
	2O$_0$	Wild oat and little seed canary grass	Wheat and common lambs' quarter	Broad-leaved dock
	L6	Broad-leaved dock	Wheat, wild oat, and common lambs' quarter	Little seed canary grass
	B11	Common lambs' quarter	Wheat	Wild oat, little seed canary grass, and broad-leaved dock
–	T23	–	Wheat	Wild oat, little seed canary grass, broad-leaved dock, and common lambs' quarter
	T31	–	Broad-leaved dock	Wheat, wild oat, little seed canary grass, and common lambs' quarter

4. Discussion

The study of diverse forms of soil-inhabiting microorganisms and their activities may be helpful in resolving many agricultural, environmental, and ecological issues created by unsustainable farming practices. The invasion of weeds in crops reduces their yields, and farmers adopt unsustainable and unhealthy practices to reduce the losses of their crops. The harmful impacts of tillage and chemicals have been established. Therefore, the present study explored an alternative, inexpensive, sustainable, and environmentally and ecologically safe technique for weed control in crops. It was aimed at finding out the natural mechanisms of rhizobacteria, which function to limit the growth of weeds, alleviate the biotic stress of weeds on crops, and produce a vigorous crop stand. Strengthening such natural processes through augmentation, inoculation, or other processes is required for the development of biological weed control in crops. This may help us to resolve the above-mentioned issues created by conventional control practices [11].

The rhizosphere inhabiting bacteria, which release phytotoxic metabolites in the rhizosphere and result in germination/growth reduction of weeds, are called allelopathic bacteria [18]. The present study is the pioneering work executed in Pakistan, which is aimed at searching such rhizobacteria with their novel characteristics to develop biological weed control. The probability of the existence of such bacteria has been speculated in the rhizosphere of weeds and crops, which are growing together over many years or where the weed invasions occur more frequently [28]. Therefore, we collected the samples of weeds and wheat from areas/fields across the District of Faisalabad, Pakistan, where the weed invasions were more frequent. The findings of this work support the above-mentioned finding of Schippers et al. [28]. They also reported that growth inhibitory rhizobacteria grew, strengthened, and increased their activities in the agricultural crops, where a single crop is grown year after year. It resulted in the reduction of yields of crops. They reported the increase in cyanogenic bacteria and cyanide production in the rhizosphere of potatoes when this crop was continuously grown over a field for 3 years. Their findings increased the importance of crop rotation.

We isolated 393 strains of rhizobacteria from the rhizosphere of five weeds and wheat in this study. These strains were passed through a comprehensive screening process based on the production of phytotoxic metabolites in vitro, suppression of indicator bacteria and plants, in vivo suppression of weeds, and their effects on wheat crops. The protocols followed for these purposes obtained support from the findings of Bakker and Schipper [22], Kremer and Souissi [29], and Kremer [24]. The first test conducted on these strains was the qualitative production of HCN. It was considered a major substance responsible for the growth inhibition of some plants by Kremer and Souissi [29]. This study obtained 22.6% of strains (89) to have produced cyanide at various levels. The distribution of cyanogenic strains in different weeds and wheat was also variable. This was synonymous with the findings of Kremer and Souissi [29]. The proportion of cyanogenic strains in their study (32%) was, however, higher than in our study. This difference might be due to differences in agro-ecological conditions and prevalent agricultural practices. Zeller et al. [19] found that the sensitivity of different weeds and crops to cyanide was variable, and the cyanogenic bacteria might cause suppression of some weeds without imparting harmful effects on the accompanying crop in certain cases. They applied various levels of cyanide to five weeds and wheat and reported that this characteristic of rhizobacteria might be used for the selective suppression of three seeds (C. jacea, G. mollugo, and H. murinum), invading the wheat crop without disturbing the growth of wheat.

The cyanogenic strains of our study were further tested for the production of toxic metabolites using the indicator of sensitive bacteria (E. coli strain K12). The relevance of this assay for the screening of rhizobacteria weed control agents was reported by Kremer et al. [30]. We got 21.3% of the cyanogenic strains to suppress the growth of sensitive bacteria. As all the cyanogenic strains did not suppress the growth of sensitive bacteria in this study, one may speculate that the strains inhibiting the growth of bacteria may also have possessed the characteristics of production of some other toxic compounds along with cyanide. This assay indicated that the strains inhibiting the growth of sensitive bacteria might be producing multiple growth inhibitory compounds, collectively termed as antimetabolites, and could be more suitable for testing on weeds and wheat in the next screening studies.

Nineteen strains, obtained from the above screening procedures, were tested on sensitive plant species, i.e., lettuce. The effects of these strains on the growth of the seedlings of lettuce were variable. Some strains inhibited, some promoted, and others remained ineffective. Hence, all the strains inhibiting the growth of E. coli did not inhibit the growth of lettuce in our study. This finding agreed with Kremer et al. [30]. Kremer and Kennedy [31] also reported the growth reduction of lettuce by such rhizobacteria. As the strains tested on lettuce were all cyanogenic in nature, Zermane et al. [32] also reported mixed effects of cyanogenic rhizobacteria on lettuce. The non-inhibition of lettuce by some strains may be due to the non-host interactions, where these strains needed to grow with their host in order to express their characteristics [33]. There also exist differences in the metabolic functions of *E. coli* and lettuce, the former being a prokaryote, and the latter being a eukaryotic plant species. There may also be the difference of compounds, causing antibiosis against bacteria and plants. The results obtained in our study reflected the release of diverse types of metabolites and their functions by these strains, which affected the growth of bacteria and plants. For similar reasons, we tested all the above-mentioned strains on weeds and wheat in the further screening process. This decision in our study has grounds in Souissi and Kremer [34]. They reported the reduction in the growth of weeds by those strains of rhizobacteria, which did not reduce the growth of lettuce. In other words, the growth reduction of lettuce and weeds by rhizobacteria could not be correlated in their study.

Stability or consistency in the characteristics of strains of our study may be evident from the above studies. It increased our reliance on these strains for further studies regarding their effects on weeds and wheat. We found all type of effects of the strains on weeds and wheat, i.e., there were strains inhibitory to all the weeds and wheat, suppressive to one or more weeds and wheat, suppressive to one or more weeds but not to wheat, and suppressive to one or more weeds but promoted the growth of wheat. This array of responses by the strains of allelopathic bacteria has multiple applications if further studies on their characterization and response under natural conditions are carried out. These may be developed for application to control weeds and strengthen crop in poor agricultural systems (selective strains) and control weeds in non-agricultural systems (non-selective strains). The reasons for selectivity may be a difference of tolerance to toxic metabolites in weeds and crop, release of toxic metabolites by these strains only in the rhizosphere of their host plants, the difference in availability of substrates required for the production of toxic metabolites in the rhizosphere of weeds and wheat, a difference of survival, colonization, and establishment in the rhizosphere of weeds and wheat, and difference of mechanisms in the rhizosphere of host and non-host plants [19,20,35]. The findings of our study became more evident when the strains were further characterized by the production of indole-3-acetic acid, exopolysaccharides, siderophores, catalases, chitinases, oxidases, and P solubilization. The most prominent strains were identified as pseudomonads. The effects of the five most efficient strains on weeds and wheat were tested under axenic conditions in Abbas et al. [36]. The strains inhibiting one or more weeds and promoting wheat may be more successful for weed control under natural conditions. These may strengthen the weak crop plants, increase their competitive ability, and, hence, increase the scale of weed control by allelopathic bacteria. The non-selective strains inhibitory to wheat may be tested for their effects on other crops to explore opportunities for their application in other cropping systems. The efforts on augmentation of effects of allelopathic bacteria under natural conditions may be helpful to realize the dream of biological weed control. The strains of allelopathic bacteria obtained from this study can be further tested for their effects on weeds and wheat under field conditions. Further efforts may be required to improve their efficiency of weed control under natural conditions. Application methods of allelopathic bacteria may also be needed to be optimized. This will produce a bioherbicide for the control of weeds in an environmentally friendly and sustainable manner.

5. Conclusions

The rhizosphere of five weeds and wheat, growing in areas of high weed invasion, was explored for the allelopathic bacteria. A large collection of strains of rhizobacteria was passed through a comprehensive screening process for this purpose. We got 22.6% strains cyanogenic in nature, 21.3% of which (19 strains) inhibited the growth of sensitive bacteria. These strains were applied to lettuce, which showed mixed effects. These strains were later tested on four weeds and wheat. We got strains inhibitory to all these weeds (eight for the broad-leaved dock, seven for wild oat, eight for little seed canary grass, and three for common lambs' quarter). They reduced the dry matter of these weeds from

23.1 to 68.1%, 38.5 to 80.2%, 16.5 to 69.4%, and 27.5 to 50.0%, respectively. Only five of these strains were inhibitory to wheat; the others either remained neutral (five strains) or improved the growth of wheat (nine strains). These strains offer opportunities for the development of biological weed control.

Author Contributions: Conceptualization: M.N., Z.A.Z.; Data curation: T.A., A.M., S.A.; Formal analysis and software: A.M., M.S.A., M.S.E.; Investigation: M.N., Z.A.Z., Methodology: T.A., A.M., and M.S.E.; Supervision: M.N., Z.A.Z., Writing—original draft: T.A.; Writing—review and editing: M.N., Z.A.Z., M.S.A. All authors have read and agreed to the published version of the manuscript.

Acknowledgments: This research was financially supported by the Higher Education Commission of Pakistan under the Indigenous Ph.D. Fellowship Program for 5000 scholars Phase II Batch II (Grant # 213-65052-2AV2-105. The authors would like to extend their sincere appreciation to the Researchers Supporting Project number (RSP-2020/173), King Saud University, Riyadh, Saudi Arabia.

References

1. Anonymous. *Pakistan Economic Survey*; Economic Adviser, Government of Pakistan, Ministry of Finance: Rawalpindi, Pakistan, 1966. Available online: http://www.irispunjab.gov.pk/StatisticalReport/Pakistan%20Economic%20Surveys/Economic%20Survey%201996-97.pdf (accessed on 24 September 2020).

2. Anonymous. *Economic Survey of Pakistan 2015–16*; Economic Advisor's Wing, Finance Division, Government of Pakistan: Islamabad, Pakistan, 2016. Available online: http://www.finance.gov.pk/survey_1516.html (accessed on 24 September 2020).

3. Oerke, C.E. Centenary review on crop losses to pests. *J. Agric. Sci.* **2006**, *144*, 31–43. [CrossRef]

4. Shad, R.A. Weeds and weed control. In *Crop Production*; Nazir, S., Bashir, E., Bantel, R., Eds.; National Book Foundation: Islamabad, Pakistan, 2015; pp. 175–204.

5. Ghorbani, R.; Leifert, C.; Seel, W. Biological control of weeds with antagonistic plant pathogens. *Adv. Agron.* **2005**, *86*, 191–225.

6. Gaines, T.A.; Duke, S.O.; Morran, S.; Rigon, C.A.G.; Tranel, P.J.; Kupper, A.; Dayan, F.E. Mechanisms of evolved herbicide resistance. *J. Biol. Chem.* **2020**, *295*, 10307–10330. [CrossRef] [PubMed]

7. Quimby, P.C.; King, L.R.; Grey, W.E. Biological control as a means of enhancing the sustainability of crop/land management systems. *Agric. Ecosyst. Environ.* **2002**, *88*, 147–152. [CrossRef]

8. Geiger, F.; Bengtsson, J.; Berendse, F.; Weisser, W.W.; Emmerson, M.; Morales, M.B.; Ceryngier, P.; Liira, J.; Tschantke, T.; Winqvist, C.; et al. Persistent negative effects of pesticides on biodiversity and biological control potential on European farmland. *Basic Appl. Ecol.* **2010**, *11*, 97–105. [CrossRef]

9. Pimentel, D. Environmental and economic costs of the application of pesticides primarily in the United States. *Environ. Dev. Sustain.* **2005**, *7*, 229–252. [CrossRef]

10. Alavanja, M.C.R.; Hoppin, J.A.; Kamel, F. Health effects of chronic pesticide exposure: Cancer and neurotoxicity. *Annu. Rev. Public Health* **2004**, *25*, 155–197. [CrossRef]

11. Abbas, T.; Zahir, Z.A.; Naveed, M.; Kremer, R.J. Limitations of existing weed control practices necessitate the development of alternative techniques based on biological approaches. *Adv. Agron.* **2017**, *147*, 239–280.

12. Quimby, P.C.; Birdsall, J.L. Fungal agents for biological control of weeds: Classical and augmentative approaches. In *Novel Approaches to Integrated Pest Management*; Reuveni, R., Ed.; CRC Press: Boca Raton, FL, USA, 1995; pp. 293–308.

13. Boyette, C.D.; Hoagland, R.E. Bioherbicidal potential of a strain of *Xamthomonas* spp. for control of common cocklebur (*Xanthium strumarium*). *Biocontrol. Sci. Technol.* **2013**, *23*, 183–196. [CrossRef]

14. Coombs, E.M.; Clark, J.K.; Piper, G.L.; Cofrancesco, A.F., Jr. *Biological Control of Invasive Plants in the United States*; Oregon State University Press: Corvallis, OR, USA, 2004.

15. Denslow, J.S.; D'Antonio, C.M. After biocontrol: Assessing indirect effects of insect releases. *Biol. Control* **2005**, *35*, 307–318. [CrossRef]

16. Farooq, M.; Bajwa, A.A.; Cheema, S.A.; Cheema, Z.A. Application of allelopathy in crop production. *Int. J. Agric. Biol.* **2013**, *15*, 1367–1378.

17. Kobayashi, K. Factors affecting phytotoxic activity of allelochemicals in soil. *Weed Biol. Manag.* **2004**, *4*, 1–7. [CrossRef]

18. Kremer, R.J. The role of allelopathic bacteria in weed management. In *Allelochemicals: Biological Control of Plant Pathogens and Diseases*; Inderjit, M.K.G., Ed.; Springer: Dordrecht, The Netherlands, 2006; pp. 143–156.

19. Zeller, S.L.; Brandl, H.; Schmid, B. Host-plant selectivity of rhizobacteria in a crop/weed model system. *PLoS ONE* **2007**, *2*, 846–858. [CrossRef]

20. Kennedy, A.C.; Johnson, B.N.; Stubbs, T.L. Host range of a deleterious rhizobacterium for biological control of downy brome. *Weed Sci.* **2001**, *49*, 792–797. [CrossRef]

21. King, E.; Ward, M.; Raney, D. Two simple media for the demonstration of pycyanin and Xuorescein. *J. Lab. Clin. Med.* **1954**, *44*, 301–307. [PubMed]

22. Bakker, A.W.; Schipper, B. Microbial cyanide production in the rhizosphere in relation to potato yield reduction and *Pseudomonas* spp. mediated plant growth stimulation. *Soil Biol. Biochem.* **1987**, *19*, 451–457. [CrossRef]

23. Gasson, M.J. Indicator technique for antimetabolic toxin production by phytopathogenic species of *Pseudomonas*. *Appl. Environ. Microbiol.* **1980**, *39*, 25–29. [CrossRef] [PubMed]

24. Kremer, R.J. Interactions between the plants and microorganisms. *Allelopath. J.* **2013**, *31*, 51–70.

25. Abd-Alla, M.H.; Morsy, F.M.; El-Enany, A.E.; Ohyama, T. Isolation and characterization of a heavy metal resistant isolate of *Rhizobium leguminosarum* bv. *viciae* potentially applicable for biosorption of Cd^{+2} and Co^{+2}. *Int. Biodeterior. Biodegrad.* **2012**, *67*, 48–55. [CrossRef]

26. Steel, R.G.D.; Torrie, J.H.; Dicky, D.A. *Principles and Procedures of Statistics—A Biometrical Approach*, 3rd ed.; McGraw Hill Book International Co.: Singapore, 1997.

27. Matloob, A.; Safdar, M.E.; Abbas, T.; Aslam, F.; Khaliq, A.; Tanveer, A.; Rehman, A.; Chadhar, A.R. Challenges and prospects for weed management in Pakistan: A review. *Crop Prot.* **2020**, *134*, 104724. [CrossRef]

28. Schippers, B.; Bakker, A.W.; Bakker, P.A.H.M. Interactions of deleterious and beneficial rhizosphere microorganisms and the effect of cropping practices. *Annu. Rev. Phytopathol.* **1987**, *25*, 339–358. [CrossRef]

29. Kremer, R.J.; Souissi, T. Cyanide production of rhizobacteria and potential for suppression of weed seedling growth. *Curr. Microbiol.* **2001**, *43*, 182–186. [CrossRef] [PubMed]

30. Kremer, R.J.; Begonia, M.F.T.; Stanley, L.; Lanham, E.T. Characterization of rhizobacteria associated with weed seedlings. *Appl. Environ. Microbiol.* **1990**, *56*, 1649–1655. [CrossRef] [PubMed]

31. Kremer, R.J.; Kennedy, A.C. Rhizobacteria as biocontrol agents of weeds. *Weed Technol.* **1996**, *10*, 601–609. [CrossRef]

32. Zermane, N.; Souissi, T.; Kroschel, J.; Sikora, R. Biocontrol of broomrape (*Orobanche crenata* Forsk. and *Orobanche foetida* Poir.) by *Pseudomonas fluorescens* isolate Bf7-9 from the faba bean rhizosphere. *Biocontrol Sci. Technol.* **2007**, *17*, 483–497. [CrossRef]

33. Weiland, G.W.; Neumann, R.; Backhaus, H. Variation of microbial communities in soil, rhizosphere and rhizoplane in response to crop species, soil type and crop development. *Appl. Environ. Microbiol.* **2001**, *67*, 5849–5854. [CrossRef]

34. Souissi, T.; Kremer, R.J. A rapid microplate callus bioassay for assessment of rhizobacteria for biocontrol of leafy spurge (*Euphorbia esula* L.). *Biocontrol Sci. Technol.* **1998**, *8*, 83–92. [CrossRef]

35. Owen, A.; Zdor, R. Effect of cyanogenic rhizobacteria on the growth of velvetleaf (*Abutilon theophrasti*) and corn (*Zea mays* L.) in autoclaved soil and the influence of supplemental glycine. *Soil Biol. Biochem.* **2001**, *33*, 801–809. [CrossRef]

36. Abbas, T.; Zahir, Z.A.; Naveed, M. Bioherbicidal activity of allelopathic bacteria against weeds associated with wheat and their effects on growth of wheat under axenic conditions. *BioControl* **2017**, *62*, 719–730. [CrossRef]

Agricultural Utilization of Unused Resources: Liquid Food Waste Material as a New Source of Plant Growth-Promoting Microbes

Waleed Asghar, Shiho Kondo, Riho Iguchi, Ahmad Mahmood⑩ and Ryota Kataoka *

Department of Environmental Sciences, Faculty of Life and Environmental Sciences, University of Yamanashi, Kofu, Yamanashi 400-8510, Japan; waleedasghar978@gmail.com (W.A.); g19lr004@yamanashi.ac.jp (S.K.); lustiness_17@yahoo.co.jp (R.I.); ahmadmahmood91@gmail.com (A.M.)
* Correspondence: rkataoka@yamanashi.ac.jp

Abstract: Organic amendment is important for promoting soil quality through increasing soil fertility and soil microbes. This study evaluated the effectiveness of using liquid food waste material (LFM) as a microbial resource, by analyzing the microbial community composition in LFM, and by isolating plant growth-promoting bacteria (PGPB) from the material. High-throughput sequencing of LFM, collected every month from May to September 2018, resulted in the detection of >1000 bacterial operational taxonomic units (OTUs) in the LFM. The results showed that *Firmicutes* was abundant and most frequently detected, followed by *Proteobacteria* and *Actinobacteria*. Of the culturable strains isolated from LFM, almost all belonged to the genus *Bacillus*. Four strains of PGPB were selected from the isolated strains, with traits such as indole acetic acid production and 1-aminocyclopropane-1-carboxylic acid deaminase activity. Lettuce growth was improved via LFM amendment with PGPB, and *Brassica rapa* showed significant differences in root biomass when LFM amendment was compared with the use chemical fertilizer. Field experiments using LFM showed slight differences in growth for *Brassica rapa*, lettuce and eggplant, when compared with the use of chemical fertilizer. LFM is a useful microbial resource for the isolation of PGPB, and its use as fertilizer could result in reduced chemical fertilizer usage in sustainable agriculture.

Keywords: bacterial community composition; liquid food waste materials (LFM); plant growth-promoting bacteria (PGPB); plant growth-promoting (PGP) traits

1. Introduction

Plant growth-promoting microorganisms (PGPM) are broadly accepted to enhance crop production [1]. Plant growth-promoting microorganisms enhance plant growth and development through a variety of functions, encompassing the increase of macro-nutrient availability to the host plant by assembly of growth-promoting chemicals [2], nitrogen fixation [3], solubilization of inorganic phosphate and mineralization of organic phosphate [4], production of different types of phytohormones-like organic compounds [5,6] and biological control of phytopathogens by synthesizing antibiotics and/or competing with harmful microorganisms [7,8]. Therefore, the continuous use of PGPM could lead to it replacing pesticides and chemical fertilizers [9].

On another front, the overuse of chemical fertilizers and continuous agricultural activities results in the deterioration of soil quality [10,11]. The associated loss of soil health, fertility and nutrient status leads to continuous input requirements. Crop nutrition needs can be met through the provision of inorganic as well as organic fertilizers and biofertilizers. Overreliance on inorganic fertilizers stretches the economics of the farming community, and also leads to consumption of available non-renewable nutrient resources, compromises the potential plant-beneficial microbiome, and can have a severe environmental impact [12]. In contrast, the concerns around the use of organic fertilizers include that they are bulky, slow release, have inconsistent composition and can spread weed seeds, among other things. Therefore, sustainable solutions must be sought for crop production, while focusing on the utilization of all available resources. Organic waste production, which can be animal- or plant-based, including food leftovers, vegetable and fruit peels and market refuse, is a worldwide issue, and its disposal and treatment is increasingly important in developing countries [13]. The large amount of this waste that is produced is a major economic, social and environmental challenge [14], which is associated with extensive handling costs. There are great potential benefits to recycling and reusing this material in agriculture. Recent efforts have led to up to a 25% reduction in food waste in some parts of the world, however, in Japan, although food waste legislation has helped to reduce the volume of food waste produced, more needs to be done in addressing this issue, as reviewed by [15]. General waste, other than that generated by food processing industries and households, contains about 60% organic matter [16]. Hence, the separation of organic matter from general waste streams should be targeted, and treated as a resource rather than a problem [11,17].

Organic waste contains fatty acids, proteins and carbohydrates [18,19] among other constituents, which can be utilized as a source of crop nutrition. The application of organic waste materials in agriculture has been reported to reduce runoff, improve soil structure and increase soil biological activity [20]. In addition, some research has showed that local effective microorganisms (LEM) are a beneficial inoculant for the nitrogen mineralization of organic materials [21,22]. Therefore, the better management of organic waste materials could lead to preservation of soil quality and sustainable crop production [9]. Previous studies have explored the potential of the utilization of food and organic wastes in domestic, agricultural and industrial applications [11,23–25]. Among the variety of waste processing and manipulation procedures prior to their application in agriculture, most have had associated physical, chemical or biological problems. In the effective utilization of food waste, quick manipulation, easy operation and little or no reduction in the nutritional composition of the waste products are all considered publicly acceptable, and could increase the waste's potential for wide application and dissemination. Under this scenario, nutrient retention can be ensured, and minimal damage to the plant-beneficial microbes present in the food waste would be achieved. A food waste recycling facility started operating in 2014 in Kai-City, Yamanashi Prefecture, Japan, which collects food waste from school restaurants in the vicinity, processes the waste using lactic acid fermentation, and supplies the final product in liquid form to farmers (Kai City Biomass [26]). The food waste recycling facility has a structure divided into four phases. In the first phase, food wastes and water, along with an inoculum of microorganisms, such as lactic acid bacteria, are added, and the mixture is agitated and gradually moved to phases 2, 3 and 4. During that time, the pH drops to 3 and the temperature rises above 50 °C to promote fermentation. This liquid food waste material (LFM) has been used as a crop nutrient source by many farmers in the area. Although LFM has been mainly employed for use in agriculture and/or for energy production, the microbiological potential of the plant growth-promoting microbes in LFM has not been studied, to the best of our knowledge. In this study, we explored the microbial community composition of the final form of the recycled waste materials, and studied the ecology of those microorganisms, while also investigating the plant growth-promoting traits of the culturable bacterial isolates.

2. Materials and Methods

2.1. Liquid Food Waste Materials

The LFM was obtained in March 2017 and March 2018 for pot and field experiments, respectively, and in May, June, July, August and September 2018 for bacterial composition analysis (and in October for isolation of microbes), from the Biomass Center at Kai City, Yamanashi, Japan. A portion (50 mL) of the material each month was stored at −80 °C for DNA extraction and high-throughput sequencing.

The Kai City facility produces LFM from residues of local school-provided lunches using the lactic acid fermentation process at the rate of approximately 90 L/day^{-1}. The total carbon and nitrogen of the final form of fertilizer were 31.7% and 1.41%, respectively. The C/N ratio of LFM was 22.4. In addition, the pH of LFM was 3.42 because of the lactic acid fermentation process. Electrical conductivity (EC), nitrate–nitrogen (NO_3^--N), ammonium-N (NH_4^+-N) and available phosphate (Trough-P) were at the values of 6.65 mS/cm^{-1}, 0.95 mg/L^{-1}, 14.8 mg/L^{-1} and 0.69 mg/L^{-1}, respectively.

2.2. Assay of Liquid Food Waste Material (LFM) Utilization

An incubation experiment was carried out to assess the mineralization of NO_3^--N from the LFM according to a modified Soil Environmental Analytical Method, 1997. A total of 100 mL of the LFM material was weighed and mixed with 300 g of soil obtained from University of Yamanashi (UofY) Research Farm (hereinafter referred to as UofY farm soil); soil type is gray lowland soil (pH 6.79 ± 0.33; EC (mS/cm^{-1}) 0.11 ± 0.08; NO_3^--N 23.1 ± 3.13 mg/kg^{-1}; available phosphate 421 ± 86.8 mg/kg^{-1}). The pots were covered by aluminium foil and incubated for 14 weeks at 25 °C. Since rapeseed cake is used as an organic fertilizer, it was used as a control for nitrogen release after application to soil. The NO_3^--N content was measured via the alkali reduction diazo dye method (Soil Environmental Analytical Method, 1997).

2.3. Isolation of Bacteria from LFM

Bacterial isolation from LFM was performed through the dilution plating technique. A total of 1 mL of LFM and 4.0 mL of sterile distilled water was placed in a test tube and mixed thoroughly using a vortex mixer. Subsequently, 50-μL dilutions were taken from the first tube and spread onto Reasoner's 2A agar (R2A) media (Eiken Chemical Co. Ltd., Tochigi, Japan) using a disposable spreader; plates of each dilution were incubated at 25 °C for 3 days. Colonies appearing after 3 days were re-streaked until a single pure colony type per plate was achieved.

2.4. DNA Extraction and PCR Amplification for Culturable Bacteria

DNA was extracted from the isolated strains using the ZR Fungal/Bacterial DNA MiniPrep Kit™ (Zymo Research Corp., Irvine, CA, USA). 16S rRNA gene sequencing was carried out for identification of the strains. Extracted DNA from isolated strains was mixed with prior to PCR amplification. The universal primers 341F (5′-CCTACGGGAGGCAGCAG-3′) and 1378R (5′-TGTGCAAGGAGCAGGGAC-3′) were used to amplify the 16S rRNA gene on a TaKaRa PCR Thermal Cycler Dice® Series Gradient (Takara, Shiga, Japan). The PCR amplification conditions were as follows: 95 °C for 5 min, followed by 30 cycles of 94 °C for 30 s, 58 °C for 30 s, and 72 °C for 1 min, followed by a final extension at 72 °C for 7 min [27]. The amplification mixture for PCR (total volume: 25 μL) contained 1 μL of DNA template, 1 μL of each primer, 9.5 μL of sterilized distilled water and 12.5 μL of GoTaq Green Master mix (Promega, Madison, WI, USA). The amplification products (5 μL) were subjected to electrophoresis on a 1% (w/v) agarose gel in tris-acetate-ethylenediaminetetraacetic acid buffer at 100 V for 25 min, and visualized by GelRed™ staining (1:20,000 dilution; Biotium, Fremont, CA, USA). The DNA sequences obtained were compared with those previously reported in the DNA Data Bank of Japan (http://blast.ddbj.nig.ac.jp/), and the nearest neighbor was noted. The sequences of numbers 2, 4, 6 and 11 were submitted to DNA Data Bank of Japan (DDBJ).

2.5. High-Throughput DNA Sequencing

DNA was isolated from the stored LFM samples using the FastDNA™ Spin Kit for Soil (MP Biomedicals Japan, Tokyo, Japan). The DNA concentration was measured using a nano-spectrophotometer and DNA was diluted to 1 ng/μL^{-1} using sterile water. The V4 region of the 16S rRNA gene was amplified using the primers 515F (5′-GTGCCAGCMGCCGCGGTAA-3′) and 806R (5′-GGACTACHVGGGTWTCTAAT-3′) with additional barcode sequences. All PCR reactions were carried out with Phusion® High-Fidelity PCR Master Mix (New England Biolabs

Japan Inc., Tokyo, Japan). The quality and quantity of PCR products was assessed by mixing equal volumes of a loading buffer (containing SYBR green) with PCR products and electrophoresing the samples on 2% (w/v) agarose gel. Samples with a bright main strip between 400 and 450 bp were chosen for further experiments. The PCR products were mixed in equal density ratios. Thereafter, the mixed PCR products were purified with a Qiagen Gel Extraction Kit (Qiagen, Hilden, Germany). The libraries—250 bp paired-end reads generated with NEBNext® UltraTM DNA Library Prep Kit for Illumina (New England Biolabs Japan Inc.,Tokyo, Japan) and quantified via Qubit and quantitative PCR—were sequenced on an Illumina HiSeq 2500 platform. Quality control was performed at each step of the procedure. Paired-end reads were assigned to samples based on their unique barcode and truncated by cutting off the barcode and primer sequence. Paired-end reads were merged using FLASH (V1.2.7, http://ccb.jhu.edu/software/FLASH/) [28]. Quality filtering of the raw tags was performed under specific filtering conditions to obtain high-quality clean tags [29] according to the QIIME (V1.7.0, http://qiime.org/scripts/split_libraries_fastq.html) quality control process [30]. The tags were compared with the reference database (Gold database, http://drive5.com/uchime/uchime_download.html) using the UCHIME algorithm (UCHIME Algorithm, http://www.drive5.com/usearch/manual/uchime_algo.html) [31] to detect chimera sequences (http://www.drive5.com/usearch/manual/chimera_formation.html). Next, the chimera sequences were removed [32], and the effective tags were finally obtained. Sequence analysis was performed via Uparse software (Uparse v7.0.1001 http://drive5.com/uparse/) using all the effective tags [33]. Sequences with ≥97% similarity were assigned to the same operational taxonomic units (OTUs). A representative sequence for each OTU was screened for further annotation. For each representative sequence, Mothur software was used against the small subunit rRNA database of SILVA (http://www.arb-silva.de/) [34] for species annotation at each taxonomic rank (Threshold: 0.8–1) [35]. The phylogenetic relationship of the representative sequences of all OTUs was obtained by using MUSCLE software (Version 3.8.31, http://www.drive5.com/muscle/) for rapid comparison of multiple sequences [36]. The abundance of OTUs was normalized using a standard sequence number corresponding to the sample with the least sequences. Subsequent analyses were all performed based on this output normalized data.

The reads were submitted to the DDBJ Sequence Read Archive (https://www.ddbj.nig.ac.jp/dra/index-e.html) under Bioproject, and are available under accession number DRA010367.

2.6. Plant Growth-Promoting Traits of Isolates

Indole acetic acid production: The isolated strains were tested for indole-3-acetic acid (IAA) production. Cultures of each isolate were grown at 25 °C for 4 days in IAA production media (2 g beef extract, 3 g $CaCO_3$, 30 g glucose, pH 7 in 1 L of distilled water) with or without 1 mM (final concentration) tryptophan. The cultures were centrifuged at 10,000 g for 10 min. IAA production was measured in 300 μL of supernatant using 1.2 mL of Salkowski's reagent [37,38]; absorbance was measured at 535 nm in a spectrophotometer, and the concentration was estimated from a standard curve. Control/blank samples were prepared without bacterial inoculation.

1-aminocyclopropane-1-carboxylic acid (ACC) deaminase and nitrogen fixation: DNA was extracted using ZR Bacterial/Fungal DNA MiniPrep Kit™ (Zymo Research Corp., Irvine, CA, USA). The PCR amplification conditions for ACC deaminase and nifH genes were as follows: 1 cycle of 95 °C for 5 min, then 30 cycles of 94 °C for 30 s, 55 °C for 30 s, and 72 °C for 1 min, followed by a final extension at 72 °C for 7 min using a T100™ Thermal Cycler (Bio-rad, Hercules, CA, USA). The PCR mixture (total volume: 25 μL) contained 1 μL of DNA template, 1 μL of 10 mM primers (Po1F (5′ TGCGAYCCSAARGCBGACTC 3′) and Po1R (5′ ATS GCC ATCATY TCR CCG GA 3′) [39] for nifH genes; ACCF (5′ GCCAARCGBGAVGACTGCAA 3′) and ACCR (5′ TGCATSGAYTTGCCYTC 3′) [40] for ACC deaminase), 12.5 μL of GoTaq® Green Master Mix and 9.5 μL sterilized distilled water. The PCR amplification products were checked via 1.0% of agarose gel electrophoresis, staining and visualization.

Siderophore production: A slightly modified Chrome Azurol S (CAS) method was used for determination of siderophore production by bacterial isolates [41,42]. A total of 100 mL of medium was prepared as follows: 7.3 mg of hexadecyl trimetly ammonium bromide, 6.04 mg of CAS, 3.04 g of piperazine-1,4-bis(2-ethanesulfonic acid) and 1 mL of 1 mM $FeCl_3 \cdot 6H_2O$. A quantity of 10 mL of the siderophore production medium was applied over the surface of agar plates containing cultivated microorganisms. The blue CAS agar changed to light yellow or orange if siderophores were produced by the bacteria; the siderophore production was evaluated by the following index: + color change, − no color change and ++ color change detected over the entire medium.

Phosphate solubilization: The medium developed by Pikovskaya [43] was used for qualitative estimation of calcium phosphate solubilization by the isolates. Selected strains were inoculated into the media and incubated for 7 days at 25 °C. Zones of clearance around the bacterial colonies were indicative of phosphate solubilization; the results were compiled on the basis of the following index: − No clear zone, ± detectable clear zone but very weak activity, + detectable clear zone.

2.7. Pot Experiments

Two pot experiments were conducted using UofY farm soil. The first compared selected bacterial isolates with an uninoculated control to evaluate the role of specific isolates, and the second compared LFM with an untreated control and a fertilizer control to determine the role of LFM in plant growth promotion. For the first experiment, 11 bacterial isolates, that were selected based on PGP traits, were compared with an uninoculated control in a *Lactuca sativa* var. crispa (lettuce) growth trial. Pots (size: 100 cm^2; Fujiwara Seisakusho, Ltd., Tokyo, Japan) were filled with 300 g of soil (dry weight) and soil moisture was maintained at 60% of water holding capacity daily. A suspension of each bacterial strain (grown for 48 h (stationary phase) at 25 °C with shaking in PDB medium) was applied to the pots while the same volume of uninoculated PDB was applied as the control. Subsequently, a lettuce seedling germinated on a petri dish was transferred to each pot. The plants were grown for five weeks, harvested, and the dry weight of aboveground and belowground parts was recorded after being put into the dry oven set at 60 °C. From the results of the pot experiment using lettuce, four isolates (numbers 2, 4, 6 and 11), which showed the maximum growth-enhancement of lettuce, were selected (data not shown), and then these strains were tested under similar growth conditions for *Brassica campestris* (brassica) and the same parameters were recorded. In the second experiment, the response to LFM was compared with that of chemical fertilizer and control treatments. To achieve this goal, 100 mL of LFM and chemical fertilizer (HYPONeX Japan Corp., Ltd., Osaka, Japan. Liquid Fertilizer, N:P:K = 6:10:5) was mixed with soil to achieve a final concentration of 200 mg/kg^{-1} soil N, while there was no amendment in the control pots. Similar growing conditions and parameters were recorded as for the first experiment.

2.8. Field Experiment Using LFM

A field experiment was conducted comparing chemical fertilizer with LFM at the University of Yamanashi Research Farm (35°60′39.5″ N, 138°57′82.9″ E). The field experimental plots (4 m × 2 m) were treated with LFM and chemical fertilizer. In this field, the chemical fertilizer plots have been continuously treated with chemical fertilizer, and cow compost was applied every two years in all subplots until the year before the study. Soil chemical properties were as follows: pH (H_2O) 7.0, EC 0.12 (mS/cm^{-1}), ex-Ca 2940 mg/kg^{-1}, ex-Mg 874 mg/kg^{-1}, ex-K 381 mg/kg^{-1}, CEC 14 (cmolc/kg^{-1}), Trough-P 344 mg/kg^{-1}, NH_4-N 5.8 mg/kg^{-1}, NO_3-N 59.4 mg/kg^{-1}. Two replicates were prepared for each of the test vegetables: *Brassica rapa* var., *Lactuca sativa* var. crispa (lettuce) and *Solanum melongena* (eggplant). LFM was input at the rate of 200 kg/ha^{-1}, 200 kg/ha^{-1} and 75 kg/ha^{-1} to the final concentration of soil N for brassica, lettuce and eggplant, respectively, whereas chemical fertilizer input was 200 kg/ha^{-1} soil N for all crops. *Brassica rapa* and lettuce were planted at 20 to 30 plants/plot, and eggplant was cultivated at 9 plants/plot. *Brassica rapa* was grown for 29 days and plant height was measured upon harvest. Lettuce was grown for 56 days; plant height and dry weight of the edible part were measured at harvest. The eggplants were harvested when the fruits grew to a suitable size (around 120 g/fruit); the quantity and weight of the fruits were measured. Fruit harvest began on 12 August 2018 and continued until 29 September 2018.

2.9. Statistical Analysis

Analysis of variance (ANOVA) was carried out to determine the statistical effects of treatments in Statistix 8.0 (Analytical Software, Tallahassee, FL, USA) followed by pairwise comparison of treatment means using Tukey's honestly significant difference (HSD) test and multiple comparisons through Dunnett's test. For the veracity of sequencing data analysis, raw data was merged and filtered to obtain clean data. Effective data was used for operational taxonomic unit (OTUs) clustering. The clustering analysis was applied, and a clustering tree was constructed to study the similarities among different samples. The unweighted pair–group method with arithmetic means (UPGMA) Clustering was performed as a type of hierarchical clustering method to interpret the distance matrix using average linkage, and was conducted using QIIME software (Version 1.7.0).

3. Results

3.1. Isolation of Plant Growth Promoting Bacteria (PGPB)

The bacterial isolation from the LFM was performed via standard methods, through the serial dilution plating technique. Various different strains appeared in the media. The number of culturable bacteria in the LFM was 3.5×10^4 colony forming units/mL^{-1} of LFM. After isolation, 31 strains were randomly selected, and 11 out of those 31 strains were examined for plant growth-promoting (PGP) traits. The sequence of strain numbers 2, 4, 6 and 11 (approximately 940 nt; GenBank accession No. LC553393, LC553394, LC553395, LC553396) was compared with other bacterial nucleotide sequences in GenBank. All strains exhibited a high sequence similarity with *Bacillus* spp.

3.2. Identification of Culturable Bacteria

All 31 isolates that were identified belonged to genus *Bacillus* (Figure S2); these were type A—closely related to *Bacillus velezensis* strain FZB42 (frequency: 3.3%), type B—closely related to *Bacillus amyloliquefaciens* strain MPA 1034 (frequency: 56.7%), type C—closely related to *Bacillus vallismortis* strain NRRL B-14890 (frequency: 26.7%), type D—closely related to *Bacillus subtilis* subsp. *inaquosorum* strain BGSC 3A28 (frequency: 3.3%), type E—closely related to *Bacillus wiedmannii* strain FSL W8-0169 (frequency: 3.3%), type F—closely related to *Bacillus velezensis* strain NTGB-29 (frequency: 3.3%), and type G—closely related to *Bacillus vallismortis* strain DSM 11031 (frequency: 3.3%).

3.3. PGP Traits

Because of the importance of microbial IAA production in influencing the root architecture and initial plant growth, IAA production was examined for 14 of the 31 isolates. Strain numbers 2, 4, 6 and 11 were positive for IAA production with tryptophan (Table 1). Amplification of the *nifH* gene confirmed N fixation potential in strain number 6, whereas amplification of the ACC deaminase gene was positive for all four selected strains (Table 1). Only strain number 11 showed zones of clearance on the Pikovskaya agar plates, indicating the phosphate solubilization ability of this strain (Table 1). Strain number 11 was also positive for siderophore production, with complete color change from blue to yellow, when compared with other non-siderophore-producing strains and the control that had a negative reaction.

Table 1. Plant growth-promoting traits, where + indicates the possession of the following trait, and − indicates the lack of the trait.

Strain No.	Indole-3-Acetic Acid (IAA)	Phosphate Solubilization	Nitrogen Fixation	1-Aminocyclopropane-1-Carboxylic Acid Deaminase	Siderophore
1	−	−	−	−	−
2	+	−	−	+	−
3	−	−	−	−	+
4	+	−	−	+	−
5	-	−	−	+	−
6	+	−	+	+	−
7	−	+	−	−	−
8	−	+	−	−	−
9	−	−	-	+	+
10	−	+	−	−	−
11	+	+	−	+	+

3.4. High-Throughput DNA Sequencing of LFM

In total, 192,355 reads were obtained; the average number of observed species per sample was 1013 ± 170 (max: 1306, min: 892), and the coefficient of variation was 0.17. The average bacterial composition was shown via the integration of the clustering results and the relative abundance of each sample by phylum (Figures 1a,b and S3). *Proteobacteria* were most frequently detected, followed by *Firmicutes* and *Actinobacteria*. The species composition at the phylum level was different in August when compared with that from the other months (Figure 1a). At the genus level, when the top 10 genera were compared between different months, the genera composition of the LFM in May was different from that from the other months (Figures 1a,b and S3). This is because there were few *Lactobacillus* spp. at this time, and the fermentation was in the early stages.

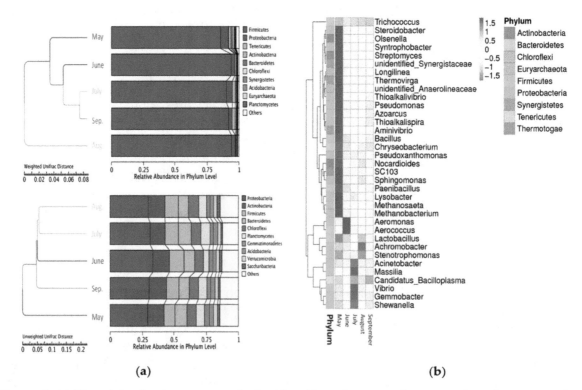

(a)   (b)

Figure 1. (a) Unweighted pair–group method with arithmetic means cluster tree based on unweighted and weighted unifrac distance. This was displayed with the integration of clustering results and the relative abundance of each sample by phylum. (b) The top 10 taxa at the genus level were selected to form the distribution histogram of relative abundance.

3.5. Incubation Study of LFM Utilization

The release of NO_3^--N and phosphate was used to determine the potential for nutrient provision from LFM. The release of NO_3^--N started from 14 days in the LFM, and approximately 200 mg/kg^{-1} of NO_3^--N had accumulated by the end of 14 weeks of incubation (Figure S1a). Phosphate availability followed the same trend as that of NO_3^--N release (Figure S1b). Available phosphate was rapidly released from rapeseed cake, whereas no phosphate was detected from LFM until week 2. From week 3, the availability of phosphates increased slightly in the LFM.

3.6. Pot Experiments

A pot experiment was conducted to examine the effect of selected strains on the growth of brassica and lettuce. In the initial experiment on lettuce, 11 strains were tested; 4 of these strains showed high activity when compared with the control. Therefore, strains 2, 4, 6 and 11 were tested on brassica, and showed significant differences in their growth-promoting effect (Figure 2). A pot experiment was also conducted to assess the effect of LFM on the growth of both brassica and lettuce. There was no significant difference in the growth characteristics of lettuce between the LFM and chemical fertilizer treatments (Figure 3a); however, brassica exhibited significant differences in its root biomass with LFM amendment (Figure 3b, Tukey's HSD, $p < 0.05$).

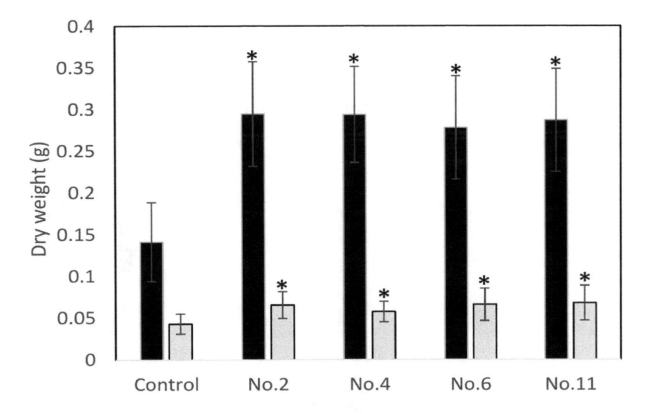

Figure 2. Growth response of *Brassica campestris* with plant growth-promoting bacteria selected in this study. Control ($n = 7$), strain number 2 ($n = 10$), strain number 4 ($n = 10$), strain number 6 ($n = 7$), strain number 11 ($n = 10$), Dunnett test ($p < 0.05$). The vertical bar indicates the standard error. * indicates significance differences between treatments when compared with the control ($p < 0.05$).

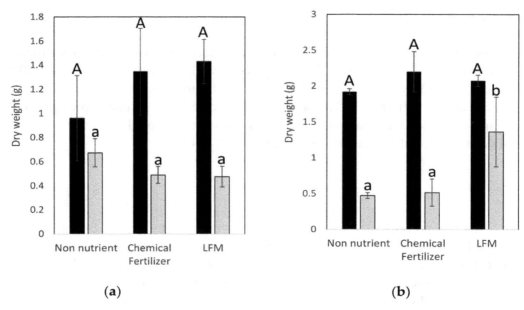

Figure 3. Pot experiment testing lettuce (**a**) and *Brassica rapa* (**b**) growth using liquid food waste materials (LFM). Values presented are means and standard error (*n* = 3). Closed bar and Gray bar mean edible part and root, respectively. Treatments of the same crop with different letters are significantly different by Tukey's HSD (*p* < 0.05).

3.7. Field Experiments

Field experiments were conducted to assess the effect of LFM on the growth of *Brassica rapa*, lettuce and eggplant. The growth of *Brassica rapa* and lettuce in the field was similar to that in the pot experiment. The heights achieved by *Brassica rapa* were 33.0 ± 0.67 cm and 32.5 ± 0.78 cm, with LFM and chemical fertilizer, respectively. The lettuce grown with LFM amendment was slightly larger than that grown with chemical fertilizer, but not significantly so (Figure 4a). Eggplant also grew slightly better with LFM than with chemical fertilizer, but the differences were not significant (Figure 4b,c for eggplant).

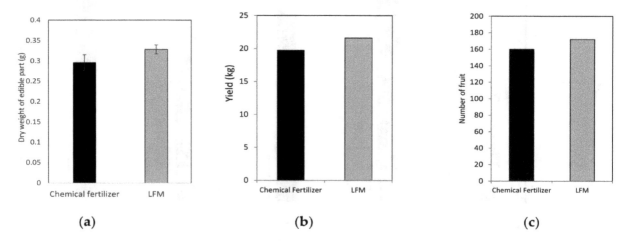

Figure 4. Field experiment testing the effect of liquid food waste materials (LFM) on the growth of Lettuce (**a**) and eggplant (**b,c**).

4. Discussion

This study showed that useful microorganisms, such as PGPB, were present in LFM produced from the recycling of unused resources. This is an important finding that leads to the promotion of recycling, and also indicates that unused resources are useful as microbial resources. LFM can be used as a fertilizer, and has other positive effects on vegetable growth. However, there are limitations to the plant growth-enhancement functions of LFM. If we apply isolated strains to LFM in order to promote a plant growth function more effective than that of the original LFM, some supplementary nutrients are needed to produce an effect. This point still remains an issue.

High-throughput sequencing analysis of the bacterial community composition in the LFM revealed substantial differences between the sampling months May and August. A similar bacterial community composition was observed for the samples from June, July and September. The May sample was still in the early stages of fermentation, and there was no school-provided lunch in August; therefore, LFM was stored in the tank, which might help to explain the differences in the community composition of these samples.

Most of the bacteria isolated from LFM were *Bacillus* spp., which is a spore-forming bacteria. This was because the LFM pH was reduced to 3.0 through the process of lactic acid fermentation, and the temperature exceeded 50 °C. However, this result was obtained because we used R2A media for isolation, and we detected the family of *Lactobacillaceae* in the LFM through high-throughput sequencing (Figure S3). The most frequently isolated *Bacillus* spp. strains were *Bacillus amyloliquefaciens* (56.7%) and *Bacillus vallismortis* (26.7%). *Bacillus* species are known to produce dormant spores [44], and enact an anti-pathogen activity through the assembly of non-ribosomal cyclic lipopeptides [45]. In addition, *Bacillus* species are considered PGPB because of their potential for antibiotic production, biofilm formation on the plant root surface, and production of plant hormones [46,47]. Furthermore, seed treatment with *Bacillus* species has been shown to significantly enhance shoot fresh and dry weight, as well as plant height, in various crops [48,49]. From the results of this study, >1000 bacterial OTUs were identified in LFM; therefore, it might be possible to isolate other useful strains, other than *Bacillus* spp., under a range of isolation conditions, including increased pH.

Organic materials slowly release nutrients, but they are still a promising alternative to chemical fertilizers, as their application can reduce nutrient leaching, volatilization and problems of toxicity [50]. In the present study, LFM was used as an alternative to chemical fertilizers in order to investigate the release of NO_3^--N and available phosphate (Figure S1). Low amounts of available phosphate were released from LFM during the incubation study because of the low total phosphate concentration in LFM. LFM released NO_3^--N from the third week of incubation, and released approximately 200 kg/ha^{-1} NO_3^--N during the 13 weeks of incubation. Moreover, the biomass richness of soil fertilized with LFM was higher than that treated with chemical fertilizer [51], and LFM did not change the soil pH after treatment through our study.

The growth-promoting effect of strain number 11, with an IAA-producing ability and an ACC deaminase activity, was confirmed in a pot experiment; growth was significantly promoted with inoculation by strain number 11, when compared with the control (Figure 2). The growth of lettuce in a pot was the same with both LFM and chemical fertilizer. For *Brassica rapa*, the growth of the edible (aboveground) part was the same with both LFM and chemical fertilizer, whereas the root biomass was significantly increased with LFM, when compared with chemical fertilizer (Figure 3b). These results indicate that the PGPB in LFM contributed to the increase in root biomass of *Brassica rapa*. A field experiment was conducted to assess the effect of LFM on the growth of *Brassica rapa*, lettuce and eggplant. Although the N input by LFM was less than half that of the chemical fertilizer, the growth of the eggplant with LFM was the same as that with chemical fertilizer. The yield with LFM was higher than, but not significantly different from, that achieved with chemical fertilizer. These results indicate the abundance and activity of PGPB in LFM, and their efficacy in supporting eggplant growth under the conditions tested. LFM could be a viable alternative to commercially available chemical fertilizers, without an adverse effect on soil and vegetable growth. A previous study showed the positive effect of PGPB inoculation on vegetable growth and yield [52].

All the selected strains that showed growth-enhancement in the pot experiment had an IAA-producing ability (Table 1). IAA is a type of plant hormone that promotes root elongation and enhances root growth. Many PGPBs with the IAA-production ability have been isolated in previous studies [1,53]. Furthermore, all strains that were positive for IAA production also showed ACC deaminase activity (strain numbers 2, 4, 6 and 11). This suggests that IAA production and ACC deaminase activity contribute greatly to enhancing the plant growth in our isolated strains, while *Caulobacter* sp. had a negative impact on plant growth, even though it produced higher levels of IAA [54]. In addition, previous studies have shown that PGP microbes and PGPB can promote plant growth indirectly or directly, through the production of ACC deaminase and through reducing the ethylene level in the developing plants through the roots [52,55], by generating plant growth hormones like IAA [56]. It is likely that ACC deaminase and IAA production promote root growth in a similar fashion [57,58]. Of the selected strains, only strain number 11 showed phosphorus solubilizing potential and siderophore production (Table 1). Phosphate solubilization is effective in soils with low available phosphoric acid, and siderophore production chelates the iron in soils with high pH to help plant uptake [52,59]. However, the detailed mechanism of plant growth-enhancement is complex, and further investigation is needed [54,60].

The selection of PGPB strains from LFM was important to confirm the positive effect of the inoculants on plant growth, and to optimize their application for maximum impact on vegetable crops. The main aim of this study was to reduce the commercial use of chemical fertilizers, by utilizing LFM as an alternative fertilizer with the maximum impact on crop growth and soil, and minimal environmental impact. Further investigation into LFM use as an organic fertilizer should evaluate any adverse impact of its application to the soil environment.

Supplementary Materials:
Figure S1: Nitrate nitrogen (**a**) and available phosphate (**b**) release from organic materials, ● rapeseed oil cake, ▲ Liquid food waste material (LFM). Figure S2: 31 strains isolated from liquid food waste material belong to genus *Bacillus*. Type A closely related to *Bacillus velezensis* strain FZB42, type B closely related to *Bacillus amyloliquefaciens* strain MPA 1034, type C closely related to *Bacillus vallismortis* strain NRRL B-14890, type D closely related to *Bacillus subtilis* subsp. inaquosorum strain BGSC 3A28, type E closely related to *Bacillus wiedmannii* strain FSL W8-0169, type F closely related to *Bacillus velezensis* strain NTGB-29, and type G closely related to *Bacillus vallismortis* strain DSM 11031. Figure S3: Top 30 bacterial compositions, in different taxonomic levels.

Author Contributions: Conceptualization, R.K.; methodology, S.K., R.I., R.K., W.A., A.M.; writing—original draft preparation, R.K., W.A., A.M.; writing—review and editing, R.K.; visualization, R.K., W.A.; supervision, R.K.; project administration, R.K.; funding acquisition, R.K. All authors have read and agreed to the published version of the manuscript.

Acknowledgments: The authors are grateful to Mr. Yuta KOBAYASHI for supporting to manage the field conditions. We also thank Kai city for their many supports.

References

1.   Wang, J.; Li, R.; Zhang, H.; Wei, G.; Li, Z. Beneficial bacteria activate nutrients and promote wheat growth under conditions of reduced fertilizer application. *BMC Microbiol.* **2020**, *20*, 1–12.

2.   Khan, N.; Bano, A.; Rahman, M.A.; Guo, J.; Kang, Z.; Babar, M.A. Comparative physiological and metabolic analysis reveals a complex mechanism involved in drought tolerance in chickpea (*Cicer arietinum* L.) induced by PGPR and PGRs. *Sci. Rep.* **2019**, *9*, 1–19.

3.   Lin, L.; Li, Z.; Hu, C.; Zhang, X.; Chang, S.; Yang, L.; Li, Y.; An, Q. Plant growth-promoting nitrogen-fixing enterobacteria are in association with sugarcane plants growing in Guangxi, China. *Microbes Environ.* **2012**, *27*, 391–398. [CrossRef] [PubMed]

4.   Zaheer, A.; Malik, A.; Sher, A.; Qaisrani, M.M.; Mehmood, A.; Khan, S.U.; Ashraf, M.; Mirza, Z.; Karim, S.; Rasool, M. Isolation, characterization, and effect of phosphate-zinc-solubilizing bacterial strains on chickpea (*Cicer arietinum* L.) growth. *Saudi J. Biol. Sci.* **2019**, *26*, 1061–1067. [CrossRef]

5.   Sivasankari, B.; Anandharaj, M.; Daniel, T. Effect of PGR producing bacterial strains isolated from vermisources on germination and growth of *Vigna unguiculata* (L.) Walp. *J. Biochem. Technol.* **2014**, *5*, 808–813.

6.   Grobelak, A.; Kokot, P.; Hutchison, D.; Grosser, A.; Kacprzak, M. Plant growth-promoting rhizobacteria as an alternative to mineral fertilizers in assisted bioremediation-sustainable land and waste management. *J. Environ. Manag.* **2018**, *227*, 1–9. [CrossRef]

7.   Sun, G.; Yao, T.; Feng, C.; Chen, L.; Li, J.; Wang, L. Identification and biocontrol potential of antagonistic bacteria strains against Sclerotinia sclerotiorum and their growth-promoting effects on Brassica napus. *Biol. Control.* **2017**, *104*, 35–43. [CrossRef]

8.   Etesami, H.; Emami, S.; Alikhani, H.A. Potassium solubilizing bacteria (KSB): Mechanisms, promotion of plant growth, and future prospects A review. *J. Soil Sci. Plant. Nutr.* **2017**, *17*, 897–911. [CrossRef]

9.   Padmini, O.S.; Wuryani, S.; Aryani, R. Application of Organic Fertilizer and Plant Growth-Promoting Rhizobacteria (PGPR) to Increase Rice Yield and Quality. In *ICoSI: Proceedings of the 2nd International Conference on Sustainable Innovation 2014*; Taufik, T., Prabasari, I., Rineksane, I.A., Yaya, R., Widowati, R., Rosyidi, S.A.P., Riyadi, S., Harsanto, P., Eds.; Springer: Berlin/Heidelberg, Germany, 2017; pp. 3–11.

10.   Komatsuzaki, M.; Ohta, H. Soil management practices for sustainable agro-ecosystems. *Sustain. Sci.* **2007**, *2*, 103–120. [CrossRef]

11.   Mahmood, A.; Iguchi, R.; Kataoka, R. Multifunctional food waste fertilizer having the capability of Fusarium-growth inhibition and phosphate solubility: A new horizon of food waste recycle using microorganisms. *Waste Manag.* **2019**, *94*, 77–84. [CrossRef]

12.   Mahmood, A.; Kataoka, R. Metabolite profiling reveals a complex response of plants to application of plant growth-promoting endophytic bacteria. *Microbiol. Res.* **2020**, *234*, 126421. [CrossRef] [PubMed]

13.   Sinha, R.K.; Herat, S.; Bharambe, G.; Patil, S.; Bapat, P.; Chauhan, K.; Valani, D. Human waste-as potential resource: Converting trash into treasure by embracing the 5 r's philosophy for safe and sustainable waste management. *Environ. Res. J.* **2009**, *13*, 111–171.

14.   Gustavsson, J.; Cederberg, C.; Sonesson, U.; Van Otterdijk, R.; Meybeck, A. *Global Food Losses and Food Waste*; FAO: Rome, Italy, 2011.

15.   Liu, C.; Hotta, Y.; Santo, A.; Hengesbaugh, M.; Watabe, A.; Totoki, Y.; Allen, D.; Bengtsson, M. Food waste in Japan: Trends, current practices and key challenges. *J. Clean. Prod.* **2016**, *133*, 557–564. [CrossRef]

16.   Lin, C.S.K.; Pfaltzgraff, L.A.; Herrero-Davila, L.; Mubofu, E.B.; Abderrahim, S.; Clark, J.H.; Koutinas, A.A.; Kopsahelis, N.; Stamatelatou, K.; Dickson, F. Food waste as a valuable resource for the production of chemicals, materials and fuels. Current situation and global perspective. *Energy Environ. Sci.* **2013**, *6*, 426–464. [CrossRef]

17.   Baroutian, S.; Munir, M.T.; Sun, J.Y.; Eshtiaghi, N.; Young, B.R. Rheological characterisation of biologically treated and non-treated putrescible food waste. *Waste Manag.* **2018**, *71*, 494–501. [CrossRef]

18.   Yan, S.; Li, J.; Chen, X.; Wu, J.; Wang, P.; Ye, J.; Yao, J. Enzymatical hydrolysis of food waste and ethanol production from the hydrolysate. *Renew. Energy* **2011**, *36*, 1259–1265. [CrossRef]

19.   Leung, C.C.J.; Cheung, A.S.Y.; Zhang, A.Y.-Z.; Lam, K.F.; Lin, C.S.K. Utilisation of waste bread for fermentative succinic acid production. *Biochem. Eng. J.* **2012**, *65*, 10–15. [CrossRef]

20.   Marinari, S.; Masciandaro, G.; Ceccanti, B.; Grego, S. Influence of organic and mineral fertilisers on soil biological and physical properties. *Bioresour. Technol.* **2000**, *72*, 9–17. [CrossRef]

21.   Ney, L.; Franklin, D.; Mahmud, K.; Cabrera, M.; Hancock, D.; Habteselassie, M.; Newcomer, Q.; Dahal, S. Impact of inoculation with local effective microorganisms on soil nitrogen cycling and legume productivity using composted broiler litter. *Appl. Soil Ecol.* **2020**, *154*, 103567. [CrossRef]

22.   Ney, L.; Franklin, D.; Mahmud, K.; Cabrera, M.; Hancock, D.; Habteselassie, M.; Newcomer, Q. Examining trophic-level nematode community structure and nitrogen mineralization to assess local effective microorganisms' role in nitrogen availability of swine effluent to forage crops. *Appl. Soil Ecol.* **2018**, *130*, 209–218. [CrossRef]

23.   Truong, L.; Morash, D.; Liu, Y.; King, A. Food waste in animal feed with a focus on use for broilers. *Int. J. Recycl. Org. Waste Agric.* **2019**, *8*, 417–429. [CrossRef]

24.   Torres-León, C.; Ramírez-Guzman, N.; Londoño-Hernandez, L.; Martinez-Medina, G.A.; Díaz-Herrera, R.; Navarro-Macias, V.; Alvarez-Pérez, O.B.; Picazo, B.; Villarreal-Vázquez, M.; Ascacio-Valdes, J.; et al. Food waste and byproducts: An opportunity to minimize malnutrition and hunger in developing countries. *Front. Sustain. Food Syst.* **2018**, *2*. [CrossRef]

25.   Girotto, F.; Alibardi, L.; Cossu, R. Food waste generation and industrial uses: A review. *Waste Manag.* **2015**, *45*, 32–41. [CrossRef]

26. Center, K.-C.B. Food waste recycle. Available online: https://www.city.kai.yamanashi.jp/kurashi_tetsuduki/ gomi_kankyo_pet/kankyo/3704.html (accessed on 27 May 2020).

27. Mahmood, A.; Takagi, K.; Ito, K.; Kataoka, R. Changes in endophytic bacterial communities during different growth stages of cucumber (*Cucumis sativus* L.). *World J. Microbiol. Biotechnol.* **2019**, *35*, 104. [CrossRef]

28. Magoč, T.; Salzberg, S.L. FLASH: Fast length adjustment of short reads to improve genome assemblies. *Bioinformatics* **2011**, *27*, 2957–2963. [CrossRef]

29. Bokulich, N.A.; Subramanian, S.; Faith, J.J.; Gevers, D.; Gordon, J.I.; Knight, R.; Mills, D.A.; Caporaso, J.G. Quality-filtering vastly improves diversity estimates from Illumina amplicon sequencing. *Nat. Methods* **2013**, *10*, 57–59. [CrossRef]

30. Caporaso, J.G.; Kuczynski, J.; Stombaugh, J.; Bittinger, K.; Bushman, F.D.; Costello, E.K.; Fierer, N.; Peña, A.G.; Goodrich, J.K.; Gordon, J.I.; et al. QIIME allows analysis of high-throughput community sequencing data. *Nat. Methods* **2010**, *7*, 335. [CrossRef]

31. Edgar, R.C.; Haas, B.J.; Clemente, J.C.; Quince, C.; Knight, R. UCHIME improves sensitivity and speed of chimera detection. *Bioinformatics* **2011**, *27*, 2194–2200. [CrossRef]

32. Haas, B.J.; Gevers, D.; Earl, A.M.; Feldgarden, M.; Ward, D.V.; Giannoukos, G.; Ciulla, D.; Tabbaa, D.; Highlander, S.K.; Sodergren, E.; et al. Chimeric 16S rRNA sequence formation and detection in Sanger and 454-pyrosequenced PCR amplicons. *Genome Res.* **2011**, *21*, 494–504. [CrossRef]

33. Edgar, R.C. UPARSE: Highly accurate OTU sequences from microbial amplicon reads. *Nat. Methods* **2013**, *10*, 996. [CrossRef]

34. Wang, Q.; Garrity, G.M.; Tiedje, J.M.; Cole, J.R. Naïve Bayesian Classifier for Rapid Assignment of rRNA Sequences into the New Bacterial Taxonomy. *Appl. Environ. Microbiol.* **2007**, *73*, 5261–5267. [CrossRef] [PubMed]

35. Quast, C.; Pruesse, E.; Yilmaz, P.; Gerken, J.; Schweer, T.; Yarza, P.; Peplies, J.; Glöckner, F.O. The SILVA ribosomal RNA gene database project: Improved data processing and web-based tools. *Nucleic Acids Res.* **2013**, *41*, D590–D596. [CrossRef] [PubMed]

36. Edgar, R.C. Muscle: Multiple sequence alignment with high accuracy and high throughput. *Nucleic Acids Res.* **2004**, *32*, 1792–1797. [CrossRef]

37. Acuña, J.; Jorquera, M.; Martínez, O.; Menezes-Blackburn, D.; Fernández, M.; Marschner, P.; Greiner, R.; Mora, M. Indole acetic acid and phytase activity produced by rhizosphere bacilli as affected by pH and metals. *J. Soil Sci. Plant. Nutr.* **2011**, *11*, 1–12.

38. Patten, C.L.; Glick, B.R. Role of Pseudomonas putida indoleacetic acid in development of the host plant root system. *Appl. Environ. Microbiol.* **2002**, *68*, 3795–3801. [CrossRef]

39. Poly, F.; Ranjard, L.; Nazaret, S.; Gourbière, F.; Monrozier, L.J. Comparison of nifH gene pools in soils and soil microenvironments with contrasting properties. *Appl. Environ. Microbiol.* **2001**, *67*, 2255–2262. [CrossRef]

40. Jha, B.; Gontia, I.; Hartmann, A. The roots of the halophyte Salicornia brachiata are a source of new halotolerant diazotrophic bacteria with plant growth-promoting potential. *Plant. Soil* **2012**, *356*, 265–277. [CrossRef]

41. Schwyn, B.; Neilands, J. Universal chemical assay for the detection and determination of siderophores. *Anal. Biochem.* **1987**, *160*, 47–56. [CrossRef]

42. Pérez-Miranda, S.; Cabirol, N.; George-Téllez, R.; Zamudio-Rivera, L.; Fernández, F. O-CAS, a fast and universal method for siderophore detection. *J. Microbiol. Methods* **2007**, *70*, 127–131. [CrossRef]

43. Pikovskaya, R. Mobilization of phosphorus in soil in connection with vital activity of some microbial species. *Mikrobiologiya* **1948**, *17*, 362–370.

44. Piggot, P.J.; Hilbert, D.W. Sporulation of Bacillus subtilis. *Curr. Opin. Microbiol.* **2004**, *7*, 579–586. [CrossRef] [PubMed]

45. Zhang, B.; Qin, Y.; Han, Y.; Dong, C.; Li, P.; Shang, Q. Comparative proteomic analysis reveals intracellular targets for bacillomycin L to induce Rhizoctonia solani Kühn hyphal cell death. *Biochim. Biophys. Acta* **2016**, *1864*, 1152–1159. [CrossRef] [PubMed]

46. Santos, C.A.; Nobre, B.; da Silva, T.L.; Pinheiro, H.; Reis, A. Dual-mode cultivation of Chlorella protothecoides applying inter-reactors gas transfer improves microalgae biodiesel production. *J. Biotechnol.* **2014**, *184*, 74–83. [CrossRef] [PubMed]

47. Rahman, A.; Uddin, W.; Wenner, N.G. Induced systemic resistance responses in perennial ryegrass against M

agnaporthe oryzae elicited by semi-purified surfactin lipopeptides and live cells of B acillus amyloliquefaciens. *Mol. Plant. Pathol.* **2015**, *16*, 546–558. [CrossRef]

48. Yuan, J.; Ruan, Y.; Wang, B.; Zhang, J.; Waseem, R.; Huang, Q.; Shen, Q. Plant growth-promoting rhizobacteria strain Bacillus amyloliquefaciens NJN-6-enriched bio-organic fertilizer suppressed Fusarium wilt and promoted the growth of banana plants. *J. Agric. Food Chem.* **2013**, *61*, 3774–3780. [CrossRef] [PubMed]

49. Gowtham, H.; Murali, M.; Singh, S.B.; Lakshmeesha, T.; Murthy, K.N.; Amruthesh, K.; Niranjana, S. Plant growth promoting rhizobacteria-Bacillus amyloliquefaciens improves plant growth and induces resistance in chilli against anthracnose disease. *Biol. Control.* **2018**, *126*, 209–217. [CrossRef]

50. Machado, D.L.M.; Lucena, C.C.d.; Santos, D.d.; Siqueira, D.L.d.; Matarazzo, P.H.M.; Struiving, T.B. Slow-release and organic fertilizers on early growth of Rangpur lime. *Rev. Ceres* **2011**, *58*, 359–365. [CrossRef]

51. Cheong, J.C.; Lee, J.T.; Lim, J.W.; Song, S.; Tan, J.K.; Chiam, Z.Y.; Yap, K.Y.; Lim, E.Y.; Zhang, J.; Tan, H.T. Closing the food waste loop: Food waste anaerobic digestate as fertilizer for the cultivation of the leafy vegetable, xiao bai cai (*Brassica rapa*). *Sci. Total Environ.* **2020**, *715*, 136789. [CrossRef]

52. Sheirdil, R.A.; Hayat, R.; Zhang, X.-X.; Abbasi, N.A.; Ali, S.; Ahmed, M.; Khattak, J.Z.K.; Ahmad, S. Exploring potential soil bacteria for sustainable wheat (*Triticum aestivum* L.) production. *Sustainability* **2019**, *11*, 3361. [CrossRef]

53. Zhang, Z.; Zhou, W.; Li, H. The role of GA, IAA and BAP in the regulation of in vitro shoot growth and microtuberization in potato. *Acta Physiol. Plant.* **2005**, *27*, 363–369. [CrossRef]

54. Berrios, L.; Ely, B. Plant growth enhancement is not a conserved feature in the Caulobacter genus. *Plant. Soil* **2020**, *449*, 81–95. [CrossRef]

55. Dey, R.; Pal, K.; Bhatt, D.; Chauhan, S. Growth promotion and yield enhancement of peanut (*Arachis hypogaea* L.) by application of plant growth-promoting rhizobacteria. *Microbiol. Res.* **2004**, *159*, 371–394. [CrossRef] [PubMed]

56. Mishra, M.; Kumar, U.; Mishra, P.K.; Prakash, V. Efficiency of plant growth promoting rhizobacteria for the enhancement of *Cicer arietinum* L. growth and germination under salinity. *Adv. Biol Res.* **2010**, *4*, 92–96.

57. Glick, B.R.; Cheng, Z.; Czarny, J.; Duan, J. Promotion of Plant Growth by ACC Deaminase-Producing Soil Bacteria. In *New Perspectives and Approaches in Plant Growth-Promoting Rhizobacteria, Research*; Bakker, P.A.H.M., Raaijmakers, J.M., Bloemberg, G., Höfte, M., Lemanceau, P., Cook, B.M., Eds.; Springer: Berlin/Heidelberg, Germany, 2007; pp. 329–339.

58. Pandey, S.; Gupta, S. ACC deaminase producing bacteria with multifarious plant growth promoting traits alleviates salinity stress in French bean (*Phaseolus vulgaris*) plants. *Front. Microbiol.* **2019**, *10*, 1506.

59. Vansuyt, G.; Robin, A.; Briat, J.-F.; Curie, C.; Lemanceau, P. Iron acquisition from Fe-pyoverdine by Arabidopsis thaliana. *Mol. Plant. Microbe Interact.* **2007**, *20*, 441–447. [CrossRef]

60. Khan, N.; Bano, A.; Babar, M.A. Metabolic and physiological changes induced by plant growth regulators and plant growth promoting rhizobacteria and their impact on drought tolerance in *Cicer arietinum* L. *PLoS ONE* **2019**, *14*, e0213040. [CrossRef] [PubMed]

Co-Inoculation of Rhizobacteria and Biochar Application Improves Growth and Nutrients in Soybean and Enriches Soil Nutrients and Enzymes

Dilfuza Jabborova [1,2,3,*], Stephan Wirth [3], Annapurna Kannepalli [2],
Abdujalil Narimanov [1], Said Desouky [4], Kakhramon Davranov [5], Riyaz Z. Sayyed [6],
Hesham El Enshasy [7,8,9], Roslinda Abd Malek [7], Asad Syed [10] and Ali H. Bahkali [10]

[1]   Laboratory of Medicinal Plants Genetics and Biotechnology, Institute of Genetics and Plant Experimental Biology, Uzbekistan Academy of Sciences, Tashkent Region, Kibray 111208, Uzbekistan; narimanov63@list.ru
[2]   Division of Microbiology, ICAR-Indian Agricultural Research Institute, Pusa, New Delhi 110012, India; annapurna96@gmail.com
[3]   Leibniz Centre for Agricultural Landscape Research (ZALF), D-15374 Müncheberg, Germany; swirth@zalf.de
[4]   Botany and Microbiology Department, Faculty of Science, Al-Azhar University, Cairo 11651, Egypt; dr_saidesouky@yahoo.com
[5]   Institute of Microbiology, Academy of Sciences of Uzbekistan, Tashkent 100128, Uzbekistan; k-davranov@mail.ru
[6]   Department of Microbiology, PSGVP Mandal's, Arts, Science & Commerce College, Shahada 425409, Maharashtra, India; sayyedrz@gmail.com
[7]   Institute of Bioproduct Development (IBD), Universiti Teknologi, Malaysia (UTM), Skudai, Johor Bahru 81310, Malaysia; henshasy@ibd.utm.my (H.E.E.); roslinda@ibd.utm.my (R.A.M.)
[8]   School of Chemical and Energy Engineering, Faculty of Engineering, Universiti Teknologi, Malaysia (UTM), Skudai, Johor Bahru 81310, Malaysia
[9]   City of Scientific Research and Technology Application, New Burg Al Arab, Alexandria 21934, Egypt
[10] Department of Botany and Microbiology, College of Science, King Saud University, P.O. Box 2455, Riyadh 11451, Saudi Arabia; asadsayyed@gmail.com (A.S.); abahkali@ksu.edu.sa (A.H.B.)
*   Correspondence: dilfuzajabborova@yahoo.com

Abstract: Gradual depletion in soil nutrients has affected soil fertility, soil nutrients, and the activities of soil enzymes. The applications of multifarious rhizobacteria can help to overcome these issues, however, the effect of co-inoculation of plant-growth promoting rhizobacteria (PGPR) and biochar on growth andnutrient levelsin soybean and on the level of soil nutrients and enzymes needs in-depth study. The present study aimed to evaluate the effect of co-inoculation of multifarious *Bradyrhizobium japonicum* USDA 110 and *Pseudomonas putida* TSAU1 and different levels (1 and 3%) of biochar on growth parameters and nutrient levelsin soybean and on the level of soil nutrients and enzymes. Effect of co-inoculation of rhizobacteria and biochar (1 and 3%) on the plant growth parameters and soil biochemicals were studied in pot assay experiments under greenhouse conditions. Both produced good amounts of indole-acetic acid; (22 and 16 μg mL^{-1}), siderophores (79 and 87%SU), and phosphate solubilization (0.89 and 1.02 99 g mL^{-1}). Co-inoculation of *B. japonicum* with *P. putida* and 3% biochar significantly improved the growth and nutrient content ofsoybean and the level of nutrients and enzymes in the soil, thus making the soil more fertile to support crop yield. The results of this research provide the basis of sustainable and chemical-free farming for improved yields and nutrients in soybean and improvement in soil biochemical properties.

Keywords: biochar; *Bradyrhizobium japonicum*; *Pseudomonas putida*; plant growth; plant nutrients; soil enzymes; soil nutrients; soybean

1. Introduction

The global climate scenario is experiencing a drastic depletion of soil nutrients due to various anthropogenic activities, burning of fossil fuel, and excess use of agrochemicals [1]. Applications of plant-growth promoting rhizobacteria (PGPR) and biochar have been advocated as an effective, cheap, and sustainable approach for the replenishment of crop health, crop nutrients, and soil nutrients and enzymes and for improving and sustaining soil fertility [2]. Furthermore, these amendments have a positive impact on the growth [3], development, and yield of several crops [4,5]. Various reports claimed that the application of plant growth-promoting rhizobacteria (PGPR) and biochar improves plant growth, plant nutrients, and physicochemical properties of soil [6–8]. Moreover, such applications of biochar also keep a check onatmospheric CO_2 levels [9] and, thus, contribute todecrease global warming effects [10], while the use of PGPR to increase soil fertility and plant nutrients will help to reduce the doses of agrochemicals in the field [11].

A wide variety of symbiotic bacteria, such as *Rhizobium* sp. and *B. japonicum*, etc., have been reported to promote seed germination, the growth of root and shoot, andthe level of nutrients in soybean and also improve soil biochemical properties [4,5].Rhizobia-legumes symbiosis plays a vital role in increasing crop yields, reducing the use of inorganic nitrogen fertilizers and improving soil fertility [12]. Rhizobial species are commonly used as inoculants in various parts of the world for improving the yield of legumes. Co-inoculation with multifarious *Bradyrhizobium* sp. and *Pseudomonas* sp. improves plant growth, plant, and soil nutrients and enzymes through the production of siderophores [13], phytohormones [14], enzymes [15], exopolysaccharide [16], stress tolerance [17], and phosphate solubilization [18–23], etc. Thus, several studies reported increases in nodules number, nodule weight, nitrogen fixed, plant growth, and yield of legumes due to co-inoculation with plant growth promoting *Bradyrhizobium* sp. and *Pseudomonas* sp. [12–14], while the combination of biochar with PGPR further increases root length, shoot length, nodule per plant, seed number, and yield of crops [5].

The activity of PGPR bioinoculants helps in improving the level of extracellular soil enzymes that facilitates the decomposition of soil organic matter and ensures the availability of nutrients in the soil [15]. Among the soil enzymes, proteases and acid and alkaline phosphomonoesterase are the major enzymes that mediate the hydrolysis of the protein and phosphate (P) into bioavailable amino acids, organic nitrogen, and soluble P [16]. However, the activities of these enzymes are governed by many factors, such as soil properties, soil organic matter level, and the presence of organic compounds [24]. We hypothesized that co-inoculation with *B. japonicum+P. putida* and biochar would facilitate the beneficial effects on soybean plant growth, plant nutrients, and soil nutrients and enzymes.

The present study was aimed at evaluating the effects of co-inoculation of multiple plant growth-promoting traits positive in *Bradyrhizobium japonicum* USDA 110 and *Pseudomonas putida* TSAU1 and different levels (1 and 3%) of biochar on seed germination, growth parameters, and nutrient levels in soybean and the level of nutrients and enzymes in soil. The outcome of this study may provide a better way of increasing soil fertility and increasing the growth and yield of soybean. This approach has multiple dimensions; as utilization of biochar is not only a cheaper option but will also help in solving the management issues of biochar, it is expected to minimize the doses of agrochemicals and produce chemical-free food. The consortium effect of PGPR and application of biochar provide excellent benefits to the farmers as theyincur less investment and yield more crop productivity, and this organically grown crop has more demand with a good selling price.

2. Materials and Methods

2.1. Bacterial Culture, Soybean, and Biochar

B. japonicum USDA 110 and *P. putida* TSAU1 strains were collected from the culture collection of the Department of Microbiology and Biotechnology, National University of Uzbekistan, Tashkent, Uzbekistan. Soybean (*Glycine max* L. Merr.) seeds were obtained from Leibniz Centre for Agricultural Landscape Research (ZALF), Müncheberg, Germany.

The maize biochar (MBC) was collected from the Leibniz-Institute for Agriculture Engineering and Bioeconomy (ATB), Potsdam, Germany. Pyrolysis of MBC was carried out at 600 °C for 30 min and the chemical compositions of MBC were analyzed according to the method of Reibe et al. [25].

2.2. Screening for the Production of PGP Metabolites

B. japonicum USDA 110 and *P. putida* TSAU1 strains were screened for phosphate (P) solubilization on Pikovoskaya's agar and in Pikovoskaya's broth [26] for the production of indole-3-acetic acid (IAA)according to the method of Brick et al. [27], for production and estimation of siderophore according to the method of Patel et al. [28] and Payne [29], and the production and estimation of aminocyclopropane-1-carboxylate deaminase (ACCD) activity according to the method of Penrose and Glick [30]. The ACCD activity was measured as the amount of α-keto-butyrate produced per mg protein per h.

2.3. Surface Sterilization, Germination, and Bacterization of Seeds

Soybean seeds were sorted to eliminate broken, small, infected seeds and sterilized with 10% sodium hypochlorite solution for 5 min and washed three times with sterile, distilled water. Seeds were germinated in 85 mm × 15 mm tight-fitting plastic Petri dishes with 5 mL of water. *B. japonicum* USDA 110 and *P. putida* TSAU 1 broth rich in PGP metabolites were used for the inoculation of germinated seeds. Germinated seeds were first placed with sterile forceps into bacterial suspension (5×10^6 CFU g^{-1}) for 10 min before planting, were air-dried, and then planted in plastic pots containing 400 g sandy loamy soil.

2.4. Experimental Design

The effect of rhizobacteria on the growth of soybean was studied in pot experiments in a greenhouse at ZALF, Müncheberg, Germany during July 2015. All the experiments were carried out in a randomized block design (RBD) with three replications. Experimental treatments included un-inoculated control (soil without biochar and soil with two levels of biochar (1 and 3%)), inoculation with *B. japonicum* USDA 110 (soil without biochar and soil with two levels of biochar (1 and 3%)), and co-inoculation with *B. japonicum* USDA 110 and *P. putida* TSAU 1 strains (soil without biochar and soil with two levels of biochar (1 and 3%)). The plants were grown in greenhouse conditions at 24 °C during the day and 16 °C at night for 30 days.

2.5. Measurement of Plant Growth Parameters and Plant Nutrients

Plants harvested after 30 days were subjected to the measurement of seed germination rate, root length, shoot length, root dry weight, shoot dry weight, and the number of nodules per plant of soybean. Plant nutrients, such as nitrogen (N), phosphorus (P), potassium (K), magnesium (Mg), sodium (Na), and calcium (Ca) were estimated from crushed plant tissue with an inductively coupled plasma optical emission spectrometer (ICP-OES; iCAP 6300 Duo, Thermo Fischer Scientific Inc., Waltham, MA, USA) via Mehlich-3 extraction [30]. The nitrogen and phosphorus contents of root and shoot were determined from dried powdered biomass. For nitrogen estimation, 1 g of plant biomass was digested with 10 mL concentrated H_2SO_4 and 5 g catalyst mixture in the digestion tube. The mixture was allowed to cool and then processed for distillation. The distillate was collected and titrated with H_2SO_4 blank (without leaf). Total nitrogen was calculated from the blank and sample titer reading [31]. For the estimation of P content, plant P was extracted with 0.5 N $NaHCO_3$ (pH8.5)and treated with ascorbic acid in an acidic medium [32]. The intensity of blue color produced was measured and the amount of P was calculated from the standard curve of P. For the estimation of potassium content of plant biomass, 25 mL of ammonium acetate solution was added in 5 g of the biomass sample, the content was shaken for 5 min and filtered, and the amount of K from the filtrate was measured [33]. For the estimation of Na, Mg, and Ca, 1 g of plant extract was mixed with 80 mL of 0.5 N HC1 for 5 min at 25 °C followed by measurement of concentrations of these elements in the filtrate [34].

2.6. Analysis of Soil Nutrient and Soil Enzymes

The rootsoil (10 g) of experimental pots was air-dried soil, shaken with 100 mL ammonium acetate (0.5 M) for 30 min to effectively displace the available nutrients, and adhered to soil minerals. The soil organic carbon (SOC), nitrogen (N), phosphate (P), and potassium (K) content of soil were determined by the dry combustion method according to the method of Sims [35] and Nelson and Sommers [36] using a CNS analyzer (TruSpec, Leco Corp., St. Joseph, MI, USA). For this purpose, 10 mL of 1 N $K_2Cr_2O_7$ and 20 mL of concentrated H_2SO_4 was added in 1g soil, mixed thoroughly and diluted with 200 mL of distilled water followed by the addition of 10 mL each of H_3PO_4 and sodium fluoride. The resulting solution was used for the elemental analysis. Blank (without soil) served as control. Soil Organic Carbon (SOC) of soil sample was calculated with the help of blank and sample titer reading.

The acid and alkaline phosphomonoesterase activities were assayed according to the method of Tabatabai and Bremner [37].Moist soil (0.5 g) was placed in a 15 mL vial, and 2 mL of modified universal buffer (MUB) (pH 6.5 for the acid phosphatase assay or pH 11 for the alkaline phosphatase assay) and 0.5 mL of p-nitrophenyl phosphate substrate solution (0.05 M) were added to the vial, sequentially. The assay and control batches were replicated 3 times. The concentration of p-nitrophenol (p-NP) produced in the assays of acid and alkaline phosphomonoesterase activities were calculated from a p-NP calibration curve after subtracting the absorbance of the control at 400 nm. Protease activity was assayed according to the method of Ladd and Butler [38]. For this, 0.5 g of soil was weighed into a glass vial, and 2.5 mL of phosphate buffer (0.2 M, pH of 7.0) and 0.5 mL of N-benzoyl-L-arginine amide (BAA) substrate solution (0.03 M) were added. The ammonium released was calculated by relating the measured absorbance at 690 nm.

2.7. Statistical Analyses

All the experiments were performed in three replicates and the average of triplicate was considered. Experimental data were analyzed with the StatView Software (SAS Institute, Cary, NC, USA, 1998) using ANOVA. The significance of the effect of treatment was determined by the magnitude of the p-value ($p < 0.05 < 0.001$).

3. Results

3.1. Analysis of Maize Biochar

Analysis of pyrolyzed maize biochar contained (g%) dry weight: 92.85, ash: 18.42, total C: 75.16, N: 1.65, P: 5.26, and K: 31.12 with a pH of 9.89 and electrical conductivity of 3.08.

3.2. Screening for the Production of PGP Metabolites

Both the cultures under study produced a wide variety of PGP traits. *B. japonicum* USDA 110 and *P. putida* TSAU1 produced 22 and 16 µg mL^{-1} of IAA, 79 and 87% siderophore, and 0.89 and 1.02 99 g mL^{-1} phosphate solubilization, respectively.

3.3. Measurement of Plant Growth Parameters and Plant Nutrients

The effect of rhizobacteria and biochar levels indicated a significant improvement in the seed germination rate and growth of the soybean plant treated with biochar and rhizobacteria over the control plant (without biochar treatment). The addition of different levels of biochar, inoculation of *B. japonicum* USDA 110, and *P. putida* strain TSAU 1 with biochar and without biocharshowed variable increases in the growth parameters. Addition of 3% biochar alone enhanced the seed germination by 15%, root length by 20% (Figure 1a), shoot length by 41% (Figure 1a), root dry weight by 22% (Figure 1b), and shoot dry weight by 13% (Figure 1b), as compared to the control plant (without biochar). Individual addition of *B. japonicum* USDA 110 and *P. putida* strains TSAU 1 with varying levels of biochar (1–3%) and without biochar also promoted the growth of the plant. However, a co-inoculation with *B. japonicum* USDA 110 and *P. putida* strains TSAU 1 with 3% biochar resulted in significant increasesin seed germination and plant growth attributes. Increases in seed germination by 20%, root length by 76% (Figure 1a), shoot length by 41% (Figure 1a), root dry weight by 56% (Figure 1b), shoot dry weight by 59% (Figure 1b), and number of nodules per plant by 57% (Figure 1c) were recorded over the control plant treated with 3% biochar alone.

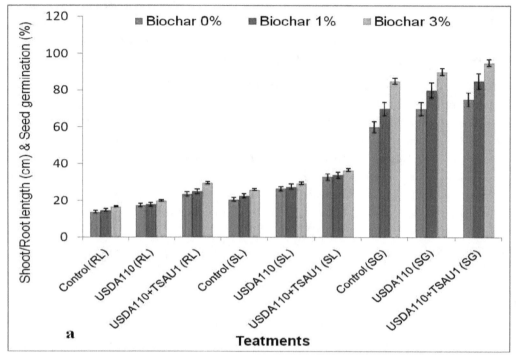

RL= Root length SL= Shoot length, SG= Seed germination *Significant at *P* (0.01)

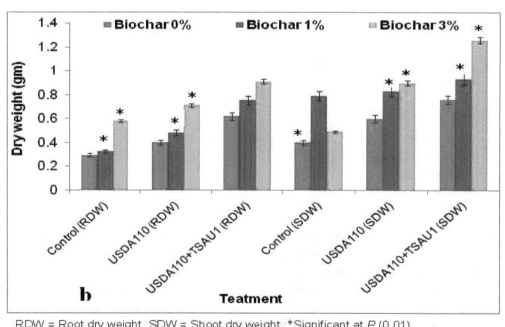

RDW = Root dry weight, SDW = Shoot dry weight, *Significant at *P* (0.01)

Figure 1. *Cont.*

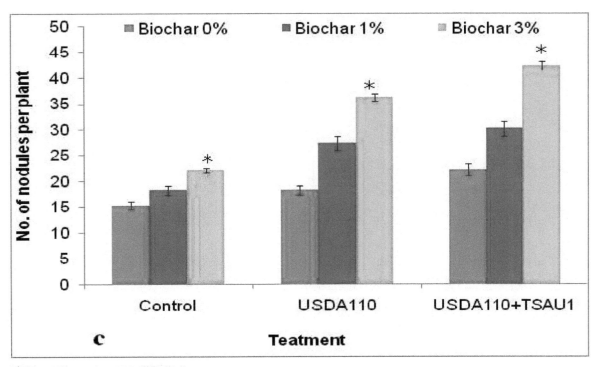

*Significant at P (0.01)

Figure 1. Effect of rhizobacteria and biochar concentrations on (**a**) root length [cm] and shoot length [cm], (**b**) dry weight of the root [g] and dry weight of the shoot [g], and (**c**) number of nodules. Plant growth parameters were measured after 30 days of growth of plant growth under greenhouse conditions.* = values significant at p 0.01.

Analysis of nutrients in a soybean plant (before sowing and after harvesting) revealedthat treatments with 1 and 3% biochar improved the content of total N, P, K, Mg, Na, and Ca in the plant. The inoculation of *B. japonicum* USDA 110 alone (0% biochar) increased N content by 36%, P content by 8.3%, K content by 5.6%, Mg content by 4.8%, Na content by 30%, and Ca content by 2.88%. However, the co-inoculation of *B. japonicum* USDA 110 and *P.putida* TSAU1 with 3% biochar showed a significant improvement in N content by 62.85%, P content by 7.42, K content by 76.85%, Mg content by 5.14%, Na content by 20%, and Ca content by 28%, as compared to the control (without biochar) (Table 1).

Table 1. Effect of rhizobacteria and biochar levels on plant nutrients.

Biochar Application	Treatments	N (%)	P (%)	K (%)	Mg (%)	Na (%)	Ca (%)
0%	Control	1.75 + 0.01	0.24 + 0.01	1.40 + 0.04	0.39 + 0.10	0.02 + 0.00	0.82 + 0.03
	TSAU1	2.00 + 0.02 *	0.25 + 0.04	1.41 + 0.02	0.43 + 0.02	0.06 + 0.01 *	0.91 + 0.03
	USDA 110	2.39 + 0.02 *	0.26 + 0.04	1.49 + 0.02	0.47 + 0.02	0.08 + 0.01 *	1.07 + 0.03
	USDA110+TSAU1	2.60 + 0.02 *	0.27 + 0.02	2.09 + 0.15 *	0.62 + 0.01 *	0.09 + 0.01 *	1.17 + 0.01 *
1%	Control	1.77 + 0.02	0.27 + 0.03	2.33 + 0.02	0.66 + 0.02	0.03 + 0.01	0.95 + 0.03
	USDA 110	2.51 + 0.02 *	0.28 + 0.02	2.52 + 0.04	0.52 + 0.02	0.07 + 0.01 *	1.25 + 0.03
	USDA+TSAU 1	2.64 + 0.02 *	0.32 + 0.02	2.33 + 0.03	0.68 + 0.02	0.13 + 0.04 *	1.21 + 0.02
3%	Control	1.91 + 0.02	0.28 + 0.01	2.41 + 0.02	0.64 + 0.02	0.03 + 0.03	1.09 + 0.02
	USDA 110	2.27 + 0.01	0.37 + 0.01	3.64 * + 0.01	0.48 + 0.01	0.02 + 0.01	0.99 + 0.01
	USDA+TSAU1	2.85 * + 0.01	0.35 + 0.01 *	3.72 * + 0.01	0.39 + 0.01	0.03 + 0.01	1.04 + 0.01

Values are the average of three replicates ± values are standard deviations. Plant nutrient contents were measured after 30 days of growth of plant under greenhouse conditions. * = values significant at p 0.01.

3.4. Estimation of Soil Nutrient Content and Soil Enzymes

Analysis of soil nutrient content revealed that the inoculation of soybean with *B. japonicum* USDA 110 alone (3% biochar) increased N content by 73%, P content by 173%, and K content by 17%, as compared to the control of 3% biochar. *B. japonicum* USDA 110 alone (3% biochar) significantly enhanced the N content by 98% and K content by 117%, as compared to the control without biochar (Table 2).

Table 2. Effect of rhizobacteria and biochar levels on soil nutrients.

Biochar Application	Treatments	SOC (%)	Total N (%)	P (mg)	K (mg)
0%	Control	21.09 ± 0.01	0.080 ± 0.01	4.29 ± 0.03	2.95 ± 0.02
	TSAU1	23.06 ± 0.01	0.082 ± 0.01	4.43 ± 0.03	3.05 ± 0.02
	USDA 110	27.08 ± 0.01	0.083 ± 0.01	4.60 ± 0.02 *	3.27 ± 0.03 *
	USDA+TSAU1	29.04 ± 0.02 *	0.094 ± 0.8 *	4.88 ± 0.02 *	5.58 ± 0.03 *
1%	Control	25.09 ± 0.01	0.091 ± 0.01	4.22 ± 0.03	4.83 ± 0.02
	USDA 110	29.06 ± 0.01	0.101 ± 0.02 *	6.14 ± 0.01 *	5.44 ± 0.01 *
	USDA+TSAU1	32.07 ± 0.8 *	0.164 ± 0.03 *	16.67 ± 0.05 *	5.68 ± 0.02 *
3%	Control	25.09 ± 0.01	0.094 ± 0.01	6.02 ± 0.01	5.35 ± 0.03
	USDA 110	33.05 ± 0.01	0.163 ± 0.01 *	16.47 ± 0.01 *	6.30 ± 0.01 *
	USDA+TSAU1	41.08 ± 0.01 *	0.170 ± 0.01 *	18.33 ± 0.01 *	8.49 ± 0.01 *

Values are the average of three replicates. ± values are standard deviations. * = values significant at p 0.01. Soil nutrient contents were measured after 30 days of growth of plant under greenhouse conditions.

The lowest level of these elements was evident in the soil without biochar treatment. The highest values of SOC, N, P, and K were observed in soil amended with 3% biochar and co-inoculation with *B. japonicum* USDA 110 and *P. putida* TSAU1 vis-à-vis the lowest value found in soil with *B. japonicum* USDA 110 and *P. putida* TSAU1 alone or in combination but without biochar and soil with no bioinoculants and no biochar treatments (Table 2).

Co-inoculation of soybean with of *B. japonicum* USDA 110 and *P. Putida* TSAU 1 strains enhanced nutrient contents of soil compared to all other treatments. The combination with *B. japonicum* USDA 110 and *P. putida* TSAU 1 (3% biochar) significantly increased N content by 80%, P content by 204%, and K content by 58% compared to the control of 3% biochar. When co-inoculated with *B. japonicum* USDA 110 and *P. putida* TSAU 1 (3% biochar)the N content rose by 11% and K content by 35% compared to variants inoculated with *B. japonicum* USDA 110 alone.

The addition of biochar to soil increased the activity of soil protease and acid and alkaline phosphomonoesterase. Substantial increases of 25.05%, 21.02%, and 23.02% in the activities of protease and acid and alkaline phosphomonoesterase, respectively, were evident due to the co-inoculation of *B. japonicum* USDA 110 and *P. Putida* TSAU1 (0% biochar). A combination of this treatment with 1% biochar further improved the activities of these enzymes. However, the activities of these enzymes were significantly improved due to the co-inoculation of *B. japonicum* USDA 110 and *P. Putida* TSAU 1 with 3% biochar. 2-fold, 1.52-fold, and 1.25-fold increases in the activities of protease and acid and alkaline phosphomonoesterase, respectively, were evident due to co-inoculation with two bioinoculants and 3% biochar (Table 3).

Table 3. Effect of rhizobacteria and biochar levels on soil enzymes.

Biochar Application	Treatments	Protease Activity ($\mu g\ NH_4^+$-$N\ g^{-1}h^{-1}$)	Acid Phosphomonoesterase Activity ($\mu g\ pNPg^{-1}1h^{-1}$)	Alkaline Phosphomonoesterase Activity ($\mu g\ pNPg^{-1}r^{-1}$)
0%	Control	19.2 ± 0.05	650.3 ± 30.1	300.1 ± 16.3
	TSUA1	20.1 ± 0.05	697.1 ± 20.1	317.1 ± 12.3
	USDA 110	23.5 ± 0.10	703.3 ± 34.5	365.6 ± 18.1
	USDA+TSAU 1	25.8 ± 0.19 *	780.6 ± 38.8 *	380.2 ± 20.4 *
1%	Control	21.4 ± 0.07	766.3 ± 35.7	370.5 ± 19.5
	USDA 110	25.8 ± 0.20 *	820.9 ± 45.3 *	425.3 ± 21.6 *
	USDA+TSAU 1	27.7 ± 0.18 *	940.6 ± 43.2 *	482.2 ± 20.8 *
3%	Control	24.3 ± 0.09	810.3 ± 37.6	420.6 ± 19.5
	USDA 110	28.5 ± 0.11 *	911.8 ± 46.3 *	483.5 ± 21.2 *
	USDA+TSAU 1	30.8 ± 0.15 *	1020.4 ± 48.6 *	535.7 ± 25.2 *

Values are the average of three replicates. $\pm$ values are standard deviations. * = values significant at p 0.01. Soil enzyme levels were measured after 30 days of inoculation of PGPR and biochar.

4. Discussion

4.1. Screening for the Production of PGP Metabolites

PGPR is known to produce a wide variety of plant-beneficial metabolites that help in plant nutrition and the overall vigor of the plant [39–42]. Production of IAA, siderophore, and P solubilization have been reported in various species of *Bradyrhizobium*, including *B. japonicum* [42–45] and *P. putida* [46,47]. Sayyed et al. [48] reported the production of siderophores from *P. fluorescence* NCIM5096 isolated from the groundnut field rhizosphere. Shaikh et al. [49] reported the production of siderophore from *P. aeruginosa* isolated from the banana field rhizosphere. Pandya et al. [50] reported the production of siderophore and phytohormones, such as IAA and gibberellins in *Pseudomonas* sp. *Rhizobium* sp., and *Azotobacter* sp. isolated from the sugarcane field rhizosphere. Theyobserved higher yields of phytohormones in *Pseudomonas* sp., as compared to the other isolates. Wani et al. [40] reported the production of siderophore in soil bacterium *P. aeruginosa* RZS9. They claimed a further increase in siderophore yield following the optimization of the process by a statistical approach. Jabborova et al. [14] reported the production of siderophore, IAA, and enzymes, such as protease, cellulose, lipase, P solubilization, and antifungal activity in nine endophytic PGPR strains. Sayyed et al. [51] reported the production of copious amounts of siderophore in *P. fluorescence* NCIM 5096 and *P. putida* NCIM2847.

4.2. Measurement of Plant Growth Parameters and Plant Nutrients

An increase in seed germination is due to the phytohormone production, while plant growth promotion during the symbiotic association is due to the nitrogen and other nutrients supplied by the bacterial symbiont. Sayyed et al. [48] reported plant growth-promoting effects of siderophore producing *P. fluorescence* NCIM5096 in wheat and groundnut. Wani et al. [40] reported the plant growth-promoting effects and antifungal-activities production of siderophore producing *P. aeruginosa*. Pandya et al. [50] reported that the inoculation of siderophore and phytohormone producing *Pseudomonas* sp., *Rhizobium* sp., and *Azotobacter* sp. promoted growth in wheat. Jabborova et al. [14] found that inoculation of siderophore, IAA, and enzymes producing P-solubilizing endophytic PGPR strains promoted the growth of medicinal plants. Sayyed et al. [13] observed growth promotion in wheat due to the inoculation of siderophore-producing *P. fluorescence* NCIM 5096 and *P.putida* NCIM2847.

Masciarelli et al. [45] reported a significant increase in the number of root nodules in soybean due to inoculation with *B. japonicum*. Egamberdieva et al. [23] reported the synergistic effect of co-inoculation of *B. japonicum* and *P. putida* to be more effective in increasing nodulation in soybean.Several researchers reported that biochar increased plant growth, nodule number, and yield in different crops [3,5,47]. Pandit et al. [7] claimed that the application of 3% biochar promoted the growth of maize. Uzoma et al. [52] recorded a significant increase in the productivity of biocharrized maize, as compared to a control under sandy soil conditions. Increased growth, more nodulation, and improved yield of soybean after the application of biochar were also reported by Iijima et al. [53].

The addition of organically rich biochar and inoculation with PGPR plays a vital role in increasing the soil microbial activity that provides more nutrition to the plant [54]. Egamberdieva et al. [55] reported significant ($p < 0.05$) increases in N, P, K, and Mg contents in chickpea plants treated with *Mesorhizobium ciceri* and biochar. It has been reported that the biochar amendment improves the water-holding capacity of soil [56], which increases the availability of minerals and nutrients [55]. Shen et al. [57] reported the positive effect of biochar amendment on the plant uptake of plant nutrients. Prendergast et al. [58] claimed that the addition of biochar can induce changes in nutrient availability and may provide additional N, P, K, Mg, Na, Ca. Shen et al. [57] observed an increase in P uptake in plants due to the application of biochar. Egamberdieva et al. [55] observed a significant increase in K content in chickpea roots and shoots treated with *M. ciceri* and biochar. Wang et al. [59] observed similar results and claimed an increasing level of K and Mg uptake in soybean due to the addition of bamboo biochar. Ma et al. [60] reported a positive effect of co-inoculation of *B. japonicum* and biochar on N and other nutrient contents in soybean root and shoot biomass. An increase in N content may be due to the positive impact of biochar on the nodule number that contributes more N to the shoot and root biomass.

4.3. Estimation of Soil Nutrients and Soil Enzymes

Since biochar is an organically rich amendment, its addition is expected to increase the level of soil nutrients. Egamberdievaet al. [55] reported a two-fold rise in SOC, N, P, K, and Mg concentrations in soil amended with biochar, and a three-fold increase in these nutrients in the soil treated with biochar and inoculation with *M. ciceri*. Similar results were reported by Wang et al. [61]. An increase in the soil's organic carbon and other nutrients can also be correlated with increased mineralization due to increased enzyme activity. A linear relationship between soil nutrients and the activities of soil enzymes involved in mineralization has been proposed by Ouyang et al. [62]. Fall et al. [63] reported significant ($p < 0.05$) increases in SOC, available N, soluble P, and total nitrogen upon the application of biochar at a higher rate (12 t ha^{-1}). They also recorded an increase in rice rhizospheric carboxylate secretions. Głodowska et al. [6] suggested a combination of biochar and *B. japonicum* strain 532 C, which significantly increased the number of nodules and the growth of soybean. The combination with biochar and *B. japonicum* resulted in enhanced nodulation, nodule biomass, and shoot biomass of soybean [63]. Numerous studies have shown that biocharapplication increases the nutrient contents of plants and soil and improves soil fertility [7,62–64]. Egamberdieva et al. [55] found that inoculation of *B. japonicum* USDA 110 halophilic *P. putida* TSAU1 promoted growth, protein content, nitrogen, and phosphorus uptake and improved the root-system architecture of soybean. Their results indicated that the synergistic effect of co-inoculation of these two strains significantly improved plant growth, nitrogen, phosphorus contents, and contents of soluble leaf proteins as compared with the inoculation with *B. japonicum* USDA 110 alone or the control.

Masciarelli et al. [45] found that co-inoculation of soybean plants with *B. Amyloliquefaciens* subsp. Plantarum and *B. japonicum* showed significant improvement in plant growth parameters and nodulation. They found that inoculation of *B. amyloliquefaciens* subsp. Plantarum with *B. japonicum* enhanced the ability of *B.japonicum* to colonize host plant roots and increase the number of nodules. Phosphomonoesterase (E.C. 3.1.3.2) in the soil is either of plant-root or microbial origin. It plays a major role in P solubilization in soils and in making P available to plants [40]. Acid phosphomonoesterase is dominant in acidic soil, while alkaline phosphomonoesterase occurs in the alkaline soil. The presence of these enzymes and their level in the soil is directly related to the extent of P solubilization and, hence, the amount of soluble P in the soil. Non-nitrogen fixers, such as *Pseudomonas* sp. assimilate nitrogen through the decomposition of protein–nitrogen to low molecular nitrogenous compounds and increase the soil nitrogen and, thus, soil fertility. Extracellular proteases enter the soil via microbial production.

Co-inoculation of *B. japonicum* and *P.putida* along with the application of biochar has been reported to enhance the activities of a wide variety of enzymes in soil [60]. The increase in activities of soil enzyme may be due to increased microbial activity as a result of the addition of consortium of organisms and the addition of biochar that contains good amounts of carbon, nitrogen, and minerals to support cell proliferation and, therefore, enzyme activities [60]. Egamberdieva et al. [55] demonstrated a 2-fold increase in protease and a 40% increase in acid phosphomonoesterase activity due to the addition of biochar. The positive effect on the activities of the soil enzymes can be attributed to the stimulating effect of biochar on microbial activity [63]. The enhancement in the soil enzyme activities due to rhizobial inoculation was also observed by Fall et al. [63]. Ouyang et al. [62] reported that the addition of biochar increases the activities of soil enzymes and attributed this increased enzyme activity to the availability of nutrients and increased microbial activities brought by the addition of biochar to the soil. Egamberdieva et al. [55] and Ma et al. [60] also reported the positive effect of increasing the level of biochar on protease activity. Oladele [64] reported a significant ($p < 0.05$) increase in soil enzymes, such as invertase, alkaline phosphatase, urease, and catalase as a result of the higher application of biochar. It has been reported that with the amendment of more biochar, more soil proteins adhere to the surfaces of biochar pores, make the protein (substrate) unavailable in the soil, and cause a decrease in protease activity [22]. However, we report increased protease activity with an increase in the biochar amendment to the soil.

5. Conclusions

The application of biochar positively affects the growth and nodulation of soybean by increasing nutrient contents, such as N, P, and K in soil. Inoculation with *B. japonicum* USDA 110 alone increased the number of nodules, the length and dry weight of roots, and the length of shoots of soybean, as compared to the control. *B. japonicum* enhanced the total N content, P content, and K content of the soil, as compared to controls with biochar and without biochar, respectively. Co-inoculation with *B. japonicum* USDA 110 and *P. putida* TSAU 1 significantly increased the growth of soybean, nutrient contents in soybeanand soil, and activities of soil protease and acid and alkaline monophosphoeserase, as compared to the control. However, the combined application of *B. japonicum* USDA 110 and *P. putida* TSAU 1 and biochar (3%) showed pronounced positive effects on growth and vigor of soybean, nutrient levels in plant biomass and soil, and activities of soil enzymes. Thus, the co-inoculation with rhizobia and application of biochar offers the best eco-friendly and chemical-freestrategy for the sustainable increase in the yield and replenishment of nutrients in soybean and soil and increase in soil biochemical properties. In general, consortia of PGPR and biochar application improves plant growth, contents of plant and soil nutrients, and soil enzyme activities, which influence soil nutrient retention, nutrient availability, and improve crop growth.

The present study demonstrates that application of *B.japonicum* alone has the capacity to improve soybean growth, nutrient contents, and improve soil biochemical properties, however, the co-inoculation of this symbiont along with *P.putida* has a more positive effect on plant growth and soil biochemicals, and co-inoculation of these rhizobia in combination with biochar possesses the capacity to significantly improve the growth and nutrient contents in soybean as well as nutrients and enzyme activities in soil. However, to claim the bio-efficacy potential of the co-inoculation of rhizobacteria and application of biochar needs multiple field studies over the season and in different agro-climatic zones.

Author Contributions: D.J. designed the experiments and wrote the manuscript; S.W. and K.D. supervised the work; A.N. and S.D. performed the methodology; R.Z.S., S.W., H.E.E., and A.S. interpreted the data and edited the paper; A.K. and R.A.M. performed analysis, and A.H.B. conceived the funding. All authors have read and approved the paper. All authors have read and agreed to the published version of the manuscript.

Acknowledgments: The authors extend their appreciation to The Researchers Supporting Project Number (RSP-2020/15), King Saud University, Riyadh, Saudi Arabia. We thank colleagues at Leibniz Centre for Agricultural Landscape Research (ZALF), Müncheberg, Germany for providing necessary support laboratory and greenhouse facilities, namely the Experimental Field Station and the Central Laboratory.

References

1.   Sarfraz, R.; Hussain, A.; Sabir, A.; Fekih, I.B.; Ditta, A.; Xing, S. Role of biochar and plant growth promoting rhizobacteria to enhance soil carbon sequestration—A review. *Environ. Monit. Assess.* **2019**, *191*, 251. [CrossRef] [PubMed]

2.   Zhang, A.; Bian, R.; Pan, G.; Cui, L.; Hussain, Q.; Li, L.; Zheng, J.; Zheng, J.; Zhang, X.; Han, X. Effects of biochar amendment on soil quality, crop yield and greenhouse gas emission in a Chinese rice paddy: A field study of 2 consecutive rice growing cycles. *Field Crops Res.* **2012**, *127*, 153–160. [CrossRef]

3.   Asai, H.; Samson, B.K.; Stephan, H.M.; Songyikhangsuthor, K.; Homma, K.; Kiyono, Y.; Inoue, Y.; Shiraiwa, T.; Horie, T. Biochar amendment techniques for upland rice production in Northern Laos: 1. Soil physical properties, leaf SPAD and grain yield. *Field Crops Res.* **2009**, *111*, 81–84. [CrossRef]

4. Saxena, J.; Rana, G.; Pandey, M. Impact of addition of biochar along with *Bacillus* sp. on growth and yield of French beans. *Sci. Hortic.* **2013**, *162*, 351–356. [CrossRef]

5. Agboola, K.; Moses, S. Effect of biochar and cowdung on nodulation, growth and yield of soybean (*Glycine max* L. Merrill). *Int. J. Agric. Biosci.* **2015**, *4*, 154–160.

6. Głodowska, M.; Schwinghamer, T.; Husk, B.; Smith, D. Biochar based inoculants improve soybean growth and nodulation. *Agric. Sci.* **2017**, *8*, 1048–1064. [CrossRef]

7. Pandit, N.; Mulder, J.; Hale, S.; Martinsen, V.; Schmidt, H.; Cornelissen, G. Biochar improves maize growth by alleviation of nutrient stress in a moderately acidic low-input Nepalese soil. *Sci. Total Environ.* **2018**, *625*, 1380–1389. [CrossRef]

8. Chen, H.; Ma, J.; Wei, J.; Gong, X.; Yu, X.; Guo, H.; Zhao, Y. Biochar increases plant growth and alters microbial communities via regulating the moisture and temperature of green roof substrates. *Sci. Total Environ.* **2018**, *635*, 333–342. [CrossRef]

9. Laird, D. The charcoal vision: A win–win–win scenario for simultaneously producing bioenergy, permanently sequestering carbon, while improving soil and water quality. *Agron. J.* **2008**, *100*, 178–181. [CrossRef]

10. Hossain, M.; Strezov, V.; Chan, K.; Nelson, P. Agronomic properties of wastewater sludge biochar and bioavailability of metals in production of cherry tomato (*Lycopersicone sculentum*). *Chemosphere* **2010**, *78*, 1167–1171. [CrossRef]

11. Zhang, H.; Prithiviraj, B.; Charles, T.; Driscoll, B.; Smith, D. Low-temperature tolerant *Bradyrhizobium japonicum* strains allowing improved nodulation and nitrogen fixation of soybean in a short season (cool spring) area. *Eur. J. Agron.* **2003**, *19*, 205–213. [CrossRef]

12. Egamberdieva, D.; Jabborova, D.; Wirth, S. Alleviation of salt stress in legumes by co-inoculation with *Pseudomonas* and *Rhizobium*. In *Plant-Microbe Symbiosis: Fundamentals and Advances*; Springer: Singapore, 2013; pp. 291–303. [CrossRef]

13. Sayyed, R.; Gangurde, N.; Patel, P.; Josh, S.; Chincholkar, S. Siderophore production by *Alcaligenes faecalis* and its application for growth promotion in *Arachis hypogaea*. *Indian J. Biotechnol.* **2010**, *9*, 302–307.

14. Jabborova, D.; Annapurna, K.; Fayzullaeva, M.; Sulaymonov, K.; Kadirova, D.; Jabbarov, Z.; Sayyed, R. Isolation and characterization of endophytic bacteria from ginger (*Zingiberofficinale* Rosc.). *Ann. Phytomed.* **2020**, *9*, 116–121. [CrossRef]

15. Buckley, S.; Allen, D.; Brackin, R.; Jämtgård, S.; Näsholm, T.; Schmidt, S. Microdialysis as an in situ technique for sampling soil enzymes. *Soil Biol. Biochem.* **2019**, *135*, 20–27. [CrossRef]

16. Sayyed, R.; Patel, P.; Shaikh, S. Plant growth promotion and root colonization by EPS producing *Enterobacter* sp. RZS5 under heavy metal contaminated soil. *Indian J. Exp. Biol.* **2015**, *53*, 116–123.

17. Sagar, A.; Riyazuddin, R.; Shukla, P.; Ramteke, P.; Sayyed, R. Heavy metal stress tolerance in *Enterobacter* sp. PR14 is mediated by plasmid. *Indian J. Exp. Biol.* **2020**, *58*, 115–121.

18. Elkoca, E.; Kantar, F.; Sahin, F. Influence of nitrogen-fixing and phosphorus solubilizing bacteria on the nodulation, plant growth, and yield of chickpea. *J. Plant Nutr.* **2007**, *31*, 157–171. [CrossRef]

19. Jabborova, D.; Davranov, K. Effect of phosphorus and nitrogen concentrations on root colonization of soybean (*Glycine max* L.) by *Bradyrhizobium japonicum* and *Pseudomonas putida*. *Int. J. Adv. Biotechnol. Res.* **2015**, *6*, 418–424.

20. Egamberdieva, D.; Jabborova, D.; Berg, G. Synergistic interactions between *Bradyrhizobium japonicum* and the endophyte *Stenotrophomonas rhizophila* and their effects on growth, and nodulation of soybean under salt stress. *Plant Soil* **2016**, *405*, 35–45. [CrossRef]

21. Egamberdieva, D.; Wirth, S.; Jabborova, D.; Räsänen, L.; Liao, H. Coordination between *Bradyrhizobium* and *Pseudomonas* alleviates salt stress in soybean through altering root system architecture. *J. Plant Interact.* **2017**, *12*, 100–107. [CrossRef]

22. Jabborova, D.; Enakiev, Y.; Davranov, K.; Begmatov, S. Effect of co-inoculation with *Bradyrhizobiumjaponicum* and *Pseudomonas putida* on root morph-architecture traits, nodulation and growth of soybean in response to phosphorus supply under hydroponic conditions. *Bulg. J. Agric. Sci.* **2018**, *24*, 1004–1011.

23. Egamberdieva, D.; Jabborova, D.; Wirth, S.; Alam, P.; Alyemeni, M.; Ahmad, P. Interaction of magnesium with nitrogen and phosphorus modulates symbiotic performance of soybean with *Bradyrhizobiumjaponicum* and its root architecture. *Front. Microbiol.* **2018**, *9*, 1.

24. Holik, L.; Vranová, V. Proteolytic activity in meadow soil after the Application of phytohormones. *Biomolecules* **2019**, *9*, 507. [CrossRef] [PubMed]

25. Reibe, K.; Götz, K.; Roß, C.; Döring, T.; Ellmer, F.; Ruess, L. Impact of quality and quantity of biochar and hydrochar on soil Collembola and growth of spring wheat. *Soil Biol. Biochem.* **2015**, *83*, 84–87. [CrossRef]

26. Nautiyal, C. An efficient microbiological growth medium for screening phosphate solubilizing microorganisms. *FEMS Microbiol. Lett.* **1999**, *170*, 265–270. [CrossRef] [PubMed]

27. Bric, J.; Bostock, R.; Silverstone, S. Rapid in situ assay for indoleacetic acid production by bacteria immobilized on a nitrocellulose membrane. *Appl. Environ. Microbiol.* **1991**, *57*, 535–538. [CrossRef]

28. Patel, P.; Shaikh, S.; Sayyed, R. Modified chrome azurol S method for detection and estimation of siderophores having affinity for metal ions other than iron. *Environ. Sustain.* **2018**, *1*, 81–87. [CrossRef]

29. Payne, S. Detection, isolation, and characterization of siderophores. In *Methods in Enzymol*; Elsevier: Amsterdam, The Netherlands, 1994; pp. 329–344. [CrossRef]

30. Penrose, D.; Glick, B. Methods for isolating and characterizing ACC deaminase–containing plant growth–promoting rhizobacteria. *Physiol. Plant.* **2003**, *118*, 10–15. [CrossRef]

31. Labconco, C. *A Guide to Kjeldahl Nitrogen Determination Methods and Apparatus*; Labconco Corporation: Houston, TX, USA, 1998.

32. Upadhyay, A.; Sahu, R. Determination of total nitrogen in soil and plant. In *Laboratory Manual on Advances in Agro-Technologies for Improving Soil, Plant and Atmosphere Systems*; CAFT: Jabalpur, India, 2012; pp. 18–19.

33. Upadhyay, A.; Sahu, R. Determination of potassium in soil and plant. In *Laboratory Manual on Advances in Agro-Technologies for Improving Soil, Plant and Atmosphere Systems*; CAFT: Jabalpur, India, 2012; pp. 23–35.

34. Sahrawat, K. Determination of calcium, magnesium, zinc and manganese in plant tissue using a dilute HCl extraction method. *Commun. Soil Sci. Plant Anal.* **1987**, *18*, 947–962. [CrossRef]

35. Sims, J. Soil test phosphorus: Principles and methods. Methods of Phosphorus Analysis for Soils, Sediments, Residuals and Waters. *Southern Coop. Ser. Bull.* **2009**, *408*, 9–19.

36. Nelson, D.; Sommers, L. Total carbon, organic carbon, and organic matter. *Methods Soil Anal. Part 2 Chem. Microbiol. Prop.* **1983**, *9*, 539–579. [CrossRef]

37. Tabatabai, M.; Bremner, J. Use of p-nitrophenyl phosphate for assay of soil phosphatase activity. *Soil Biol. Biochem.* **1969**, *1*, 301–307. [CrossRef]

38. Ladd, J.; Butler, J. Short-term assays of soil proteolytic enzyme activities using proteins and dipeptide derivatives as substrates. *Soil Biol. Biochem.* **1972**, *4*, 19–30. [CrossRef]

39. Patel, P.; Shaikh, S.; Sayyed, R. Dynamism of PGPR in bioremediation and plant growth promotion in heavy metal contaminated soil. *Indian J. Exp. Biol.* **2016**, *54*, 286–290.

40. Wani, S.; Shaikh, S.; Sayyed, R. Statistical-based optimization and scale-up of siderophore production process on laboratory bioreactor. *3Biotech* **2016**, *6*, 69.

41. Saxena, B.; Rani, A.; Sayyed, R.; El-Enshasy, H.A. Analysis of Nutrients, Heavy Metals and Microbial Content In Organic and Non-Organic Agriculture Fields of Bareilly Region-Western Uttar Pradesh, India. *Biosci. Biotechnol. Res. Asia* **2020**, *17*, 399–406. [CrossRef]

42. Sagar, A.; Sayyed, R.; Ramteke, P.; Sharma, S.; Najat Marraiki Elgorban, A.; Syed, A. ACC deaminase and antioxidant enzymes producing halophilic *Enterobacter* sp. PR14 promotes the growth of rice and millets under salinity stress. *Plant Physiol. Mol. Biol.* **2020**. [CrossRef]

43. Seneviratne, M.; Gunaratne, S.; Bandara, T.; Weerasundara, L.; Rajakaruna, N.; Seneviratne, G.; Vithanage, M. Plant growth promotion by *Bradyrhizobium japonicum* under heavy metal stress. *S. Afr. J. Bot.* **2016**, *105*, 19–24. [CrossRef]

44. Mubarik, N.; Mahagiani, I.; Wahyudi, A. Production of IAA by Bradyrhizobium sp. *World Acad. Sci. Eng. Technol.* **2013**, 152–155. [CrossRef]

45. Masciarelli, O.; Llanes, A.; Luna, V. A new PGPR co-inoculated with *Bradyrhizobium japonicum* enhances soybean nodulation. *Microbiol. Res.* **2014**, *169*, 609–615. [CrossRef]

46. Meliani, A.; Bensoltane, A.; Benidire, L.; Oufdou, K. Plant growth-promotion and IAA secretion with *Pseudomonas fluorescens* and *Pseudomonas putida*. *Res. Rev. J. Bot. Sci.* **2017**, *6*, 16–24.

47. Genesio, L.; Miglietta, F.; Baronti, S.; Vaccari, F.P. Biochar increases vineyard productivity without affecting grape quality: Results from a four years field experiment in Tuscany. *Agric. Ecosyst. Environ.* **2015**, *201*, 20–25. [CrossRef]

48. Sayyed, R.; Naphade, B.; Joshi, S.; Gangurde, N.; Bhamare, H.; Chincholkar, S. Consortium of *A. feacalis* and *P. fluorescens* promoted the growth of *Arachis hypogea* (Groundnut). *Asian J. Microbiol. Biotechnol. Environ. Sci.* **2009**, *48*, 83–86.

49. Shaikh, S.; Patel, P.; Patel, S.; Nikam, S.; Rane, T.; Sayyed, R. Production of biocontrol traits by banana field *fluorescent Pseudomonads* and comparison with chemical fungicide. *Indian J. Exp. Biol.* **2014**, *52*, 917–920. [PubMed]

50. Pandya, N.; Desai, P.; Sayyed, R. Antifungal, and phytohormone production ability of plant growth-promoting rhizobacteria associated with the rhizosphere of sugarcane. *World J. Microbiol. Biotechnol.* **2011**, *13*, 112–116.

51. Sayyed, R.; Badgujar, M.; Sonawane, H.; Mhaske, M.; Chincholkar, S. Production of microbial iron chelators (siderophores) by *fluorescent Pseudomonads*. *Indian J. Biotechnol.* **2005**, *4*, 484–490.

52. Uzoma, K.; Inoue, M.; Andry, H.; Fujimaki, H.; Zahoor, A.; Nishihara, E. Effect of cow manure biochar on maize productivity under sandy soil condition. *Soil Use Manag.* **2011**, *27*, 205–212. [CrossRef]

53. Iijima, M.; Yamane, K.; Izumi, Y.; Daimon, H.; Motonaga, T. Continuous application of biochar inoculated with root nodule bacteria to subsoil enhances yield of soybean by the nodulation control using crack fertilization technique. *Plant Prod. Sci.* **2015**, *18*, 197–208. [CrossRef]

54. Lehmann, J.; Rillig, M.C.; Thies, J.; Masiello, C.A.; Hockaday, W.C.; Crowley, D. Biochar effects on soil biota–a review. *Soil Biol. Biochem.* **2011**, *43*, 1812–1836. [CrossRef]

55. Egamberdieva, D.; Li, L.; Ma, H.; Wirth, S.; Bellingrath-Kimura, S. Soil amendment with different maize biochars improves chickpea growth under different moisture levels by improving symbiotic performance with *Mesorhizobium ciceri* and soil biochemical properties to varying degrees. *Front. Microbiol.* **2019**, *10*, 2423. [CrossRef]

56. Bruun, E.; Petersen, C.; Hansen, E.; Holm, J.; Hauggaard-Nielsen, H. Biochar amendment to coarse sandy subsoil improves root growth and increases water retention. *Soil Use Manag.* **2014**, *30*, 109–118. [CrossRef]

57. Shen, Q.; Hedley, M.; Camps Arbestain, M.; Kirschbaum, M. Can biochar increase the bioavailability of phosphorus? *J. Soil Sci. Plant Nutr.* **2016**, *16*, 268–286. [CrossRef]

58. Prendergast-Miller, M.T.; Duvall, M.; Sohi, S. Localisation of nitrate in the rhizosphere of biochar-amended soils. *Soil Biol. Biochem.* **2011**, *43*, 2243–2246. [CrossRef]

59. Wang, C.; Alidoust, D.; Yang, X.; Isoda, A. Effects of bamboo biochar on soybean root nodulation in multi-elements contaminated soils. *Ecotoxicol. Environ. Saf.* **2018**, *150*, 62–69. [CrossRef] [PubMed]

60. Ma, H.; Egamberdieva, D.; Wirth, S.; Bellingrath-Kimura, S.D. Effect of biochar and irrigation on soybean-*Rhizobium* symbiotic performance and soil enzymatic activity in field rhizosphere. *Agronomy* **2019**, *9*, 626. [CrossRef]

61. Wang, Y.; Yin, R.; Liu, R. Characterization of biochar from fast pyrolysis and its effect on chemical properties of the tea garden soil. *J. Anal. Appl. Pyrolysis* **2014**, *110*, 375–381. [CrossRef]

62. Ouyang, L.; Tang, Q.; Yu, L.; Zhang, R. Effects of amendment of different biochars on soil enzyme activities related to carbon mineralisation. *Soil Res.* **2014**, *52*, 706–716. [CrossRef]

63. Fall, D.; Bakhoum, N.; NourouSall, S.; Zoubeirou, A.; Sylla, S.; Diouf, D. Rhizobial inoculation increases soil microbial functioning and gum arabic production of 13-year-old *senegaliasenegal* (L.) britton, trees in the north part of Senegal. *Front. Plant Sci.* **2016**, *7*, 1355. [CrossRef]

64. Oladele, S. Effect of biochar amendment on soil enzymatic activities, carboxylate secretions and upland rice performance in a sandy clay loam Alfisol of Southwest Nigeria. *Sci. Afr.* **2019**, *4*, e00107. [CrossRef]

Volatile Organic Compounds from Rhizobacteria Increase the Biosynthesis of Secondary Metabolites and Improve the Antioxidant Status in *Mentha piperita* L. Grown under Salt Stress

Lorena del Rosario Cappellari, Julieta Chiappero, Tamara Belén Palermo, Walter Giordano and Erika Banchio *(ID)

INBIAS Instituto de Biotecnología Ambiental y Salud (CONICET—Universidad Nacional de Río Cuarto), Campus Universitario, 5800 Río Cuarto, Argentina; lcappellari@exa.unrc.edu.ar (L.d.R.C.); jchiappero@exa.unrc.edu.ar (J.C.); tpalermo@exa.unrc.edu.ar (T.B.P.); wgiordano@exa.unrc.edu.ar (W.G.)
* Correspondence: ebanchio@exa.unrc.edu.ar.

Abstract: Salinity is a major abiotic stress factor that affects crops and has an adverse effect on plant growth. In recent years, there has been increasing evidence that microbial volatile organic compounds (mVOC) play a significant role in microorganism–plant interactions. In the present study, we evaluated the impact of microbial volatile organic compounds (mVOC) emitted by *Bacillus amyloliquefaciens* GB03 on the biosynthesis of secondary metabolites and the antioxidant status in *Mentha piperita* L. grown under 0, 75 and 100 mM NaCl. Seedlings were exposed to mVOCs, avoiding physical contact with the bacteria, and an increase in NaCl levels produced a reduction in essential oil (EO) yield. Nevertheless, these undesirable effects were mitigated in seedlings treated with mVOCs, resulting in an approximately a six-fold increase with respect to plants not exposed to mVOCs, regardless of the severity of the salt stress. The main components of the EOs, menthone, menthol, and pulegone, showed the same tendency. Total phenolic compound (TPC) levels increased in salt-stressed plants but were higher in those exposed to mVOCs than in stressed plants without mVOC exposure. To evaluate the effect of mVOCs on the antioxidant status from salt-stressed plants, the membrane lipid peroxidation was analyzed. Peppermint seedlings cultivated under salt stress and treated with mVOC showed a reduction in malondialdehyde (MDA) levels, which is considered to be an indicator of lipid peroxidation and membrane damage, and had an increased antioxidant capacity in terms of DPPH (2,2-diphenyl−1-picrylhydrazyl) radical scavenging activity in relation to plants cultivated under salt stress but not treated with mVOCs. These results are important as they demonstrate the potential of mVOCs to diminish the adverse effects of salt stress.

Keywords: mVOCs; Plant growth promoting rhizobacteria; PGPR; *Mentha piperita*; *Bacillus amyloliquefaciens* GB03; salt stress; secondary metabolites; MDA; DPPH

1. Introduction

Many aromatic plants, such as *Mentha piperita* L. (peppermint), are important sources of essential oil (EO) production. The EOs are generated and stored in glandular trichomes, where they form complex mixtures of secondary metabolites (SM) mainly composed of the volatile mono- and sesquiterpenes responsible for the characteristic aromas of various plant species [1,2]. Therefore, the quality of aromatic plants is recognized by the composition and concentration of these components for each species. Furthermore, the quantity and quality of SM is determined by environmental factors including temperature, soil quality, light intensity, and/or water availability [3].

Biotic and abiotic stresses are major constraints on crop yield, with environmental stress representing a strong restriction on increasing crop productivity as well as affecting the use of natural resources. A soil is considered to be saline when the ion concentration reaches an electrical conductivity of >4 dS m^{-1}, measured on a saturated soil at 25 °C, and consequently interferes with the growth of species of agricultural interest [4]. Salinity impacts agricultural production in most crops by affecting the physical-chemical properties of the soil and the ecological balance of the cultivated area [5]. As salinity affects many aspects of the physiology and metabolism of the plants, the presence of soluble salts in general has a negative consequence for the plant's growth by decreasing the water potential and thus restricting the absorption of water by the roots (osmotic effect). In addition, the absorption of specific saline ions leads to their accumulation in tissues in concentrations at which they can become toxic and induce physiological disorders (ionic toxicity) in the plant, with high concentrations of saline ions being able to modify the absorption of essential nutrients and leading to nutritional imbalances (nutritional effect) [6]. These effects are reflected by a decrease in germination, vegetative growth, and reproductive development [4,7].

Plant tolerance to salt stress is linked to the use of different strategies, including osmotic adjustment, the exclusion of toxic ions from the aerial part, translocation of photoassimilates to underground organs, an increased growth of the root system, and ensuring the availability of water and nutrients, among others. Furthermore, salinity can produce an accumulation of reactive oxygen species (ROS) [6], which may lead to a deterioration of photosynthetic pigments, lipid peroxidation, alterations in the selective permeability of the cell membranes, protein denaturation, and DNA mutations [8–10]. Damage of the cell membrane produces small hydrocarbons such as malondialdehyde (MDA), which is a sign of membrane cellular damage. Plants have well-described protection and repair systems that mitigate ROS damage. In addition, certain species have developed protective mechanisms that include enzymatic and non-enzymatic components [11,12].

Plant growth promoting rhizobacteria (PGPR) are beneficial microorganisms capable of colonizing the rhizosphere of plants and benefiting them both directly and indirectly [13]. It is well known that PGPR functions in different ways: synthesizing specific compounds for the plants, helping the uptake of nutrients, and protecting the plants from diseases [14–16]. In general, it has been observed that the negative effects that salinity produces in plant development can be mitigated by the use of microorganisms as inoculants, which is an alternative technology to improve the abiotic stress tolerance capacity of plants [17–21]. In this regard, considerable attention has been focused on understanding the molecular, physiological, and morphological mechanisms underlying rhizobacterial-mediated stress tolerance. In fact, the mechanisms by which these bacteria mediate abiotic stress tolerance continue to be widely studied, largely because they are difficult to elucidate [22,23].

Advances in research have revealed that certain PGPR strains are capable of emitting microbial volatile organic compounds (mVOCs) [24–28]. These compounds mainly consist of an abundant and very complex mixture of compounds, including alcohols, alkanes, alkenes, esters, ketones, sulfur, and terpenoids, characterized by their low molecular weight and high vapor pressure under normal conditions, which can vaporize significantly and enter the atmosphere. The analysis of mVOCs is a developing research area that has an effect on the applied agricultural, medical, and biotechnical applications, with a related interesting mVOC database containing available information regarding microbial volatiles having been published [29]. Recent studies have also provided new insights into the participation of mVOCs in inter- and intra-specific communication [30]. These compounds have been observed to have the ability to promote plant growth and induce systemic resistance (ISR) against pathogenic organisms, thereby improving the well-being of crops [24,27,28,31,32]. VOCs from *Paraburkholderia phytofirmans* have been shown to increase plant growth rate and tolerance to salinity, reproducing the effects of direct bacterial inoculation of roots [32]. Thus, the emission of mVOCs is currently recognized as being a very relevant aspect in microorganism–plant interactions [17,21,28,33,34].

We have previously demonstrated that both the direct inoculation of PGPR and exposure to VOCs emitted by these rhizobacteria stimulate the biosynthesis of SM and increase the biomass production in different aromatic plants [25,26,35–39]. Although there are few reports about the effects of mVOCs emitted by rhizobacteria on the SM yield of aromatic plants under conditions of abiotic stress, studies related to the emission of volatile organic compounds with biological activity by rhizobacteria is a novel area attracting increasing interest.

It should also be noted that it is necessary to examine the use of fertilizers and chemical synthesis pesticides related to the concentration of salts in the soil in order to develop sustainable agriculture, as this is key to assessing the proposal of alternative and complementary strategies. Taking this into consideration, among the possible alternatives, the use of microbial inoculants, considered to be a clean technology aligned with the principles of sustainable agriculture, becomes more relevant. Thus, the present study was founded on the hypothesis that the investigation of mVOCs with respect to the description of their biological functions and ecological roles is crucial for elucidating the mechanisms related to the control of critical biological processes in plant health and that this could also offer useful benefits to confront agronomic and environmental complications. In this present study, the aim was to explore the potential of mVOCs in ameliorating salinity effects in *M. piperita*, with an important objective of the study being to evaluate the role of mVOCs in EOs and the phenolic compound levels, as well as their function in the antioxidant status of plants grown under salt stress conditions.

2. Materials and Methods

2.1. Bacterial Strains and In Vitro Plant Treatments

2.1.1. Bacterial Cultures

Bacillus amyloliquefaciens GB03 (originally described as *Bacillus subtilis* GB03) [40] strain was grown on LB (Luria-Bertani) medium for routine use and maintained in nutrient broth with 15% glycerol at $-80\ ^\circ$C for storage. The bacterial culture was grown overnight at $30\ ^\circ$C and centrifuged at 120, washed twice in 0.9% NaCl by Eppendorf centrifugation ($4300\times g$, 10 min, $4\ ^\circ$C), re-suspended in sterile water, and adjusted to a final concentration of $\sim 10^9$ CFU/ mL for use as an inoculum.

2.1.2. Plant Micropropagation

The *M. piperita* plant is a commercially cultivated crop grown in the Traslasierra valley (Córdoba province, Argentina). Young shoots from peppermint were surface-disinfected and micropropagated, as previously described by Santoro et al. [26].

2.1.3. In Vitro Exposure to mVOCs

Single nodes from aseptically cultured plantlets were planted in sterilized glass jars (250 mL) containing 50 mL MS (Murashige and Skoog) solid media with 0.8% (*w/v*) agar and 3% (*w/v*) sucrose. Then, a small (10 mL) glass vial containing ca. 3 mL of Hoagland media with 0.8% (*w/v*) agar and 3% (*w/v*) sucrose was introduced into each jar. The small vial was inoculated with GB03 (50 μL), which served as the source of bacterial volatiles, with sterile water used in the control. Plants were exposed to mVOCs without having any physical contact with the rhizobacteria. Jars containing plants and bacteria were covered with aluminum foil, sealed with parafilm to avoid contamination, and placed in a growth chamber under controlled conditions (16/8-h light/dark cycle), temperature ($22 \pm 2\ ^\circ$C) and relative humidity ($\sim 70\%$). After 30 days, all plants were collected [38].

2.1.4. Treatments

MS media (plant growth media) and Hoagland media (bacterial growth media) were supplemented with different salt concentrations: 0, 75, and 100 mM NaCl. For each experimental set, both the plant and bacteria were grown under the same concentration of NaCl but without contact with each other. Salt level concentrations were selected based on previous observations: at lower concentrations (25 and

50 mM), plant growth was not affected, and at higher levels (125 and 150 mM), the rooting capacity decreased significantly. Experiments were repeated three times (10 jars per treatment; 1 plant/jar).

2.2. Essential Oil Extraction and Analysis

Shoot samples were individually weighed and subjected to hydrodistillation in a Clevenger-like apparatus for 40 min. The volatile fraction was collected in dichloromethane, and β-pinene (1 μL in 50 μL ethanol) was added as an internal standard (as it was previously reported, β-pinene is not present in peppermint plants [37]). The major *M. piperita* EO components, which comprise ~60% of the total oil volume, are limonene, linalool, (−) menthone, (−) menthol, and (+) pulegone. These compounds were quantified in relation to the standard added during the distillation procedure described above. The flame ionization detector (FID) response factors for each compound generated essentially equivalent areas (differences $p < 0.05$).

Chemical analyses were performed using a Perkin-Elmer Q-700 gas chromatograph (GC), equipped with a CBP−1 capillary column (30 m × 0.25 mm, film thickness 0.25 μm) and a mass selective detector. Analytical conditions were as follows: injector temperature 250 °C; detector temperature 270 °C; oven temperature programmed from 60 °C (3 min) to 240 °C at 4°/min; carrier gas = helium at a constant flow rate of 0.9 mL/min; source 70 eV. The oil components ((−) menthone, (−) menthol, and (+) pulegone) were established by comparison of the diagnostic ions (NIST 2014 library) and GC retention times with those of the respective authentic standard compounds purchased from Sigma-Aldrich [34]. GC analysis was performed using a Shimadzu GC-RIA gas chromatograph fitted with a 30 m × 0.25 mm fused silica capillary column coated with Supelcowax 10 (film thickness 0.25 μm). The GC operating conditions were as follows: injector and detector temperatures 250 °C; oven temperature programmed from 60 °C (3 min) to 240 °C at 4°/min; detector = FID; carrier gas = nitrogen at a constant flow rate of 0.9 mL/min.

2.3. Total Phenolic Content (TPC) Determination

The total phenolic content of the extract was determined by the Folin–Ciocalteu method, as previously described by Cappellari et al. [41]. The TPC were expressed in terms of μg gallic acid (a common reference compound) equivalent per g plant fresh weight using the standard curve.

2.4. Antioxidant Activity

The capacity of radical scavenging in extracts against stable DPPH• (2,2-diphenyl−1-picrylhydrazyl) was determined by the Brand-Williams et al. method [42] with minor modifications, as previously described by Chiappero et al. [43]. A calibration curve was obtained using ascorbic acid, and the scavenging capacity of the plant extracts was expressed as mM ascorbic acid equivalents (AAE) per g fresh weight (mM AEE/g FW). All experiments were performed in triplicate for each experimental unit.

2.5. Lipid Peroxidation

Lipid peroxidation was measured by quantifying the malondialdehyde (MDA) production using the thiobarbituric acid reaction. The MDA content was measured following the method of Heath and Packer [44], with some modifications, as reported by Chiappero et al. [43]. The amount of MDA was determined by its molar extinction coefficient ($155 \text{ mM}^{-1} \text{ cm}^{-1}$), which was expressed as μmol MDA/g FW (grams of fresh weight). The experiments were performed in triplicate for each experimental unit.

2.6. Statistical Analysis

Data were subjected to a two-way analysis of variance (ANOVA) (mVOcs × salt stress), followed by a comparison of multiple treatment levels with those of the control, using the post hoc Fisher LSD test. Infostat software version 2018 (Group Infostat, Universidad Nacional de Córdoba, Argentina) was used for the statistical analysis. Principal component analysis (PCA) using Infostat statistical package was conducted. The analysis of extracts shows the relationships among the treatments (mVOCs exposure and salt stress conditions) and the different variables measured (EO, TPC, lipid peroxidation (MDA), and antioxidant capacity (AAE)). At least 15 observations were used for each treatment in the multivariate dataset.

3. Results

3.1. Essential Oil

Peppermint plants subjected to salt stress showed a reduction in EO content. Plants grown under 75 or 100-mM salt concentrations and those not treated with mVOCs revealed a 50% decrease in EO yield ($p < 0.05$) (Figure 1). When plants were treated with mVOCs under control conditions, the EO content rose approximately 3.3 times compared to plants not exposed to mVOCs (Figure 1). When plants were grown under salt stress conditions and treated with mVOC, positive effects of mVOCs on EO yields were detected. The levels of EOs increased approximately 5.6 and 6.5-fold in plants grown under 75 or 100 mM and treated with mVOCs, respectively, in relation to plants subjected to salt conditions but not treated with mVOCs, with a statistically significant interaction effect between salt stress and mVOCs being found ($p < 0.05$).

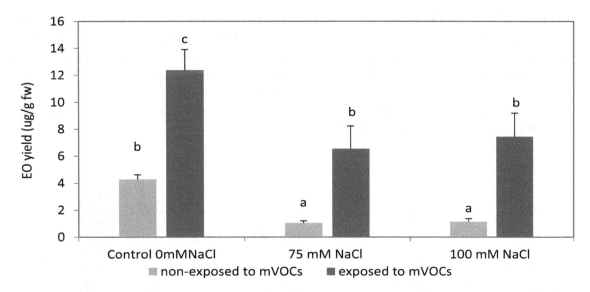

Figure 1. Essential oil yield in *Mentha piperita* plants grown under different salt concentrations (0, 75, and 100 mM NaCl) and exposed to *B. amyloliquefaciens* GB03 mVOCs (mean ± SE). Values followed by the same letter in a column are not significantly different according to Fisher's LSD test ($p < 0.05$).

Regarding the main compounds of the EOs, growing under salt stressed conditions resulted in a decrease in menthone and menthol (Table 1); although menthol content was approximately 3.5 times lower in plants grown under 75 or 100 mM concentrations and not treated with mVOCs ($p < 0.05$), the effect on menthol concentration was not statistically significant but followed the same trend as for menthone, which was significant. However, the pulegone concentration was not significantly different for control plants exposed to salt. For plants treated with mVOCs, the levels of menthone and pulegone increased approximately 2 and 3-fold, respectively, compared to those of the corresponding controls at each salinity level. However, the menthol concentration was not modified by mVOC exposure. In plants submitted to 75 mM NaCl and treated with GB03 mVOCs, the concentrations of menthone, menthol, and pulegone were approximately 6.7, 5.8, and 3.4-fold higher, respectively, in relation to plants subjected to salt conditions but not treated to mVOCs and similar to plants treated to mVOCs and not salt stressed. At 100 mM NaCl, the menthone and pulegone contents revealed the same tendency, with an increase observed in plants treated with mVOCs ($p < 0.05$), but the menthol concentration was not modified by the mVOCs (Table 1).

Table 1. Concentrations of main essential oil (EO) compounds in *Mentha piperita* grown under salt stress media (0, 75, and 100 mM NaCl) and exposed to *B. amyloliquefaciens* GB03 mVOCs emission (mean ± SE). Values are mean ± standard error (SE).

NaCl Concentration	(−)-Menthone (μg/g fw)	(−)-Menthol (μg/g fw)	(+)-Pulegone (μg/g fw)
0 mM			
control	0.99± 0.28 *b*	1.07± 0.15 *a*	1.18± 0.14 *a*
B. amyloliquefaciens GB03	2.27± 0.42 *c*	1.14± 0.23 *a*	5.29± 0.54 *c*
75 mM			
control	0.25± 0.05 *a*	0.10± 0.05 *a*	0.55± 0.12 *a*
B. amyloliquefaciens GB03	1.55± 0.17 *bc*	0.81± 0.03 *a*	2.73± 0.41 *b*
100 mM			
control	0.26± 0.05 *a*	0.22± 0.08 *a*	0.56± 0.13 *a*
B. amyloliquefaciens GB03	1.35± 0.49 *b*	0.63± 0.03 *a*	2.87± 0.79 *b*

Means followed by the same letter in a given column are not significantly different according to Fisher's LSD test ($p < 0.05$).

3.2. Total Phenolic Content

The level of TPC in plants subjected to salt stress conditions increased with the severity of the NaCl concentration ($p < 0.05$), both in plants exposed and not exposed to mVOCs. In plants grown under salt conditions (75 or 100 mM), the TPC levels rose by 15 and 50%, respectively, in relation to control plants (Figure 2). In addition, the plants subjected to both concentrations of NaC and treated with GB03 VOCs registered an increase in TPC compared to non-exposed plants ($p < 0.05$), but no statistically significant interaction effect was found ($p > 0.05$). The highest TPC concentrations were detected in plants treated with salt 100 mM and mVOCs.

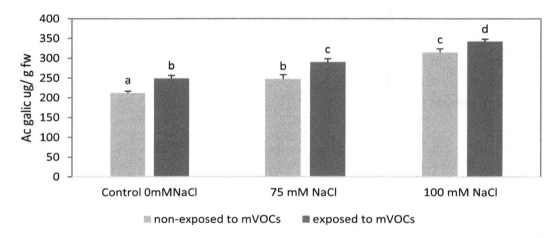

Figure 2. Total phenolic content of *Mentha piperita* plants grown under salt stress media (0, 75, and 100 mM NaCl) and exposed to *B. amyloliquefaciens* GB03 mVOCs emission (mean ± SE). Values followed by the same letter in a column are not significantly different according to Fisher's LSD test ($p < 0.05$).

3.3. Radical Scavenging Capacity

The antioxidant capacity of the DPPH• radical scavenger increased 2.6 and 3.6-fold in peppermint leaves grown under 75 and 100 mM NaCl conditions, respectively ($p < 0.05$) (Figure 3). Moreover, when plants were subjected to salt conditions and treated with mVOCs, the antioxidant capacity increased ($p < 0.05$) by 50% and 30% for 75 and 100 mM NaCl, respectively, in relation to salt stressed plants not exposed to mVOCs. The highest levels of antioxidant activity were observed when plants were exposed to VOCs and grown under 100 mM NaCl conditions, with the ascorbic acid equivalents (AAE) increasing 4.75-fold with respect to control plants (not exposed to mVOCs).

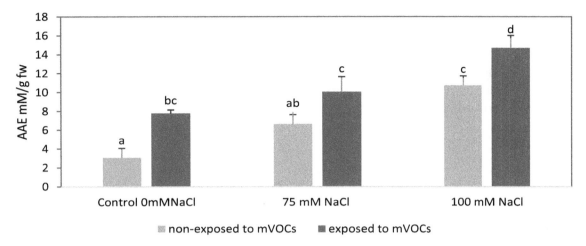

Figure 3. Antioxidant activity expressed as ascorbic acid equivalents (AAE) in *Mentha piperita* grown under salt stress media (0, 75, and 100 mM NaCl) and exposed to *B. amyloliquefaciens* GB03 mVOCs emission (mean ± SE). Values followed by the same letter in a column are not significantly different according to Fisher's LSD test ($p < 0.05$).

3.4. Lipid Peroxidation

Oxidative damage to the membrane lipids was observed due to salt stress, as shown by the MDA levels (Figure 4), with the highest MDA levels being observed ($p < 0.05$) at the higher salt concentration. The lipid peroxidation increased 1.4 and 2-fold in 75 and 100 mM NaCl treated plants, respectively, in relation to control plants. For plants treated with mVOCs and subjected to salt stress, the MDA content was approximately 25% lower than for plants stressed and not treated with mVOCs (75 and 100 mM NaCl plants).

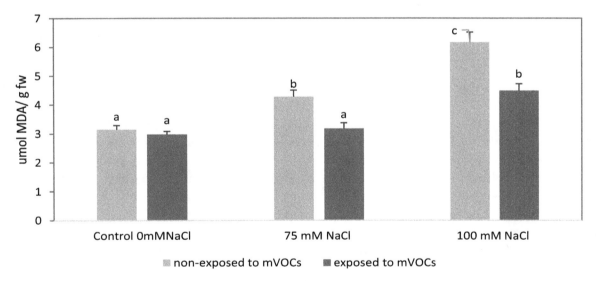

Figure 4. Malondialdehyde (MDA) content in *Mentha piperita* grown under salt stress media (0, 75, and 100 mM NaCl) and exposed to *B. amyloliquefaciens* GB03 mVOCs emission (mean ± SE). Values followed by the same letter in a column are not significantly different according to Fisher's LSD test ($p < 0.05$).

3.5. Principal Component Analysis

PCA represents a graphic image that simplifies the visualization and perception of the dataset and the variables. We used the PCA to extract and reveal the relationships among the factors (growth conditions and exposure to mVOCs) and different variables as EO, TPC, lipid peroxidation (MDA), and antioxidant capacity (AAE) in the multivariate analysis (Figure 5). The plot defined by the first two principal components was enough to explain most of the variations in the data (96.8%) and give a cophenetic correlation coefficient of 0.997. The PCA (Figure 5) showed that 100 mM NaCl (high salt concentrations) combined with exposure to mVOCs was strongly associated with TPC content and antioxidant capacity (AAE), as revealed by the circle in Figure 5. Considering the relationships among variables, a strong positive correlation (acute angle) was observed between TPC levels and AAE. There were also positive correlations found among MDA levels with no mVOC exposure and 100 mM NaCl. In addition, in PC2, positive relationships were observed between AAE, EO, and TPC with mVOC exposure.

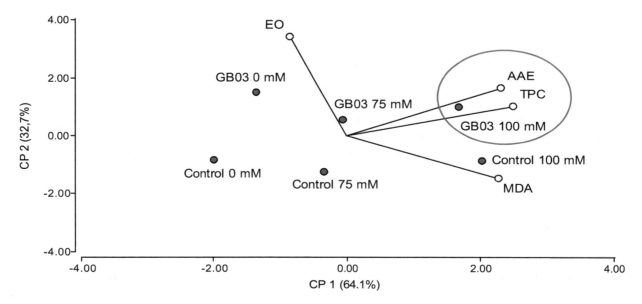

Figure 5. Principal component analysis for the physiological response of *Mentha piperita* grown under different salt stress concentrations (0, 75, and 100 mM NaCl) and *B. amyloliquefaciens* GB03 mVOCs emission. PRO: proline, TPC: total phenolic content, and MDA: lipid peroxidation were determined by estimating the amount of malondialdehyde (MDA); AEE: DPPH radical scavenging capacity.

4. Discussion

Salinity is one of the most important environmental factors diminishing plant yield, mainly in arid and semi-arid environments. The responses of plants to salt stress are intricate and affect several components, with plants having the ability to respond via signal transduction pathways by adjusting their metabolism [45,46]. These responses can differ in relation to toxic ion uptake, ion compartmentation and/or exclusion, osmotic regulation, CO_2 assimilation, photosynthetic electron transport, chlorophyll content and fluorescence, ROS generation, and antioxidant defenses [45–48].

PGPR make a significant contribution to the protection against abiotic stress through their biological activities at the rhizosphere, as exopolysaccharides production (EPS), phytohormones and 1-aminocyclopropane- 1-carboxylate (ACC) deaminase synthesis, induction of the accumulation of osmolytes and antioxidants, upregulating or downregulating the stress responsive genes, and by changes in the root morphology and volatile compounds [17–21,49,50]. In addition, in recent years, an increasing number of PGPR VOC studies have demonstrated an effect against abiotic stresses [7,38,51].

In the present study, we found that when peppermint plants were subjected to salt stress, the EO yield decreased by 50% for both concentrations evaluated (75 and 100 mM NaCl). Additionally, there was a corresponding decrease in the main compounds menthone, menthol, and pulegone. Comparable effects were reported in *M. arvensis* grown under 100, 300, and 500 mM NaCl, with a reduction of 31%, 54%, and 67%, respectively [52]. In contrast, Karray-Bouraoui et al. [53] noted an enhanced *M. pulegium* EO yield of about 2.75-fold under 50-mM salt stress conditions, with a higher density of glandular trichomes on the leaves. Furthermore, Neffati and Marzouk [54] showed that the compounds of *Coriandrum sativum* L. oil were modified by salinity and were revealed to be dependent on salt level treatment. There are contradictory reports concerning changes in EO yield in relation to salt stress. An increase in EOs and in their composition in response to low levels of salinity was reported in *Satureja hortensis* [55], in sage [56] and in thyme [57]. In contrast, other studies reported a decrease in EOs in lemon balm and in sweet marjoram [58]. Additionally, Ben Taarit et al. [59] reported that the compositions of EOs of *Salvia officinalis* were altered in moderate or high salt stress, in controls and in plants grown under 25 mM NaCl, with the major compound of the EOs being viridiflorol, whereas at higher levels (50 and 75 mM NaCl), 1, 8-cineole was predominant, and at 100 mM NaCl, manool was the principal compound.

The EO yield variations reported under abiotic stress could have resulted from the fact that their production is affected by different physiological, biochemical, metabolic, and genetic factors, which are complex to isolate from one another. In addition, the geographical, seasonal, developmental, and organ variations all contribute to EO yield, as do anatomical and hormonal factors [60–63]. The impact of salt stress on the EO levels probably was due to acclimation processes in stressed plants. Whereas in the initial stage of stress, the metabolism is severely affected, later, the acclimatization processes may reduce the secondary metabolite biosynthesis [64,65].

In the present study, the EO content in salt stressed plants treated with mVOCs showed a 5.6 and 6.5-fold increase with respect to their respective controls (plants grown under 75 or 100 mM NaCl and not treated with mVOCs, respectively), demonstrating that GB03 mVOCs have the capacity to reverse the negative effects of salinity on the EO yield. In fact, mVOCs induced salt tolerance in plants in a previous study of ours, with peppermint plants subjected to salt stress conditions and treated with GB03 VOCs having a higher shoot fresh weight, root dry weight, and total chlorophyll content compared to controls [38]. In this sense, the biosynthesis of terpenoids is affected by the primary metabolism—for example, the photosynthesis for carbon and energy supply. Factors that increase biomass production may have an impact on the relationships among the primary and secondary metabolisms, causing an increased biosynthesis of secondary metabolites [66]. Related to this, augmented plant biomass seems to lead to a larger availability of substrate for monoterpene biosynthesis [35,67].

We have also observed that abscisic acid (ABA) was not connected to salt tolerance generated in plants subjected to salt stress and treated with VOCs [38]. This observation suggests that GB03 VOCs protection against osmosis is ABA independent [68]. The jasmonic acid (JA) levels were similar in salt treated plants, when treated with mVOCs or not. In contrast, the salicylic acid (SA) levels were higher in plants subjected to salt and treated with mVOCs compared to plants subjected to salt conditions and not treated with mVOCs. SA is an important signal molecule for modulating plant responses to stress [38]. Chemical analysis using Solid Phase Microextraction (SPME) fibers of the VOC emissions from GB03 grown under salt conditions revealed the release of a total of seven components, belonging to the following four classes: hydrocarbons (cyclohexane, dodecane, undecane and hexadecane), ketones (acetoin), aldehydes (benzaldehyde), and ethers (2-butanone-3metioxy-3 methyl). The relative quantity of acetoin, the major VOC compound emitted by GB03, enhanced with salt concentration [38]. Concerning the complex profile of compounds, VOC emission is strongly affected by the collection methodology employed, the growth medium, and the density of the bacterium [50,69,70]. For instance, Farag et al. [71] identified a higher number of compounds from GB03 VOCs than Cappellari and Banchio [38], probably due to the different collection methodology used.

It has also been reported that plants treated with GB03 mVOCs and grown in a saline media accumulated less Na + through the regulation of the Na transporter. The GB03 VOCs decreased the Na level in *Arabidopsis* by decreasing Na uptake and/or increasing Na exudation [49]. Furthermore, they led to an acidification of the rhizosphere [72]. Certain bacterial VOCs activate closure of the stomata, reducing the water evaporation [73], and are also involved in biofilm formation, which maintains soil moisture content and increases drought tolerance in plants [51,74,75]. In addition, mVOCs emitted by PGPR also act as a biocontrol against several phytopathogens and trigger plant defense responses through the induction of systemic resistance (ISR) [24,71,76]. For example, the production of EOs is related to the defense response system [63], since numerous terpenes have antimicrobial activity [77]. Similarly, monoterpene synthesis is induced by herbivore feeding in *Minthostachys mollis* [78] and several plant species, suggesting that these compounds protect leaves from future attacks [67,79–81]. Consequently, as mentioned above, endogenous SA levels increased in plants cultivated under salt conditions and treated with GB03, with previous observations suggesting that the biosynthesis of *M. piperita* monoterpenes is SA and JA dependent [82].

A rise in TPC levels in different tissues under salt conditions has also been described in different plant species [83–85]. A consequence of abiotic stress is superoxide production, which leads to a detoxification mechanism. Related to this, phenolics are synthesized by many plant species for protection against abiotic stress conditions, and their levels are correlated with antioxidant activity [63,86]. Salinity stress induces metabolic and physiological reactions, as well as drastically decreasing the CO_2 uptake due to stomatal restrictions. As a consequence, the consumption of reduction equivalents (NADPH 2+) for CO_2 fixation via the Calvin cycle decreases significantly, leading to oxidative stress and an oversupply of reduction equivalents, with the metabolic processes being moved to biosynthetic activities that consume reduction equivalents. Hence, the biosynthesis of reduced compounds, such as phenols, is increased [63,85,87]. Among the SM found in *M. piperita* are phenolic compounds such as caffeic acid, rosmarinic acid, eriocitrin, and luteolin- 7-O-glucoside [88,89], with their proportion in leaves being approximately 19–23% of dry weight [90–92]. Here, we found that peppermint plants either subjected to salt conditions and/or treated with GB03 VOCs produced a positive effect on the TPC content compared to the respective control plants. Plants grown under 100 mM NaCl and treated with VOCs revealed a higher TPC content. In fact, phenolic compounds are important and powerful agents in scavenging free radicals [93–96]. The antioxidant capacity of phenolic compounds is due to their high reactivity as hydrogen or electron donors, to the particularity of the polyphenol-derived radical to stabilize and delocalize the unpaired electron, and to their capacity to chelate transition metal ions [92,97].

In a previous study, we observed that direct inoculation as well as drought stress in *M. piperita* increased TPC and phenylalanine ammonia lyase (PAL) activity, with the latter being responsible for the synthesis of phenolic compounds [41,43]. In agreement, the TPC was observed to increase in different plant species submitted to abiotic stress [86]—for example, in *T. vulgaris* subjected to drought stress [96] and in *M. pulegium* under salt stress [98]. Conversely, Rahimi et al. [99] and Alhaithloul et al. [100] described a reduction in TPC in *M piperita* plants subjected to drought stress. However, in *Tagetes minuta* plants inoculated with *P. fluorescens* WCS417r and *Azospirillum brasilense,* and in chickpea inoculated with *P. fluorescens* [101], TPC levels increased significantly [36]. Jayapala et al. [102] reported the induction of resistance against pathogens through enhancement of the activities of defense-related enzymes and a higher accumulation of TPC in chili plants inoculated with *Bacillus* sp. Furthermore, Tahir et al. [27] revealed that *Bacillus* sp. mVOCs negatively influence the development of the pathogen *R. solanacearum* by activating ISR in tobacco plants. Molecular studies have shown that resistance is the consequence of an increase in the SM levels and defense-related enzymes, including PAL.

Phenolic compounds are antioxidants that may be required for scavenging ROS and protecting the lipid membrane from oxidative stress [12]. For example, Fagopyrum *esculentum* plants grown under media with increasing salt concentrations revealed a concentration-dependent increase in the accumulation of phenolic compounds, resulting in a higher DPPH free radical scavenging potential [103]. This effect was corroborated in the present study in plants subjected to salinity environments and treated with mVOCs, which showed a heightened antioxidant capacity, as revealed by the high levels of AAE detected in the DPPH• scavenging assay and by the low amounts of MDA. The highest levels of antioxidant activity were observed when plants were grown under 100 mM NaCl and mVOC. The GB03 mVOCs decreased the MDA levels in plants subjected to salt stress, to similar levels as those in control plants. In contrast, after water deficit treatment in peppermint plants, heightened amounts of MDA, as a cell membrane damage index, were detected [99]. Additionally, peppermint growing under control conditions was revealed to be more effective in scavenging DPPH free radicals and had a higher reducing power than when exposed to drought and heat stress. This observation provides signals that tissues of peppermint subjected to heat and/or drought stress contain fewer antioxidants and reducing compounds [100].

The PCA analysis showed that plants subjected to high salt concentrations combined with exposure to mVOCs strongly affected the TPC content and antioxidant capacity (AAE). This relationship was also detected in drought-stressed peppermint plants inoculated with GB03 [43].

In plants that were inoculated and subjected to osmotic stress, similar results in MDA reduction were observed to those reported for cucumber plants inoculated with a consortium of PGPR under drought stress conditions [104], as well as those in white clover and *M. arvensis* inoculated under saline conditions [51,105]. The decrease in the leaf MDA content resulting from mVOC treatment suggests its ability to reduce the peroxidation of cell membrane lipids under salt stress and to protect the leaf cell from damage. Moreover, Gopinath et al. [106] reported in *Nicotiana tabacum* that when callus was exposed to volatile compounds from *Bacillus badius* M12 and the volatile, 2,3- butanediol, this led to increased antioxidant activity by the expression of SOD, a key antioxidant enzyme. In addition, treatment with mVOCs from GB03 and *Pseudomonas simiae* increased choline and glycine betaine biosynthesis in *Arabidopsis* [51,68]. These osmolytes have positive effects on enzyme and membrane integrity, along with adaptive roles in mediating osmotic adjustment in plants subjected to stress conditions [107]. In another investigation, 2,3-butanediol was found to induce plant production of nitric oxide (NO) and hydrogen peroxide [108], and it was reported that NO regulates antioxidant enzymes at the level of activity and gene expression [109]. At the same time, the plant hormone SA is required for plant growth under abiotic stress [7,17,73]. Finally, an increase in the SA levels was shown in peppermint plants subjected to salt stress and treated with GB03 VOCs [38].

5. Conclusions

Salt stresses affect the growth and productivity of crop plants and are detrimental to the plants, thereby reducing their yield. Thus, it is necessary to improve the technologies of abiotic stress management. In recent decades, several studies have shown that PGPR has the ability to ameliorate the negative effects of salt or water. However, only a few reports have been published on PGPR VOCs as elicitors of tolerance to abiotic stress in aromatic and medicinal plants. The GB03 VOCs have been shown to increase plant growth and chlorophyll content and lead to better morphological characteristics in *M. piperita* plants subjected to salt stress. The results shown in the present study establish that for peppermint plants grown in the laboratory under salt media, the volatiles emitted by GB03 significantly increased SM production and improved the antioxidant status. This suggests that the accumulation of SMs is a plant strategy to avoid oxidative damage caused by ROS, a direct result of salt stress. Bacterial volatiles are promising candidates for a rapid non-invasive technique to increase SM production in aromatic and medicinal crops growing under abiotic stress conditions. In addition, this is a potentially useful system for the production of SMs, which have remarkable biological activities and are often exploited as medicinal and food ingredients for therapeutic, aromatic, and culinary purposes. However, future studies are still necessary to elucidate how plants modulate and perceive PGPR VOC-elicited abiotic tolerance.

Author Contributions: L.d.R.C., J.C., and T.B.P. performed the experiments; E.B. designed the research and analyzed the data; L.d.R.C., E.B., and W.G. wrote the manuscript. All authors read, revised, and approved the final manuscript. All authors have read and agreed to the published version of the manuscript.

Acknowledgments: This research was supported by grants from the Secretaría de Ciencia y Técnica de la Universidad Nacional de Río Cuarto, Consejo Nacional de Investigaciones Científicas y Técnicas (CONICET), and Agencia Nacional de Promoción Científica y Tecnológica (ANPCyT). E.B. and W.G. are Career Members of CONICET. E.B. obtained financial support from a Georg Forster Research Fellowship of the Alexander von

Humboldt Foundation. L.C., J.C., and T.B.P. have fellowships from CONICET. The authors are grateful to Paul Hobson, native speaker, for editorial assistance.

References

1. McConkey, M.E.; Gershenzon, J.; Croteau, R.B. Developmental regulation of monoterpene biosynthesis in the glandular trichomes of peppermint. *Plant Physiol.* **2000**, *122*, 215–224. [CrossRef] [PubMed]

2. Lange, B.M.; Mahmoud, S.S.; Wildung, M.R.; Turner, G.W.; Davis, E.M.; Lange, I.; Baker, R.C.; Boydston, R.A.; Croteau, R.B. Improving peppermint essential oil yield and composition by metabolic engineering. *Proc. Natl. Acad. Sci. USA* **2011**, *108*, 16944–16949. [CrossRef] [PubMed]

3. Ramakrishna, A.; Ravishankar, G.A. Influence of abiotic stress signals on secondary metabolites in plants. *Plant Signal. Behav.* **2011**, *6*, 1720–1731. [PubMed]

4. Khan, N.; Bano, A.; Curá, J.A. Role of Beneficial Microorganisms and Salicylic Acid in Improving Rainfed Agriculture and Future Food Safety. *Microorganisms* **2020**, *8*, 1018. [CrossRef] [PubMed]

5. Upadhyay, S.K.; Singh, J.S.; Singh, D.P. Exopolysaccharide-producing plant growth-promoting rhizobacteria under salinity condition. *Pedosphere* **2011**, *21*, 214–222. [CrossRef]

6. Zhu, J.K. Plant salt tolerance. *Trends Plant Sci.* **2001**, *6*, 66–71. [CrossRef]

7. Liu, X.M.; Zhang, H. The effects of bacterial volatile emissions on plant abiotic stress tolerance. *Front. Plant Sci.* **2015**, *6*, 774. [CrossRef]

8. Mittler, R. Oxidative stress, antioxidants and stress tolerance. *Trends Plant Sci.* **2002**, *7*, 405–410. [CrossRef]

9. Mithofer, A.; Schulze, B.; Boland, W. Biotic and heavy metal stress response in plants: Evidence for common signals. *FEBS Lett.* **2004**, *566*, 1–5. [CrossRef]

10. Noctor, G.; Mhamdi, A.; Foyer, C.H. The roles of reactive oxygen metabolism in drought: Not so cut and dried. *Plant Physiol.* **2014**, *164*, 1636–1648. [CrossRef]

11. Reddy, A.R.; Chaitanya, K.V.; Vivekanandan, M. Drought-induced responses of photosynthesis and antioxidant metabolism in higher plants. *J. Plant Physiol.* **2004**, *161*, 1189–1202. [CrossRef] [PubMed]

12. Khan, N.; Bano, A.; Ali, S.; Babar, M.A. Crosstalk amongst phytohormones from planta and PGPR under biotic and abiotic stresses. *Plant Growth Reg.* **2020**, *90*, 189–203. [CrossRef]

13. Kloepper, J.W.; Lifshitz, R.; Zablotowicz, R.M. Free-living bacterial inoculation for enhancing crop productivity. *Trends Biotecnol.* **1989**, *7*, 39–49. [CrossRef]

14. Vessey, J.K. Plant growth promoting rhizobacteria as biofertilizers. *Plant Soil.* **2003**, *255*, 571–586. [CrossRef]

15. Van Loon, L.C. Plant responses to plant growth-promoting rhizobacteria. *Eur. J. Plant Pathol.* **2007**, *119*, 243–254. [CrossRef]

16. Hassan, M.; Boersma, M.; Lawaju, B.R.; Lawrence, K.S.; Liles, M.; Kloepper, J.W. Effects of secondary metabolites produced by PGPR amended with orange peel on the mortality of second-stage juveniles of *Meloidogyne incognita*. *Plant Health* **2019**, 108, Abstracts of Presentations, subsection S2.

17. Farag, M.A.; Zhang, H.; Ryu, C.M. Dynamic chemical communication between plants and bacteria through airborne signals: Induced resistance by bacterial volatiles. *J. Chem. Ecol.* **2013**, *39*, 1007–1018. [CrossRef]

18. Timmusk, S.; Islam, A.; Abd El, D.; Lucian, C.; Tanilas, T.; Ka nnaste, A.; Behers, L.; Nevo, E.; Seisenbaeva, G.; Stenström, E.; et al. Drought-tolerance of wheat improved by rhizosphere bacteria from harsh environments: Enhanced biomass production and reduced emissions of stress volatiles. *PLoS ONE* **2014**, *9*, e96086. [CrossRef]

19. Vurukonda, S.S.K.P.; Vardharajula, S.; Shrivastava, M.; SkZ, A. Enhancement of drought stress tolerance in crops by plant growth promoting rhizobacteria. *Microbiol. Res.* **2016**, *184*, 13–24. [CrossRef]

20. Enebe, M.C.; Babalola, O.O. The influence of plant growth-promoting rhizobacteria in plant tolerance to abiotic stress: A survival strategy. *Appl. Microbiol. Biotechnol.* **2018**, *102*, 7821–7835. [CrossRef]

21. Khan, N.; Zandi, P.; Ali, S.; Mehmood, A.; Adnan Shahid, M.; Yang, J. Impact of salicylic acid and PGPR on the drought tolerance and phytoremediation potential of Helianthus annus. *Front. Microbiol.* **2018**, *9*, 2507. [CrossRef] [PubMed]

22. Kumar, A.; Verma, J.P. Does plant-Microbe interaction confer stress tolerance in plants: A review? *Microbiol. Res.* **2018**, *207*, 41–52. [CrossRef] [PubMed]

23. Liu, H.; Brettell, L.E.; Qiu, Z.; Singh, B.K. Microbiome-Mediated Stress Resistance in Plants. *Trends Plant Sci.* **2020**, *25*, 733–743. [CrossRef] [PubMed]

24. Ryu, C.M.; Farag, M.A.; Hu, C.H.; Reddy, M.S.; Kloepper, J.W.; Pare, P.W. Bacterial volatiles induce systemic resistance in Arabidopsis. *Plant Physiol.* **2004**, *134*, 1017–1026. [CrossRef]

25. Banchio, E.; Xie, X.; Zhang, H.; Paré, P.W. Soil bacteria elevate essential oil accumulation and emissions in sweet basil. *J. Agric. Food Chem.* **2009**, *57*, 653–657. [CrossRef]

26. Santoro, M.; Zygadlo, J.; Giordano, W.; Banchio, E. Volatile organic compounds from rhizobacteria increase biosynthesis of essential oils and growth parameters in peppermint (*Mentha piperita*). *Plant Physiol. Biochem.* **2011**, *49*, 1077–1082. [CrossRef]

27. Tahir, H.A.; Gu, Q.; Wu, H.; Raza, W.; Hanif, A.; Wu, L.; Colman, M.V.; Gao, X. Plant Growth Promotion by Volatile Organic Compounds Produced by *Bacillus subtilis* SYST2. *Front. Microbiol.* **2017**, *8*, 171. [CrossRef]

28. Naseem, H.; Ahsan, M.; Shahid, M.A.; Khan, N. Exopolysaccharides producing rhizobacteria and their role in plant growth and drought tolerance. *J. Basic Microbiol.* **2018**, *58*, 1009–1022. [CrossRef]

29. Lemfack, M.C.; Gohlke, B.-O.; Toguem, S.M.T.; Preissner, S.; Piechulla, B.; Preissner, R. mVOC 2.0: A database of microbial volatiles. *Nucleic Acids Res.* **2018**, *46*, 1261–1265. [CrossRef]

30. Kanchiswamy, C.N.; Malnoy, M.; Maffei, M.E. Chemical diversity of microbial volatiles and their potential for plant growth and productivity. *Front. Plant Sci.* **2015**, *6*, 151. [CrossRef]

31. Lee, B.; Farag, M.A.; Park, H.B.; Kloepper, J.W.; Lee, S.H.; Ryu, C.M. Induced Resistance by a long-chain bacterial volatile: Elicitation of plant systemic defense by a CVolatile Produced by *Paenibacillus polymyxa*. *PLoS ONE* **2012**, *7*, e48744. [CrossRef] [PubMed]

32. Khan, N.; Bano, A. Rhizobacteria and Abiotic Stress Management. In *Plant Growth Promoting Rhizobacteria for Sustainable Stress Management*; Springer: Singapore, 2019; pp. 65–80.

33. Hossain, M.J.; Ran, C.; Liu, K.; Ryu, C.M.; Rasmussen-Ivey, C.R.; Williams, M.A.; Hassan, M.K.; Choi, S.-K.; Jeong, H.; Newman, M.; et al. Deciphering the conserved genetic loci implicated in plant disease control through comparative genomics of *Bacillus amyloliquefaciens* subsp. plantarum. *Front. Plant Sci.* **2015**, *6*, 631. [CrossRef] [PubMed]

34. Ledger, T.; Rojas, S.; Timmermann, T.; Pinedo, I.; Poupin, M.J.; Garrido, T.; Richter, P.; Tamayo, J.; Donoso, R. Volatile-mediated effects predominate in *Paraburkholderia phytofirmans* growth promotion and salt stress tolerance of *Arabidopsis thaliana*. *Front. Microbiol.* **2016**, *7*, 1838. [CrossRef] [PubMed]

35. Banchio, E.; Bogino, P.; Santoro, M.V.; Torres, L.; Zygadlo, J.; Giordano, W. Systemic induction of monoterpene biosynthesis in *Origanum x majoricum* by soil bacteria. *J. Agric. Food Chem.* **2010**, *58*, 650–654. [CrossRef] [PubMed]

36. Cappellari, L.; Santoro, M.V.; Nievas, F.; Giordano, W.; Banchio, E. Increase of secondary metabolite content in marigold by inoculation with plant growth-promoting rhizobacteria. *Appl. Soil Ecol.* **2013**, *70*, 16–22. [CrossRef]

37. Cappellari, L.R.; Santoro, M.V.; Reinoso, H.; Travaglia, C.; Giordano, W.; Banchio, E. Anatomical, morphological, and phytochemical effects of inoculation with plant growth-promoting rhizobacteria on peppermint (*Mentha piperita*). *J. Chem. Ecol.* **2015**, *41*, 149–158. [CrossRef] [PubMed]

38. Cappellari, L.; Banchio, E. Microbial Volatile Organic Compounds Produced by *Bacillus amyloliquefaciens* GBAmeliorate the Effects of Salt Stress in *Mentha piperita* Principally Through Acetoin Emission. *J. Plant Growth Regul.* **2020**, *39*, 764–775. [CrossRef]

39. Santoro, M.V.; Bogino, P.C.; Nocelli, N.; Cappellari, L.; Giordano, W.; Banchio, E. Analysis of plant growth-promoting effects of fluorescent pseudomonas strains isolated from *Mentha piperita* rhizosphere and effects of their volatile organic compounds on essential oil composition. *Front. Microbiol.* **2016**, *7*, 1085. [CrossRef]

40. Choi, S.K.; Jeong, H.; Kloepper, J.W.; Ryu, C.M. Genome sequence of *Bacillus amyloliquefaciens* GB03, an active ingredient of the first commercial biological control product. *Gen Announc.* **2014**, *2*, 01092–01098. [CrossRef]

41. Cappellari, L.R.; Chiappero, J.; Santoro, M.; Giordano, W.; Banchio, E. Inducing phenolic production and volatile organic compounds emission by inoculating *Mentha piperita* with plant growth-promoting rhizobacteria. *Sci. Hortic.* **2017**, *220*, 193–198. [CrossRef]

42. Brand-Williams, W.; Cuvelier, M.E.; Berset, C.L.W.T. Use of a free radical method to evaluate antioxidant activity. *LWT Food Sci. Technol.* **1995**, *28*, 25–30. [CrossRef]

43. Chiappero, J.; Cappellari, L.; Sosa Alderete, L.G.; Palermo, T.B.; Banchio, E. Plant growth promoting rhizobacteria improve the antioxidant status in *Mentha piperita* grown under drought stress leading to an enhancement of plant growth and total phenolic content. *Ind. Crops Prod.* **2019**, *139*, 111553. [CrossRef]

44. Heath, R.L.; Packer, L. Photoperoxidation in isolated chloroplasts. Kinetics and stoichiometry of fatty acid peroxidation. *Arch. Biochem. Biophys.* **1968**, *125*, 189–198. [CrossRef]

45. Huang, G.-T.; Ma, S.-L.; Bai, L.-P.; Zhang, L.; Ma, H.; Jia, P.; Liu, J.; Zhong, M.; Guo, Z.-F. Signal transduction during cold, salt, and drought stresses in plants. *Mol. Biol. Rep.* **2012**, *39*, 969–987. [CrossRef] [PubMed]

46. Forni, C.; Duca, D.; Glick, B.R. Mechanisms of plant response to salt and drought stress and their alteration by rhizobacteria. *Plant Soil.* **2017**, *410*, 335–356. [CrossRef]

47. Khan, N.; Bano, A. Effects of exogenously applied salicylic acid and putrescine alone and in combination with rhizobacteria on the phytoremediation of heavy metals and chickpea growth in sandy soil. *Int. J. Phytoremed.* **2018**, *20*, 405–414. [CrossRef]

48. Acosta-Motos, J.R.; Ortuño, M.F.; Bernal-Vicente, A.; Diaz-Vivancos, P.; Sanchez-Blanco, M.J.; Hernandez, J.A. Plant Responses to Salt Stress: Adaptive Mechanisms. *Agronomy* **2017**, *7*, 18. [CrossRef]

49. Zhang, H.; Kim, M.S.; Sun, Y.; Dowd, S.E.; Shi, H.; Paré, P.W. Soil bacteria confer plant salt tolerance by tissue-specific regulation of the sodium transporter HKT1. *Mol. Plant-Microbe Interact.* **2008**, *21*, 737–744. [CrossRef]

50. Yang, J.; Kloepper, J.W.; Ryu, C.M. Rhizosphere bacteria help plants tolerate abiotic stress. *Trends Plant Sci.* **2009**, *14*, 1–4. [CrossRef]

51. Vaishnav, A.; Kumari, S.; Jain, S.; Varma, A.; Choudhary, D.K. Putative bacterial volatile-mediated growth in soybean (*Glycine max* L. Merrill) and expression of induced proteins under salt stress. *J. Appl Microbiol.* **2015**, *119*, 539–551. [CrossRef]

52. Bharti, N.; Barnawal, D.; Awasthi, A.; Yadav, A.; Kalra, A. Plant growth promoting rhizobacteria alleviate salinity induced negative effects on growth, oil content and physiological status in *Mentha arvensis*. *Acta Physiol. Plant.* **2014**, *36*, 45–60. [CrossRef]

53. Karray-Bouraoui, N.; Ksouri, R.; Falleh, H.; Rabhi, M.; Grignon, C.; Lachaal, M. Effects of environment and development stage on phenolic content and antioxidant activities of Tunisian *Mentha pulegium* L. *J. Food Biochem.* **2009**, *34*, 79–89. [CrossRef]

54. Neffati, M.; Marzouk, B. Changes in essential oil and fatty acid composition in coriander (*Coriandrum sativum* L.) leaves under saline conditions. *Ind. Crops Prod.* **2008**, *28*, 137–142. [CrossRef]

55. Baher, Z.F.; Mirza, M.; Ghorbanli, M.; Rezaii, M.B. The influence of water stress on plant height, herbal and essential oil yield and composition in *Satureja hortensis* L. *Flavour Frag. J.* **2002**, *17*, 275–277. [CrossRef]

56. Hendawy, S.F.; Khalid, K.A. Response of sage (*Salvia officinalis* L.) plants to zinc application under different salinity levels. *J. Appl. Sci. Res.* **2005**, *1*, 147–155.

57. Ezz El-Din, A.A.; Aziz, E.E.; Hendawy, S.F.; Omer, E.A. Response of *Thymus vulgaris* L. to salt stress and Alar (B9) in newly reclaimed soil. *J. Appl. Sci. Res.* **2009**, *5*, 2165–2170.

58. Shalan, M.N.; Abdel-Latif, T.A.T.; Ghadban, E.A. Effect of water salinity and some nutritional compounds of the growth and production of sweet marjoram plants (*Marjorana hortensis* L.). *Egypt. J. Agric. Res.* **2006**, *84*, 959.

59. Ben Taarit, M.K.; Msaada, K.; Hosni, K.; Marzouk, B. Changes in fatty acid and essential oil composition of sage (*Salvia officinalis* L.) leaves under NaCl stress. *Food Chem.* **2010**, *9*, 951–956. [CrossRef]

60. Mehmood, A.; Hussain, A.; Irshad, M.; Hamayun, M.; Iqbal, A.; Khan, N. In vitro production of IAA by endophytic fungus Aspergillus awamori and its growth promoting activities in Zea mays. *Symbiosis* **2019**, *77*, 225–235. [CrossRef]

61. Gleadow, R.M.; Woodrow, I.E. Defense chemistry of cyanogenic *Eucalyptus cladocalyx* seedlings is affected by water supply. *Tree Physiol.* **2002**, *22*, 939–945. [CrossRef]

62. Falk, K.L.; Tokuhisa, J.G.; Gershenzon, J. The effect of sulfur nutrition on plant glucosinolate content: Physiology and molecular mechanisms. *Plant Biol.* **2007**, *9*, 573–581. [CrossRef] [PubMed]

63. Khan, N.; Bano, A.; Rahman, M.A.; Rathinasabapathi, B.; Babar, M.A. UPLC-HRMS-based untargeted metabolic profiling reveals changes in chickpea (*Cicer arietinum*) metabolome following long-term drought stress. *Plant Cell Environ.* **2019**, *42*, 115–132. [CrossRef] [PubMed]

64. Harb, A.; Awad, D.; Samarah, N. Gene expression and activity of antioxidant enzymes in barley (*Hordeum vulgare* L.) under controlled severe drought. *J. Plant Interact.* **2015**, *10*, 109–116. [CrossRef]

65. Kleinwächter, M.; Paulsen, J.; Bloem, E.; Schnug, E.; Selmar, D. Moderate drought and signal transducer induced biosynthesis of relevant secondary metabolites in thyme (*Thymus vulgaris*), greater celandine (*Chelidonium majus*) and parsley (*Petroselinum crispum*). *Ind. Crops Prod.* **2015**, *64*, 158–166. [CrossRef]

66. Pott, D.M.; Osorio, S.; Vallarino, J.G. From Central to Specialized Metabolism: An Overview of Some Secondary Compounds Derived from the Primary Metabolism for Their Role in Conferring Nutritional and Organoleptic Characteristics to Fruit. *Front. Plant Sci.* **2019**, *10*, 835. [CrossRef] [PubMed]

67. Harrewijn, P.; Van Oosten, A.M.; Piron, P.G.M. *Natural Terpenoids as Messengers: A Multidisciplinary Study of Their Production, Biological Functions and Practical Applications*; Kluwer Academic Publishers: London, UK, 2001.

68. Zhang, H.; Murzello, C.; Sun, Y.; Kim, M.S.; Xie, X.; Jeter, R.M.; Zak, J.C.; Dowd, S.E.; Paré, P.W. Choline and osmotic-stress tolerance induced in Arabidopsis by the soil microbe *Bacillus subtilis* (GB03). *Mol. Plant Microbe Interact.* **2010**, *23*, 1097–1104. [CrossRef] [PubMed]

69. Blom, D.; Fabbri, C.; Connor, E.C.; Schiestl, F.P.; Klauser, D.R.; Boller, T.; Eberl, L.; Weisskopf, L. Production of plant growth modulating volatiles is widespread among rhizosphere bacteria and strongly depends on culture conditions. *Environ. Microbiol.* **2011**, *13*, 3047–3058. [CrossRef]

70. Rath, M.; Mitchell, T.R.; Gold, S.E. Volatiles produced by *Bacillus mojavensis* RRC101 act as plant growth modulators and are strongly culture-dependent. *Microbiol. Res.* **2018**, *208*, 76–84. [CrossRef]

71. Farag, M.A.; Ryu, C.M.; Sumner, L.W.; Pare, P.W. GC-MS SPME profiling of rhizobacterial volatiles reveals prospective inducers of growth promotion and induced systemic resistance in plants. *Phytochemistry* **2006**, *67*, 2262–2268. [CrossRef]

72. Khan, N.; Ali, S.; Shahid, M.A.; Kharabian-Masouleh, A. Advances in detection of stress tolerance in plants through metabolomics approaches. *Plant Omics* **2017**, *10*, 153. [CrossRef]

73. Cho, S.M.; Kang, B.R.; Han, S.H.; Anderson, A.J.; Park, J.Y.; Lee, Y.H.; Cho, B.H.; Yang, K.Y.; Ryu, C.M.; Kim, Y.C. 2R, 3R-butanediol, a bacterial volatile produced by *Pseudomonas chlororaphis* O6, is involved in induction of systemic tolerance to drought in *Arabdopsis thaliana*. *Mol. Plant-Microbe Interact.* **2008**, *21*, 1067–1075. [CrossRef]

74. Naseem, H.; Bano, A. Role of plant growth-promoting rhizobacteria and their exopolysaccharide in drought tolerance of maize. *J. Plant Interact.* **2014**, *9*, 689–701. [CrossRef]

75. Chen, L.; Liu, Y.; Wu, G.; Njeri, K.V.; Shen, Q.; Zhang, N.; Zhang, R. Induced maize salt tolerance by rhizosphere inoculation of *Bacillus amyloliquefaciens* SQR. *Physiol. Plant.* **2016**, *158*, 34–44. [CrossRef] [PubMed]

76. Ikram, M.; Ali, N.; Jan, G.; Guljan, F.; Khan, N. Endophytic fungal diversity and their interaction with plants for agriculture sustainability under stressful condition. *Recent Pat. Food Nutr. Agric.* **2019**, *12*. [CrossRef] [PubMed]

77. Sangwan, N.S.; Farooqi, A.H.A.; Shabih, F.; Sangwan, R.S. Regulation of essential oil production in plants. *Plant Growth Regul.* **2001**, *24*, 3–21. [CrossRef]

78. Banchio, E.; Zygadlo, J.; Valladares, G. Quantitative variations in the essential oil of *Minthostachys mollis* (Kunth.) Griseb. in response to insects with different feeding habits. *J. Agric. Food Chem.* **2005**, *53*, 6903–6906. [CrossRef]

79. Hartmann, T. Plant-derived secondary metabolites as defensive chemicals in herbivorous insects: A case study in chemical ecology. *Planta* **2004**, *219*, 1–4. [CrossRef] [PubMed]

80. Freeman, B.C.; Beattie, G.A. An overview of plant defenses against pathogens and herbivores. *Plant Health Instr.* **2008**. [CrossRef]

81. Kim, Y.S.; Choi, Y.E.; Sano, H. Plant vaccination: Stimulation of defense system by caffeine production in planta. *Plant Signal Behav.* **2010**, *5*, 489–493. [CrossRef]

82. Cappellari, L.; Santoro, V.M.; Schmidt, A.; Gershenzon, J.; Banchio, E. Induction of essential oil production in *Mentha x piperita* by plant growth promoting bacteria was correlated with an increase in jasmonate and salicylate levels and a higher density of glandular trichomes. *Plant Physiol. Biochem.* **2019**, *141*, 142–153. [CrossRef]

83. Parida, A.K.; Das, A.B. Salt tolerance and salinity effects on plants: A review. *Ecotoxicol. Environ. Saf.* **2005**, *60*, 324–349. [CrossRef] [PubMed]

84. Kousar, B.; Bano, A.; Khan, N. PGPR Modulation of Secondary Metabolites in Tomato Infested with *Spodoptera litura*. *Agronomy* **2020**, *10*, 778. [CrossRef]

85. Ellenberger, J.; Siefen, N.; Krefting, P.; Schulze Lutum, J.-B.; Pfarr, D.; Remmel, M.; Schröder, L.; Röhlen-Schmittgen, S. Effect of UV Radiation and Salt Stress on the Accumulation of Economically Relevant Secondary Metabolites in Bell Pepper Plants. *Agronomy* **2020**, *10*, 142. [CrossRef]

86. Selmar, D.; Kleinwächter, M. Influencing the product quality by deliberately applying drought stress during the cultivation of medicinal plants. *Ind. Crops Prod.* **2013**, *42*, 558–566. [CrossRef]

87. Singh, D.; Prabha, R.; Meena, K. Induced accumulation of polyphenolics and flavonoids in cyanobacteria under salt stress protects organisms through enhanced antioxidant activity. *Am. J. Plant Sci.* **2014**, *5*, 726–735. [CrossRef]

88. Dorman, H.J.; Koşar, M.; Başer, K.H.; Hiltunen, R. Phenolic profile and antioxidant evaluation of *Mentha × piperita* L. (peppermint) extracts. *Nat. Prod. Commun.* **2009**, *4*, 535–542. [CrossRef]

89. Farnad, N.; Heidari, R.; Aslanipour, B. Phenolic composition and comparison of antioxidant activity of alcoholic extracts of Peppermint (*Mentha piperita*). *J. Food Meas. Charact.* **2014**, *8*, 113–121. [CrossRef]

90. McKay, D.L.; Blumberg, J.B. A review of the bioactivity and potential health benefits of peppermint tea (*Mentha piperita* L.). *Phytother. Res.* **2006**, *20*, 619–633. [CrossRef]

91. Khan, N.; Bano, A. Modulation of phytoremediation and plant growth by the treatment with PGPR, Ag nanoparticle and untreated municipal wastewater. *Int. J. Phytoremed.* **2016**, *18*, 1258–1269. [CrossRef]

92. Riachi, L.G.; De Maria, C.A.B. Peppermint antioxidants revisited. *Food Chem.* **2015**, *176*, 72–81. [CrossRef]

93. Bagues, M.; Hafsi, C.; Yahia, Y.; Souli, I.; Boussora, F.; Nagaz, K. Modulation of Photosynthesis, Phenolic Contents, Antioxidant Activities, and Grain Yield of Two Barley Accessions Grown under Deficit Irrigation with Saline Water in an Arid Area of Tunisia. *Pol. J. Environ. Stud.* **2019**, *28*, 3071–3080. [CrossRef]

94. Awika, J.M.; Rooney, L.W.; Wu, X.; Prior, R.L.; Cisneros-Zevallos, L. Screening Methods to Measure Antioxidant Activity of Sorghum (*Sorghum bicolor*) and Sorghum Products. *J. Agric. Food Chem.* **2003**, *51*, 6657–6662. [CrossRef] [PubMed]

95. Agati, G.; Tattini, M. Multiple functional roles of flavonoids in photoprotection. *New Phytol.* **2010**, *156*, 786–793. [CrossRef] [PubMed]

96. Khalil, N.; Fekry, M.; Bishr, M.; El-Zalabani, S.; Salama, O. Foliar spraying of salicylic acid induced accumulation of phenolics, increased radical scavenging activity and modified the composition of the essential oil of water stressed *Thymus vulgaris* L. *Plant Physiol. Biochem.* **2018**, *123*, 65–74. [CrossRef] [PubMed]

97. Oh, J.; Jo, H.; Cho, A.R.; Kim, S.J.; Han, J. Antioxidant and antimicrobial activities of various leafy herbal teas. *Food Control* **2013**, *31*, 403–409. [CrossRef]

98. Oueslati, S.; Karray-Bouraoui, N.; Attia, H.; Rabhi, M.; Ksouri, R.; Lachaal, M. Physiological and antioxidant responses of *Mentha pulegium* (Pennyroyal) to salt stress. *Acta Physiol. Plant.* **2010**, *32*, 289–296. [CrossRef]

99. Rahimi, Y.; Taleei, A.; Ranjbar, M. Long-term water deficit modulates antioxidant capacity of peppermint (*Mentha piperita* L.). *Sci. Hortic.* **2018**, *237*, 36–43. [CrossRef]

100. Alhaithloul, H.A.; Soliman, M.H.; Ameta, K.L.; El-Esawi, M.A.; Elkelish, A. Changes in Ecophysiology, Osmolytes, and Secondary Metabolites of the Medicinal Plants of *Mentha piperita* and *Catharanthus roseus* Subjected to Drought and Heat Stress. *Biomolecules* **2020**, *10*, 43. [CrossRef]

101. Khan, N.; Bano, A. Role of PGPR in the Phytoremediation of Heavy Metals and Crop Growth under Municipal Wastewater Irrigation. In *Phytoremediation*; Springer: Cham, Switzerland, 2018; pp. 135–149.

102. Jayapala, N.; Mallikarjunaiah, N.; Puttaswamy, H.; Gavirangappa, H.; Ramachandrappa, N.S. Rhizobacteria Bacillus spp. induce resistance against anthracnose disease in chili (*Capsicum annuum* L.) through activating host defense response. *Egypt. J. Biol. Pest Control.* **2019**, *29*, 45. [CrossRef]

103. Lim, J.H.; Park, K.J.; Kim, B.K.; Jeong, J.W.; Kim, H.J. Effect of salinity stress on phenolic compounds and carotenoids in buckwheat (*Fagopyrum esculentum* M.) sprout. *Food Chem.* **2012**, *135*, 1065–1070. [CrossRef]

104. Wang, C.J.; Yang, W.; Wang, C.; Gu, C.; Niu, D.D.; Liu, H.X.; Wang, Y.P.; Guo, J.H. Induction of drought tolerance in cucumber plants by a consortium of three plant growth-promoting rhizobacterium strains. *PLoS ONE* **2012**, *7*, e52565. [CrossRef] [PubMed]

105. Han, Q.Q.; Lü, X.P.; Bai, J.P.; Qiao, Y.; Paré, P.W.; Wang, S.M.; Zhang, J.L.; Wu, Y.N.; Pang, X.P.; Xu, W.B.; et al. Beneficial soil bacterium *Bacillus subtilis* (GB03) augments salt tolerance of white clover. *Front. Plant Sci.* **2014**, *5*, 525. [CrossRef] [PubMed]

106. Gopinath, S.; Kumaran, K.S.; Sundararaman, M.A. New initiative in micropropagation: Airborne bacterial volatiles modulate organogenesis and antioxidant activity in tobacco (*Nicotiana tabacum* L.) callus. *In Vitro Cell. Dev. Biol. Plant* **2015**, *51*, 514–523. [CrossRef]

107. Giri, J. Glycinebetaine and abiotic stress tolerance in plants. *Plant Signal. Behav.* **2011**, *6*, 1746–1751. [CrossRef]

108. Mehmood, A.; Hussain, A.; Irshad, M.; Khan, N.; Hamayun, M.; Ismail; Afridi, S.G.; Lee, I.J. IAA and flavonoids modulates the association between maize roots and phytostimulant endophytic Aspergillus fumigatus greenish. *J. Plant Interact.* **2018**, *1*, 532–542. [CrossRef]

109. Groß, F.; Durner, J.; Gaupels, F. Nitric oxide, antioxidants and prooxidants in plant defence responses. *Front. Plant Sci.* **2013**, *4*, 419. [CrossRef]

Salicylic Acid Improves Boron Toxicity Tolerance by Modulating the Physio-Biochemical Characteristics of Maize (*Zea mays* L.) at an Early Growth Stage

Muhammad Nawaz [1,*], **Sabtain Ishaq** [2], **Hasnain Ishaq** [1], **Naeem Khan** [3], **Naeem Iqbal** [1], **Shafaqat Ali** [4,5,*], **Muhammad Rizwan** [4], **Abdulaziz Abdullah Alsahli** [6] **and Mohammed Nasser Alyemeni** [6]

[1] Department of Botany, Government College University Faisalabad, Lahore 54000, Pakistan; hasnain.official1@gmail.com (H.I.); drnaeem@gcuf.edu.pk (N.I.)

[2] Department of Botany, University of Agriculture Faisalabad, Faisalabad 38000, Pakistan; sabtainishaq28@gmail.com

[3] Department of Agronomy, Institute of Food and Agricultural Sciences, University of Florida, Gainesville, FL 32611, USA; naeemkhan@ufl.edu

[4] Department of Environmental Sciences and Engineering, Government College University, Lahore 54000, Pakistan; mrazi1532@yahoo.com

[5] Department of Biological Sciences and Technology, China Medical University, Taichung 40402, Taiwan

[6] Botany and Microbiology Department, College of Science, King Saud University, P.O. Box. 2460, Riyadh 11451, Saudi Arabia; aalshenaaalshenaifi@ksu.edu.sa (A.A.A.); mnyemeni@ksu.edu.sa (M.N.A.)

* Correspondence: muhammadnawaz@gcuf.edu.pk (M.N.); shafaqataligill@yahoo.com or shafaqataligill@gcuf.edu.pk (S.A.)

Abstract: The boron (B) concentration surpasses the plant need in arid and semi-arid regions of the world, resulting in phyto-toxicity. Salicylic acid (SA) is an endogenous signaling molecule responsible for stress tolerance in plants and is a potential candidate for ameliorating B toxicity. In this study, the effects of seed priming with SA (0, 50, 100 and 150 μM for 12 h) on the growth, pigmentation and mineral concentrations of maize (*Zea mays* L.) grown under B toxicity were investigated. One-week old seedlings were subjected to soil spiked with B (0, 15 and 30 mg kg^{-1} soil) as boric acid. Elevating concentrations of B reduced the root and shoot length, but these losses were significantly restored in plants raised from seeds primed with 100 μM of SA. The B application decreased the root and shoot fresh/dry biomasses significantly at 30 mg kg^{-1} soil. The chlorophyll and carotenoid contents decreased with increasing levels of B, while the contents of anthocyanin, H_2O_2, ascorbic acid (ASA) and glycinebetaine (GB) were enhanced. The root K and Ca contents were significantly increased, while a reduction in the shoot K contents was recorded. The nitrate concentration was significantly higher in the shoot as compared to the root under applied B toxic regimes. However, all of these B toxicity effects were diminished with 100 μM SA applications. The current study outcomes suggested that the exogenously applied SA modulates the response of plants grown under B toxic conditions, and hence could be used as a plant growth regulator to stimulate plant growth and enhance mineral nutrient uptake under B-stressed conditions.

Keywords: biomass reduction; cereal crops; growth regulators; metal stress

1. Introduction

Abrupt changes in climate along with the potential abiotic and biotic stresses are serious challenges for plant growth and production worldwide [1]. Environmental stresses negatively influence the germination, growth and yield of the crop plants. The continuous yield losses caused by abiotic stresses are one of the important reasons for socioeconomic imbalance [2]. Drought reduces the yield of staple food crops throughout the world up to 70% [3], and the effects of drought and salt stress on plant growth mechanisms and patterns have been discussed [3,4]. In the last few decades, soil and water resources are being contaminated with toxic elements due to industrial revolution and urbanization together with the use of artificial fertilizers [5,6]. Increasing levels of these toxic elements are imposing harmful effects on plants, plant-dependent animals and ultimately human health [7].

Boron is an important micronutrient in many plants for their normal functioning [8]. It is also considered to be an essential element for vascular plants according to the defined criteria for essentiality. The indirect association of B with photosynthesis has been reported in crop plants—e.g., soybean [9]. However, the rate of emergence and productivity is also decreased in many plants, including tomato, maize, wheat, alfalfa and carrot under B toxicity [10]. The B toxicity significantly reduces the yield of crop plants in relatively dry areas of the world [11]. Some of the factors contributing to the elevating levels of B are the use of fertilizers, mining and irrigation [12,13]. The B-induced toxicity occurs more commonly in saline soil in semi-arid geographical zones [14]. The interplay between salt stress and B nutrition in plants has been described, with contrasting results showing antagonistic and synergistic relations even within the same plant species [15]. It has been observed that salinity increases B toxicity [16], but the interaction of salinity and B is not fully understood [17], making it an important area of research in plant physiology and ecotoxicology.

Oxidative stress may result from a deficiency or excess of B, which triggers the over-production of reactive oxygen species (ROS). The ROS and their derivatives are highly toxic agents and damage cellular membranes due to lipid peroxidation, causing protein denaturation and mutations in DNA [18]. Different nutrients such as silicon (Si) [19], zinc (Zn) [20,21], potassium (K) [22] and calcium (Ca) [23] can ameliorate B toxicity in different crop plants. The SA signal molecule [24] plays an important role in reducing the hazardous effects posed by biotic and abiotic stresses. Thus, SA has been used by many researchers to reduce the hazardous effects of different stresses such as osmotic stress [25], heat, saline and B toxicity in wheat [26].

Among the most important staple foods, maize holds an important position after wheat and rice [27]. Maize is well known for its high potential of extracting heavy metals from soil [28]. Despite this phytoextraction ability, maize is affected by various environmental stresses along with the high metal concentrations. The abiotic stress effects on maize growth and yield have been studied [29,30]. In the current study, the main objective was to assess the effects of high B toxicity under the remodeling effects of SA in terms of physio-biochemical improvements in the maize cultivar Gohar-19.

2. Results

For assessing the effects of SA on mitigating the effects of B toxicity, plants were supplied with 0, 50, 100 and 150 μM of SA. The B toxicity levels were 0, 15 and 30 mg kg^{-1} soil. Roots transport B via passive diffusion or facilitate transport [30] in the plant body through transpiration streams and it is accumulated in older shoots without being translocated [31], therefore the study parameters include both the root and shoot data of maize cv. Gohar-19.

2.1. Root and Shoot Length

The B toxicity significantly reduced the root and shoot length of maize seedlings. High B concentrations in soil inhibit the root and shoot growth due to the decreased photosynthetic activity and net plant productivity. Elevating the B concentration in soil decreased the root and shoot length up to 21.77% and 25.25%, respectively, which are significant reductions (Table 1, Figure 1). The priming of seeds with SA reduced the B toxic effects and retained the root and shoot lengths. Plant seeds that were primed with various concentrations (0, 50, 100 and 150 μM) of SA improved the root and shoot lengths. Significant increases in the root and shoot lengths were observed at 100 μM SA (Figure 2, Table 1). A 23.8% increase in root length was observed with the application of 100 μM of SA in 30 mg kg^{-1} of

B-treated plants, while a 26.7% decrease was observed in the shoot length of 30 mg kg^{-1} B-treated plants as compared with the control. The SA application at 100 μM was found to be the best treatment and caused increases in the shoot length in 30 mg kg^{-1} B-treated plants up to 31.8%.

Table 1. Effects of SA (0, 50, 100 and 150 μM) on the plant root and shoot length of maize cultivar Gohar-19 under different B toxicity levels (0, 15 and 30 mg kg^{-1}).

		Root Length (cm)		
		0 mg kg^{-1} B	15 mg kg^{-1} B	30 mg kg^{-1} B
SA	0 μM	27.1 ± 0.89 ^{c}	24.3 ± 1.05 ^{b}	21.2 ± 0.88 ^{c}
	50 μM	27.8 ± 1.01 ^{b}	24.2 ± 0.97 ^{b}	21.34 ± 1.03 ^{c}
	100 μM	29 ± 0.98 ^{a}	28.2 ± 0.87 ^{a}	26.4 ± 0.77 ^{a}
	150 μM	26.4 ± 1.12 ^{d}	23 ± 1.24 ^{c}	21.8 ± 1.02 ^{b}
		Shoot Length (cm)		
SA	0 μM	30.3 ± 0.69 ^{b}	28.5 ± 0.85 ^{b}	22.65 ± 1.25 ^{c}
	50 μM	30.2 ± 1.13 ^{b}	28.7 ± 0.77 ^{b}	22.10 ± 1.02 ^{d}
	100 μM	32.0 ± 0.99 ^{a}	30.5 ± 0.98 ^{a}	29.00 ± 0.84 ^{a}
	150 μM	28.0 ± 1.21 ^{c}	27.4 ± 0.66 ^{c}	27.00 ± 0.96 ^{b}

LSD 5% = 0.44. Values in the same column with different letters in superscript differ significantly.

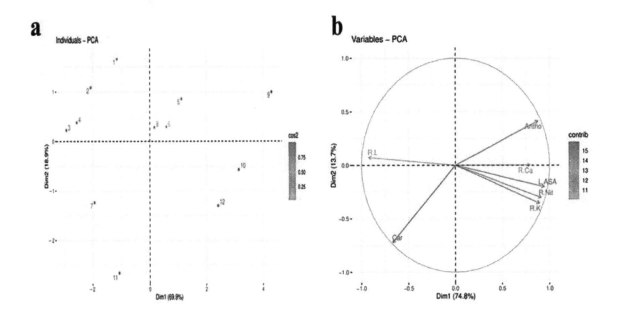

Figure 1. Score (**a**) and loading plot (**b**) of principal component analysis (PCA) on different attributes of maize cultivar Gohar-19 plants supplemented with and without SA while grown under B stress. Score plot represents the separation of treatments as T1: 0 mg B without SA; T2: 0 mg B with 50 μM SA; T3: 0 mg B with 100 μM SA; T4: 0 mg B with 150 μM SA; T5: 15 mg/kg B without SA; T6: 15 mg/kg B with 50 μM SA; T7: 15 mg/kg B with 100 μM SA; T8: 15 mg/kg B with 150 μM SA; T9: 30 mg/kg B without SA; T10: 30 mg/kg B with 50 μM SA; T11: 30 mg/kg B with 100 μM SA; T12: 30 mg/kg B with 150 μM SA. Attributes evaluated include R L = root length; Car = carotenoids; R Nit = root nitrate; R K = root potassium; R Ca = root calcium, Antho = anthocyanin; L ASA = leaf ascorbic acid.

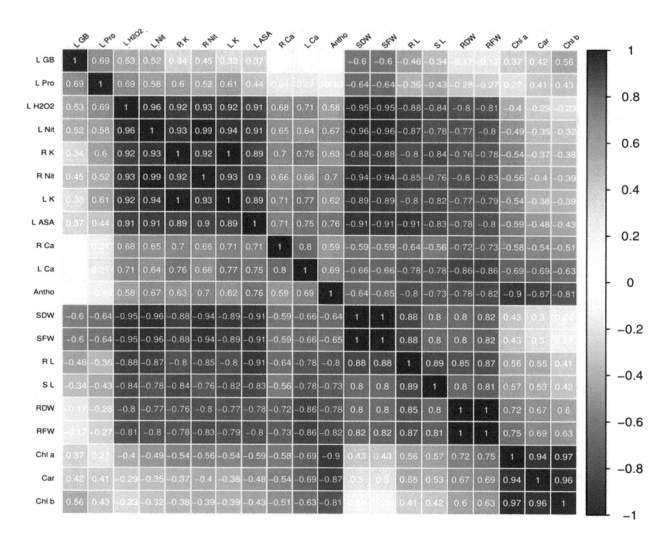

Figure 2. Correlations (r values) among the different studied parameters of maize cultivar Gohar-19 grown under different B stress levels and fertigated with and without SA. R L = root length; S L = shoot length; RFW=root fresh weight; RDW = root dry weight; SFW = shoot fresh weight; SDW = shoot dry weight; Chl a = chlorophyll a; Chl b = chlorophyll b; Car = carotenoids; Antho = anthocyanin; L ASA = leaf ascorbic acid; L H_2O_2 = leaf hydrogen peroxide; L Pro = leaf proline; L GB = leaf glycine betaine; R K = root potassium; L K = leaf potassium; R Ca = root calcium; L Ca = leaf calcium; R Nit = root nitrate; L Nit = leaf nitrate.

2.2. Plant Biomass

Plant fresh and dry biomass were also reduced in response to increasing the B treatment levels. The results obtained exhibited a positive correlation with the root and shoot lengths. Reducing and increasing patterns in plant fresh and dry biomass were observed in the root and shoot lengths. We observed 30% and 32.89% decreases in the root and shoot fresh biomass, respectively. The maximum increase in plant biomass was noted in plants raised from seeds primed with 100 µM of SA, as shown in Table 2. This gain in the plant growth biomarkers was due to the enhanced photosynthetic activity and improved antioxidant status of the plant body (Figure 2).

Table 2. Effect of SA (0, 50, 100 and 150 µM) on the plant fresh and dry weight of maize cultivar Gohar-19 under varying B toxicity levels (0, 15 and 30 mg kg^{-1}).

Treatment	Root Fresh Weight (g)	Root Dry Weight (g)	Shoot Fresh Weight (g)	Shoot Dry Weight (g)
00 µM SA + 00 mg/g^{-1} B	1.00 ± 0.89 [abc]	0.75 ± 0.21 [bc]	3.80 ± 0.33 [abc]	2.53 ± 1.01 [abcd]
00 µM SA + 15 mg kg^{-1} B	0.89 ± 0.95 [bc]	0.65 ± 0.34 [d]	3.10 ± 0.41 [abc]	2.07 ± 0.85 [abcd]
00 µM SA + 30 mg kg^{-1} B	0.70 ± 0.55 [c]	0.5 ± 0.21 [d]	2.55 ± 0.55 [c]	1.70 ± 0.33 [d]
50 µM SA + 00 mg kg^{-1} B	1.25 ± 0.45 [ab]	1.02 ± 0.35 [ab]	3.85 ± 1.01 [abc]	2.57 ± 0.65 [abcd]
50 µM SA + 15 mg kg^{-1} B	1.15 ± 0.75 [abc]	0.95 ± 0.34 [ab]	3.00 ± 0.95 [abc]	2.00 ± 0.35 [abcd]
50 µM SA + 30 mg kg^{-1} B	0.85 ± 0.65 [bc]	0.62 ± 0.32 [d]	2.70 ± 0.55 [bc]	1.80 ± 0.45 [cd]
100 µM SA + 00 mg kg^{-1} B	1.50 ± 0.76 [ab]	1.26 ± 0.22 [a]	4.30 ± 0.25 [a]	2.87 ± 0.27 [a]
100 µM SA + 15 mg kg^{-1} B	1.35 ± 0.55 [ab]	1.09 ± 0.36 [abc]	3.60 ± 0.97 [abc]	2.40 ± 0.85 [abc]
100 µM SA + 30 mg kg^{-1} B	1.15 ± 0.75 [abc]	0.9 ± 0.23 [abc]	3.00 ± 0.85 [abc]	2.00 ± 0.33 [abcd]
150 µM SA + 00 mg kg^{-1}B	1.25 ± 0.82 [ab]	0.99 ± 0.45 [abc]	3.90 ± 0.21 [abc]	2.60 ± 0.43 [abc]
150 µM SA + 15 mg kg^{-1} B	1.00 ± 071 [abc]	0.75 ± 0.35 [bc]	3.25 + 0.85 [abc]	2.17 ± 0.55 [abcd]
150 µM SA + 30 mg kg^{-1} B	0.95 ± 0.66 [bc]	0.71 ± 0.32 [c]	2.85 ± 0.79 [abc]	1.90 ± 0.65 [bcd]
LSD 5%	0.51	0.49	1.46	0.98

Values in the same column with different letters in superscript differ significantly.

2.3. Photosynthetic Pigments

Elevated B levels significantly reduced the photosynthetic pigment contents of maize seedlings. It was observed that the chl *a* contents were reduced with increasing B treatment levels. The 30 mg kg^{-1} B imposed deteriorative effects and reduced the chl *a* contents effectively (Figure 3a). An improvement in the chl *a* concentration was recorded through priming seeds with SA. Seeds primed with 100 µM of SA expressed the maximum chl *a* content, which suggests reduced toxicity effects.

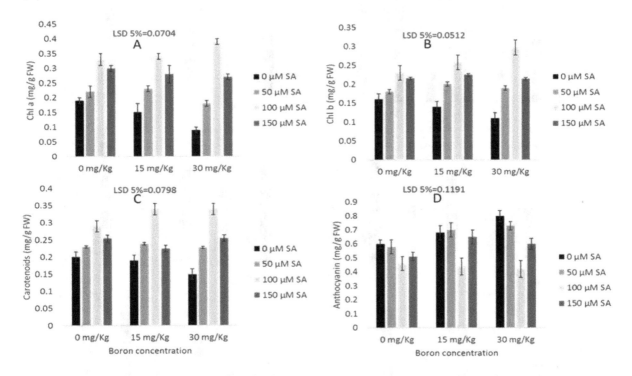

Figure 3. Effect of SA on chlorophyll a (**A**), chlorophyll b (**B**), carotenoids (**C**) and anthocyanin (**D**) contents of maize cultivar Gohar-19 under varying B toxicity levels.

The chl *b* contents were also reduced under B toxicity as compared to the control. An increase in the chl *b* contents was observed with respect to the control in the plants emerging from primed seeds. An increase of 30.4% in the chl *b* contents was observed at 100 µM SA treatment, while a non-significant increase was observed at 150 µM SA treatment as compared to the control (Figure 3b). The carotenoid contents were reduced effectively under 30 mg kg^{-1} B. The B application at 30 mg Kg^{-1} caused reductions of 52.6%, 31.3% and 45% in the chl*a*, chl*b* and carotenoids, respectively. The SA priming improved the carotenoid contents by reducing the drastic effects of B toxicity. A non-significant change in the carotenoid contents was observed in 50 and 150 μM SA primed seeds, while 100 μM SA significantly enhanced the carotenoid contents as compared to the control (Figures 2 and 3c).

2.4. Anthocyanin

The anthocyanin contents increased with increasing the levels of B toxicity. Significant increases in the anthocyanin contents were observed in plants treated with 15 and 30 mg kg^{-1} B. There was a 33.33% increase in anthocyanin contents when 30 mg kg^{-1} soil B was applied, as compared to the control. The SA priming reduced the anthocyanin contents overall, but only 100 μM of SA caused a 47.5% reduction in the anthocyanin contents (Figure 2 andFigure 3c).

2.5. Ascorbic Acid

The toxic effects of B increased the ASA contents of maize seedlings. The B treatment of 30 mg kg^{-1} significantly increased the ASA content up to 44% as compared with the control (Figure 4). Priming with SA reduced the B toxic effects. Only 100 and 150 μM of SA effectively mitigated the toxic effects on plants grown in pots containing 15 mg kg^{-1} B. However, under a high boron toxicity, only 100 μM of SA significantly reduced the ASA content up to 36% as compared to the control, as shown in Table 3, Figure 2.

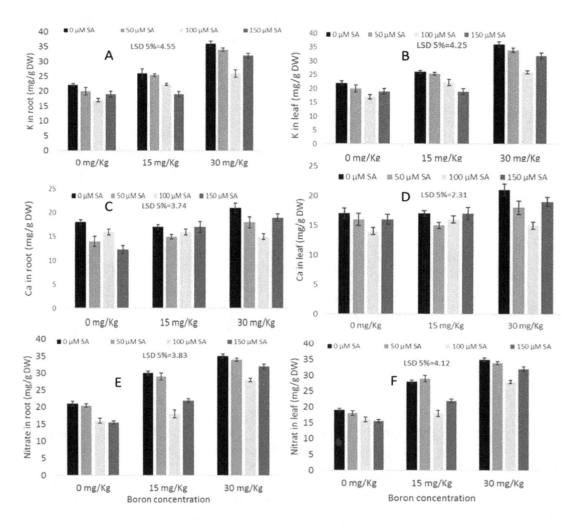

Figure 4. Effect of SA on the K (**A,B**), Ca (**C,D**) and nitrate (**E,F**) contents in the roots and leaves of maize cultivar Gohar-19 under varying B toxicity levels.

Table 3. Effect of SA (0, 50, 100 and 150 μM) on the leaf ascorbic acid, H_2O_2, proline and glycine betaine contents of maize cultivar Gohar-19 under varying B toxicity levels (0, 15 and 30 mg kg^{-1}).

Treatment	Leaf ASA (μmoles/g FW)	Leaf H_2O_2 (mg/g FW)	Leaf Proline (μMole/g FW)	Leaf GB (μg/g FW)
00 μM SA + 00 mg kg^{-1} B	210 ± 1.53	0.80 ± 0.15	32.00 ± 0.5	1.60 ± 0.05
00 μM SA + 15 mg kg^{-1} B	320 ± 1.15	1.80 ± 0.06	36.50 ± 0.5	1.80 ± 0.03
00 μM SA + 30 mg kg^{-1} B	375 ± 0.58	2.50 ± 0.10	46.00 ± 0.5	1.90 ± 0.03
50 μM SA + 00 mg kg^{-1} B	209 ± 1.00	0.70 ± 0.05	33.00 ± 0.5	1.70 ± 0.05
50 μM SA + 15 mg kg^{-1} B	300 ± 0.58	1.34 ± 0.03	37.00 ± 0.58	2.50 ± 0.06
50 μM SA + 30 mg kg^{-1} B	360 ± 0.58	2.40 ± 0.03	47.00 ± 0.50	2.50 ± 0.08
100 μM SA + 00 mg kg^{-1} B	200 ± 1.00	0.64 ± 0.02	33.00 ± 0.29	1.80 ± 0.05
100 μM SA + 15 mg kg^{-1} B	260 ± 0.58	1.00 ± 0.03	46.00 ± 0.76	2.00 ± 0.05
100 μM SA + 30 mg kg^{-1} B	240 ± 0.58	1.90 ± 0.03	58.00 ± 0.29	2.80 ± 0.05
150 μM SA + 00 mg kg^{-1} B	211 ± 1.15	0.78 ± 0.02	37.00 ± 0.29	1.90 ± 0.05
150 μM SA + 15 mg kg^{-1} B	276 ± 0.76	1.45 ± 0.05	37.00 ± 0.58	2.50 ± 0.09
150 μM SA + 30 mg kg^{-1} B	335 ± 0.29	2.20 ± 0.20	48.00 ± 0.29	2.40 ± 0.08
LSD 5%	0.51	0.49	1.46	0.98

2.6. H_2O_2 Concentration

Increasing the B toxicity enhanced the H_2O_2 content effectively. The H_2O_2 content was highly affected by 30 mg kg^{-1} B in soil. The effective treatment in term of reducing the H_2O_2 content was 100 μM SA, which decreased the H_2O_2 content up to 84% (Table 3, Figure 2).

2.7. Proline Content

The toxic effects of B significantly increased the proline contents. The B treatment of 30 mg kg^{-1} increased the proline contents up to 43.75% as compared to the control. Priming with SA remained productive in reducing the B toxic effects. Only 100 and 150 μM SA effectively mitigated the toxic effects in plants grown in pots containing 15 mg kg^{-1} B. However, under a high boron toxicity, only 150 μM SA significantly reduced the toxic effects by increasing the proline content in comparison with the control, as shown in Table 3, Figure 2.

2.8. Glycine Betaine

An outstanding improvement was noted in the leaf GB contents of plants grown under B stress. Exogenous applications of SA further increased the GB content in the leaves of plants experiencing B toxicity stress. All the treatments of SA affected the GB level, however 100 μM SA increased the GB contents up to 100% under 30 mg kg^{-1} of B treatment as compared to the control, as indicated in Table 3, Figure 2.

2.9. Potassium Content

An increase in the K contents was observed in response to the B toxicity. Applications of SA reduced the K contents and a maximum reduction of up to 27.8% was noted in the plants primed with 100 μM of SA. The K uptake and accumulation exhibited quite similar patterns in the plant root and shoot (Figures 2 and 4a,b).

2.10. Calcium Content

Boron toxicity significantly influenced the Ca accumulation in the root of the maize cultivar Gohar-19. The Ca content was reduced at a lower SA treatment level as compared to the control. With increasing the SA concentration up to 100 and 150 μM, higher increments in the Ca contents relative to the control were recorded. A total of 100 μM SA treatment was found to be effective in reducing toxic effects by lowering the Ca content up to 28.7% in the 30 mg kg^{-1} B group. The Ca accumulation in the plant leaves was reduced due to the B toxicity even at high concentrations of SA applications. No significant effects of SA on Ca contents were observed (Figures 2 and 4a,b).

2.11. Nitrate Concentration

An increase in the nitrate contents was observed with an increasing B level. Various SA concentrations (0, 50, 100 and 150 µM) reduced the nitrate content. The application of 100 µM of SA reduced the nitrate content up to 20% in the most destructive B treatment of 30 mg kg^{-1} (Figures 2 and 4a,b).

The relationship between B toxicity and the morpho-physiological attributes of maize under SA application is illustrated in Figure 2.

A Pearson correlation analysis was conducted to quantify the interactive effects of B toxicity and SA application on plant growth and biomass, chlorophyll contents, lipid peroxidation and the antioxidant and nutrient uptake of maize (Figure 2). B toxicity was negatively correlated with plant growth and biomass, photosynthetic pigments, oxidative stress and antioxidative response. Chlorophyll contents were positively correlated with plant biomass accumulation. Positive correlations were also identified among growth attributes and K, Ca and nitrate contents.

2.12. Principal Component Analysis

The combinatorial effect of B toxicity and SA application was evaluated on important attributes of maize plants by the synthesis of the score and loading plots of PCA, as presented in Figure 4. All the three applied B treatments with and without SA were successfully dispersed by the first two principal components (Figure 1a). The maximum variance among all the components was based on extracted components—i.e., PC1 (Dim1) and PC2 (Dim2), where component Dim1 contributed 69.9% while the contribution of Dim2 was 18.9% (Figure 1b).

3. Discussion

The only non-metallic element of group 13 of the periodic Table is B, which exhibits a trivalent oxidation state. Naturally, B is present in the form of borate, boric acid and borosilicate mineral. In the Earth crust, the B level varies between 1–500 mg kg^{-1} and 2–100 mg kg-1 as per the geographical region and soil composition status [32].

B has a considerable importance due to its supportive role in plant development and growth. It helps in the processes of cell division, the formation of cell wall and the elongation of cells [33]. However, B causes toxic effects at very high or very low levels. B toxicity mostly co-exists together with some other abiotic stresses—e.g., salt and drought stress [34]. A high B toxicity reduces the plant growth and other attributes.

El-Shazoly [25] conducted a study to describe the SA effects on B toxicity stress in wheat. The results of such a study were in agreement with those of the present study. The SA application also enhanced the root and shoot length, supporting the findings of the previous works [35]. It has been reported that a high level of B causes abnormal cell division in the root meristematic zone [35], hypodermis formation and suberin deposition [36], thus limiting plant growth and development. The excess of B also causes cytotoxic effects during mitosis, which in turn reduces the root and shoot biomass [37,38]. In the present study, 100 µM of SA significantly enhanced the plant biomass by mitigating the B toxicity. Sarafi et al. [39] reported that the B toxicity reduces the plant dry weight up to 48%, the number of leaves and the root dry weight in the pepper plant (Capsicum annuum). In this study, the applications of melatonin (MEL) and resveratrol (RES) were studied, where a treatment of 100 µM of RES and 1 µM of MEL effectively reversed the reductions in fresh and dry weights under B toxic effects, respectively. Eser and Aydemir [22] reported that kinetin application prevented the B-induced reductions in the plant fresh and dry weight of wheat plants under B stress. Moreover, the high B content (50 mg kg^{-1}) in soil reduced the shoot fresh and dry weight of tomato plants [40]. It has been particularized that B toxicity causes the down-regulation of the photosystem biochemical components and the inhibition of the electron transport rate [41], thus lowering the activity of carbon fixation enzymes [41,42]. High levels of B can also cause the root growth inhibition, accompanied by a decrease in plant dry weight [43]. The reduction in root growth may be due to the intense lignification of cell wall [44]. However, it has been reported that lignification is not mainly responsible for root growth inhibition, but is rather a defensive attribute for reducing B uptake [36].

The high B level (30 mg kg^{-1}) reduced the photosynthetic pigments biosynthesis. However, SA application reversed these negative effects, and the most effective treatment was 100 µM of SA. The findings of present study are in line with those of EI-Shazoly [25]. Plant growth and development are considerably dependent on photosynthetic pigments. It has been reported that the inhibition of plant growth by B stress is associated with reduced photosynthetic pigments. Indeed, the present study indicated that the biosynthesis of chlorophyll and carotenoid was negatively affected by B toxicity stress. Our results depicted a negative relationship between the biosynthesis of photosynthetic pigments and the increasing applied B stress regimes. This decline in photosynthetic pigments might be owing to H_2O_2 accumulation, which damages the photosynthetic reaction centers. Papadakis et al. [45] reported that one of the possible reasons for the reduction in photosynthetic activity in plants grown under excess of B was the structural damage of thylakoids. In general, SA, being a versatile molecule, interacts with other hormones to promote the induction of enzymes and antioxidants to alleviate the toxic effects of stress [46].

Regarding the mineral contents, Kaya and Ashraf [46] described that B toxicity significantly reduces the N, K and Ca contents in tomato. However, nitric oxide application induced the level of minerals and minimized the B toxicity effects. EI-Shazoly [25] described that a low level of boron (3 mg kg^{-1} soil) does not affect the K content in wheat plants, however a high level could decrease it. Moreover, the Ca level was reduced due to the B toxicity (3 mg kg^{-1} soil), but increased upon SA and thiamin application.

High levels of B increased the anthocyanin contents in sweet basil (*Ocimum basilicum* L.) plants, indicating possible stress responses or poor nutrient mobilization from the plant root [47]. We also found that B stress elevated the anthocyanin levels in the root and shoot of maize cultivar Gohar-19. The application of SA at higher levels reduced the stress level. Additionally, the reduction in anthocyanin content in plants treated with SA predicts a reduction in stress severity.

Ascorbic acid is an important antioxidant and scavenger of ROS [48–50]. The ASA content was significantly affected by B stress in the present study. Abiotic stresses result in a higher accumulation of ASA than that of other stress. Increased ROS scavenging enzymatic and nonenzymatic activity by excessive B concentrations has already been reported in barley, chickpea, tomato and grapes [51–54].

In general, plants up-regulate the synthesis of different osmolytes in cytosol and other organelles to cope with the deleterious effects of environmental stresses. Proline and GB are considered to be key osmolytes for the osmotic adjustment. Proline, being a secondary metabolite, plays a key role in stress tolerance as an antioxidant and osmoprotectant [55]. Stress-related genes are activated by GB to detoxify ROS and protect photosynthetic machinery under stressed conditions [56]. In the current study, SA applications triggered the accumulation of proline and GB to cope with the B toxicity effects through scavenging ROS and the activation of the antioxidant defense systems.

The PCA results depicted that the application of salicylic acid had a significant ameliorative effect for B toxicity on the studied parameters of maize plants. The same effects of SA have been reported in salt-stressed sunflower plants [57]. Overall, the applied B stress exerted hazardous effects on the growth and ecophysiological attributes of maize. These results were in accordance with the findings of previous reports which have reported decreases in the growth traits of various plant species grown under environmental stress conditions [58–66]. Based on the findings of the current study, we conclude that SA applications improved the growth of B-stressed maize plants at the seedling stage through increasing the biosynthesis of photosynthetic pigments, osmolytes and antioxidants. The high level of B deteriorates photosystem II centers, as the low levels of chlorophyll and carotenoids are linked with biomass reduction caused by B toxicity. High levels of osmoprotectants such as proline may act as signaling molecules for scavenging ROS, thus stabilizing the membrane structures as well as cascading the stress-tolerant gene expression. Further studies at the molecular level may elaborate the comprehensive understanding lying behind these modulations of SA against B toxicity in maize. The induction of B toxicity tolerance in maize plants after SA application is also associated with antioxidant defense system improvement.

4. Materials and Methods

4.1. Plant Material and Experimental Design

Seeds of maize (cultivar Gohar-19) were obtained from the Maize & Millets Research Institute, Yusafwala, Sahiwal, Pakistan, and the experiments were conducted at the Department of Botany, Government College University Faisalabad, Pakistan. The seeds were surface-sterilized using a 1% sodium hypochloride solution for 5 min. Surface-sterilized seeds were thoroughly washed with distilled water and air-dried for 12 h. These seeds were soaked in 0, 50, 100 and 150 μM of SA solution for 12 h. Plastic pots were filled with 1 kg of washed and air-dried sand at the botanical garden, Government College University Faisalabad, and a 100% field capacity was maintained in pots by adding boron-free water. The experiment was carried out in a completely randomized design (CRD) with three replicates. Ten seeds were sown in each pot. After 5 days of germination, 5 seedlings were selected based on their similarity in size and vigor. B stress was applied using Nable's solution containing boric acid (H_3BO_3) by maintaining pH at 5.7. The final B concentrations of 0, 15 and 30 mg Kg^{-1} soil were maintained in each pot for one week. After one week of B stress application, the plants were harvested and stored in a freezer for further analysis.

4.2. Morphological Parameters and Plant Biomass

The root and shoot lengths were measured for individual plants using a meter scale. The root and shoot fresh and dry weight, after drying at 70 °C for 72 h, was calculated using the same weight balance.

4.3. Physiological and Biochemical Analysis

Different physio-biochemical analyses were carried out as described below.

4.3.1. Photosynthetic Pigments

Contents of chlorophyll a, b and carotenoids were determined using a 0.5 g fresh leaf sample. The Arnon [67] method with minor modifications was used for the determination of photosynthetic pigments. The collected sample was ground in 15 mL of 85% acetone and centrifuged at 10,000× g for 15 min. The absorbance of the supernatant was measured at 480, 645 and 663 nm using a spectrophotometer (Hitachi U-2001, Tokyo, Japan).

4.3.2. Anthocyanin Content

The anthocyanin content was measured as reported previously [68]. Fresh root and shoot samples (0.1 g each) were ground separately in 2 mL of 1% acidified methanol (1 mL HCL and 99 mL methanol), then the extract was heated up to 50 °C for one in a water bath. The anthocyanin content was then quantified using a spectrophotometer (Hitachi U-2001, Tokyo, Japan) at 535 nm.

4.3.3. Ascorbic Acid Content

The protocol of Mukherjee and Choudhuri [68] was followed to determine the ascorbic acid content. Root and shoot fresh samples (0.1 g) were taken and ground in 5 mL of trichloroacetic acid (TCA) using a pestle and mortar. The extract was filtered, and 4 mL of the homogenate sample was allowed to react with 2 mL of 2% dinitrophenyl hydrazine in an acidified medium. One drop of 10% thiourea (prepared in 70% ethanol) was added and the mixture was then allowed to boil at 100 °C for 15 min. The absorbance at 530 nm was recorded through UV-spectrophotometer (Hitachi U-2001, Tokyo, Japan) for calculating the ascorbic acid content.

4.3.4. H_2O_2 Content Determination

Shoot samples were extracted in a cold acetone for H_2O_2 content determination. One milliliter of extract was mixed with 1 mL of 0.1% titanium dioxide in 20% sulfuric acid and centrifuged at 8000 rpm for 15 min. The supernatant was then used to measure the absorbance at 415 nm. H_2O_2 content was calculated using a standard curve plotted in a range of 0.5–5 mM H_2O_2 and was expressed as mg g^{-1} FW.

4.3.5. Potassium Content

Dry samples of plant root and shoot (1 mg) were dissolved in 9 mL of distilled water. Flame photometer was used for the determination of potassium content, as reported previously [69].

4.3.6. Calcium Content

Dry samples of plant root and shoot (1 mg) were added to 9 mL of distilled water. Flame photometer method was used for the determination of the calcium content [69].

4.3.7. Nitrate Content

Dry root and shoot samples were dissolved in 1 mL of TCA (1.24 g TCA+ 500 mL H_2SO_4) and 1 mL of distilled water. For determining the nitrate oxide, the absorbance was recorded at 530 nm using a UV vis spectrophotometer (Hitachi U-2001, Tokyo, Japan).

5. Statistical Analysis

Collected data were subjected to an analysis of variance (ANOVA) using the statistical software Co-Stat version 6.2, Cohorts Software, 2003 (Monterey, CA, USA). The treatment means were equated by the least significant difference method (Fisher's LSD) at p value of ≤0.05 level. Before applying ANOVA, the data were standardized using means of inverse or logarithmic transformations wherever necessary. The correlations and PCA of the mean values of all variables were found using XL-STAT 2010.

Author Contributions: Conceptualization, M.N., N.K., N.I., S.A., M.R., A.A.A. and M.N.A.; Data curation, M.N. and S.I.; Formal analysis, M.N. and S.I.; Funding acquisition, A.A.A. and M.N.A.; Investigation, S.I. and N.K.; Methodology, M.N., S.I. and N.I.; Project administration, M.N., N.I. and S.A.; Resources, M.N., H.I., S.A. and A.A.A.; Software, H.I., N.I., S.A., M.R. and M.N.A.; Supervision, M.N. and S.A.; Validation, H.I.; Visualization, H.I.; Writing–original draft, M.N., N.K., M.R. and A.A.A.; Writing–review & editing, N.K., S.A., M.R. and M.N.A. All authors have read and agreed to the published version of the manuscript.

Acknowledgments: This study was financially supported by the Higher Education Commission (HEC) of Pakistan. The authors would like to extend their sincere appreciation to the Researchers Supporting Project Number (RSP-2020/236), King Saud University, Riyadh, Saudi Arabia.

References

1. Jaradat, A.; Simulated, A. Climate change deferentially impacts phenotypic plasticity and stoichiometric homeostasis in major food crops. *Emir. J. Food Agr.* **2018**, *30*, 429–442.
2. Khan, N.; Ali, S.; Shahid, M.A.; Kharabian-Masouleh, A. Advances in detection of stress tolerance in plants through metabolomics approaches. *Plant Omics* **2017**, *10*, 153. [CrossRef]
3. Mantri, N.; Patade, V.; Penna, S.; Ford, R.; Pang, E. Abiotic stress responses in plants: Present and future. In *Abiotic Stress Responses in Plants*; Springer: New York, NY, USA, 2012; pp. 1–19.

4.   Bakhsh, A. Engineering crop plants against abiotic stress: Current achievements and prospects. *Emir. J. Food Agr.* **2015**, *27*, 24–39. [CrossRef]

5.   Cheng, S. Heavy metal pollution in China: Origin, pattern and control. *Environ. Sci. Pollut. Res.* **2003**, *10*, 192–198. [CrossRef]

6.   Kelepertzis, E. Accumulation of heavy metals in agricultural soils of Mediterranean: Insights from Argolida basin, Peloponnese, Greece. *Geoderma* **2014**, *221*, 82–90. [CrossRef]

7.   Zhao, F.J.; Ma, Y.; Zhu, Y.G.; Tang, Z.; McGrath, S.P. Soil contamination in China: Current status and mitigation strategies. *Environ. Sci. Technol.* **2014**, *49*, 750–759. [CrossRef]

8.   Hussain, S.; Zhang, J.H.; Zhong, C.; Zhu, L.F.; Cao, X.C.; YU, S.M.; Jin, Q.Y. Effects of salt stress on rice growth, development characteristics and the regulating ways. *J. Integr. Agric.* **2017**, *16*, 2357–2374. [CrossRef]

9.   Liu, P.; Yang, Y.S.; Xu, G.D.; Fang, Y.H.; Yang, Y.A.; Kalin, R.M. The effect of molybdenum and boron in soil on the growth and photosynthesis of three soybean varieties. *Plant Soil Environ.* **2005**, *51*, 197–205. [CrossRef]

10. Khan, N.; Bano, A.; Babar, M.A. The stimulatory effects of plant growth promoting rhizobacteria and plant growth regulators on wheat physiology grown in sandy soil. *Arch. Microbiol.* **2019**, *201*, 769–785. [CrossRef]

11. Ayvaz, M.; Avcı, M.K.; Yamaner, C.; Koyuncu, M.; Güven, A.; Fagerstedt, K. Does excess boron affect the malondialdehyde levels of potato cultivars? *Eurasia J. Biosci.* **2013**, *7*, 47–53. [CrossRef]

12. Stiles, A.R.; Liu, C.; Kayama, Y.; Wong, J.; Doner, H.; Funston, R.; Terry, N. Evaluation of the boron tolerant grass, Puccinellia distans, as an initial vegetative cover for the phytorestoration of a boron-contaminated mining site in southern California. *Environ. Sci. Technol.* **2011**, *45*, 8922–8927. [CrossRef] [PubMed]

13. Princi, M.P.; Lupini, A.; Araniti, F.; Longo, C.; Mauceri, A.; Sunseri, F.; Abenavoli, M.R. Boron toxicity and tolerance in plants: Recent advances and future perspectives. In *Plant Metal Interaction*; Elsevier: Amsterdam, The Netherlands, 2016; pp. 115–147.

14. Kayıhan, C.; Öz, M.T.; Eyidogan, F.; Yucel, M.; Oktem, H.A. Physiological, biochemical and transcriptomic responses to boron toxicity in leaf and root tissues of contrasting wheat cultivars. *Plant Mol. Biol. Rep.* **2017**, *35*, 97–109. [CrossRef]

15. Yermiyahu, U.; Ben-Gal, A.; Keren, R.; Reid, R.J. Combined effect of salinity and excess boron on plant growth and yield. *Plant Soil* **2008**, *304*, 73–87. [CrossRef]

16. Diaz, F.J.; Grattan, S.R. Performance of tall wheatgrass (*Thinopyrum ponticum* cv. Jose) irrigated with saline-high boron drainage water: Implications on ruminant mineral nutrition. *Agric. Ecosyst. Environ.* **2009**, *131*, 128–136. [CrossRef]

17. Zhang, B.; Chu, G.; Wei, C.; Ye, J.; Li, Z.; Liang, Y. The growth and antioxidant defense responses of wheat seedlings to omethoate stress. *Pestic. Biochem. Phys.* **2011**, *100*, 273–279. [CrossRef]

18. Khan, N.; Zandi, P.; Ali, S.; Mehmood, A.; Adnan Shahid, M.; Yang, J. Impact of salicylic acid and PGPR on the drought tolerance and phytoremediation potential of Helianthus annus. *Front. Microbiol.* **2018**, *9*, 2507. [CrossRef]

19. Tavallali, V. Interactive effects of zinc and boron on growth, photosynthesis and water relations in pistachio. *J. Plant Nutr.* **2017**, *40*, 1588–1603. [CrossRef]

20. Nasim, M.; Rengel, Z.; Aziz, T.; Regmi, B.D.; Saqib, M. Boron toxicity alleviation by zinc application in two barley cultivars differing in tolerance to boron toxicity. *Pak. J. Agric. Sci.* **2015**, *52*, 151–158.

21. Samet, H.; Cikili, Y.; Dursun, S. The role of potassium in alleviating boron toxicity and combined effects on nutrient contents in pepper (*Capsicum annuum* L.). *Bulg. J. Agric. Sci.* **2015**, *21*, 64–70.

22. Siddiqui, M.H.; Al-Whaibi, M.H.; Sakran, A.M.; Ali, H.M.; Basalah, M.O.; Faisal, M.; Al-Amri, A.A. Calcium-induced amelioration of boron toxicity in radish. *J. Plant Growth Regul.* **2013**, *32*, 61–71. [CrossRef]

23. Moustafa-Farag, M.; Mohamed, H.I.; Mahmoud, A.; Elkelish, A.; Misra, A.N.; Guy, K.M.; Kamran, M.; Ai, S.; Zhang, M. Salicylic Acid Stimulates Antioxidant Defense and Osmolyte Metabolism to Alleviate Oxidative Stress in Watermelons under Excess Boron. *Plants* **2020**, *9*, 724. [CrossRef] [PubMed]

24. Maghsoudia, K.; Arvinb, M.J. Salicylic acid and osmotic stress effects on seed germination and seedling growth of wheat (*Triticum aestivum* L.) cultivars. *Plant Ecophysiol.* **2010**, *2*, 7–11.

25. El-Shazoly, R.M.; Metwally, A.A.; Hamada, A.H. Salicylic acid or thiamin increases tolerance to boron toxicity stress in wheat. *J. Plant Nutr.* **2019**, *42*, 702–722. [CrossRef]

26. Hussain, H.A.; Men, S.; Hussain, S.; Zhang, Q.; Ashraf, U.; Anjum, S.A.; Ali, I.; Wang, L. Maize Tolerance against Drought and Chilling Stresses Varied with Root Morphology and Antioxidative Defense System. *Plants* **2020**, *9*, 720. [CrossRef] [PubMed]

27. Aliyu, H.G.; Adamu, H.M. The potential of maize as phytoremediation tool of heavy metals. *Eur. Sci. J.* **2014**, *6*, 32–33.

28. Wang, B.; Liu, C.; Zhang, D.; He, C.; Zhang, J.; Li, Z. Effects of maize organ-specific drought stress response on yields from transcriptome analysis. *BMC Plant Biol.* **2019**, *19*, 335. [CrossRef]

29. Hussain, H.A.; Men, S.; Hussain, S.; Chen, Y.; Ali, S.; Zhang, S.; Zhang, K.; Li, Y.; Xu, Q.; Liao, C.; et al. Interactive effects of drought and heat stresses on morpho-physiological attributes, yield, nutrient uptake and oxidative status in maize hybrids. *Sci. Rep.* **2019**, *9*, 3890. [CrossRef]

30. Khan, N.; Bano, A. Effects of exogenously applied salicylic acid and putrescine alone and in combination with rhizobacteria on the phytoremediation of heavy metals and chickpea growth in sandy soil. *Int. J. Phytoremediation* **2018**, *16*, 405–414. [CrossRef]

31. Brown, P.H.; Bellaloui, N.; Wimmer, M.A.; Bassil, E.S.; Ruiz, J.; Hu, H.; Pfeffer, H.; Dannel, F.; Römheld, V. Boron in plant biology. *Plant Biol.* **2002**, *4*, 205–223. [CrossRef]

32. Parks, J.L.; Edwards, M. Boron in the environment. *Crit. Rev. Environ. Sci. Technol.* **2005**, *35*, 81–114. [CrossRef]

33. Blevins, D.G.; Lukaszewski, K.M. Boron in plant structure and function. *Annu. Rev. Plant Biol.* **1998**, *49*, 481–500. [CrossRef] [PubMed]

34. Zafar-ul-Hye, M.; Munir, K.; Ahmad, M.; Imran, M. Influence of boron fertilization on growth and yield of wheat crop under salt stress environment. *Soil Environ.* **2016**, *35*, 181–186.

35. Liu, D.; Jiang, W.; Zhang, L.; Li, L. Effects of boron ions on root growth and cell division of broad bean (*Vicia Faba* L.). *Isr. J. Plant Sci.* **2000**, *48*, 47–58. [CrossRef]

36. Ghanati, F.; Morita, A.; Yokota, H. Deposition of suberin in roots of soybean induced by excess boron. *Plant Sci.* **2005**, *168*, 397–405. [CrossRef]

37. Konuk, M.; Liman, R.; Cigerci, I.H. Determination of genotoxic effect of boron on *Allium cepa* root meristematic cells. *Pak. J. Bot.* **2007**, *39*, 73–79.

38. Sarafi, E.; Tsouvaltzis, P.; Chatzissavvidis, C.; Siomos, A.; Therios, I. Melatonin and resveratrol reverse the toxic effect of high boron (B) and modulate biochemical parameters in pepper plants (*Capsicum annuum* L.). *Plant Physiol. Biochem.* **2017**, *112*, 173–182. [CrossRef]

39. Eser, A.; Aydemir, T. The effect of kinetin on wheat seedlings exposed to boron. *Plant Physiol. Biochem.* **2016**, *108*, 158–164. [CrossRef]

40. Khan, N.; Bano, A.; Ali, S.; Babar, M.A. Crosstalk amongst phytohormones from planta and PGPR under biotic and abiotic stresses. *Plant Growth Regul.* **2020**, *90*, 189–203. [CrossRef]

41. Chen, M.; Mishra, S.; Heckathorn, S.A.; Frantz, J.M.; Krause, C. Proteomic analysis of *Arabidopsis thaliana* leaves in response to acute boron deficiency and toxicity reveals effects on photosynthesis, carbohydrate metabolism and protein synthesis. *J. Plant Physiol.* **2013**, *171*, 235–242. [CrossRef]

42. Turan, M.; Taban, N.; Taban, S. Effect of calcium on the alleviation of boron toxicity and localization of boron and calcium in cell wall of wheat. *Not. Bot. Horti Agrobot. Cluj-Napoca* **2009**, *37*, 99–103.

43. Ghanati, F.; Morita, A.; Yokota, H. Induction of suberin and increase of lignin content by excess boron in tobacco cells. *J. Plant Nutr. Soil Sci.* **2002**, *48*, 357–364. [CrossRef]

44. Sharma, A.; Sidhu, G.P.S.; Araniti, F.; Bali, A.S.; Shahzad, B.; Tripathi, D.K.; Brestic, M.; Skalicky, M.; Landi, M. The Role of Salicylic Acid in Plants Exposed to Heavy Metals. *Molecules* **2020**, *25*, 540. [CrossRef] [PubMed]

45. Papadakis, I.; Dimassi, K.; Bosabalidis, A.; Therios, I.; Patakas, A.; Giannakoula, A. Boron toxicity in 'Clementine' mandarin plants grafted on two root stocks. *Plant Sci.* **2004**, *166*, 539–547. [CrossRef]

46. Kaya, C.; Ashraf, M. Exogenous application of nitric oxide promotes growth and oxidative defense system in highly boron stressed tomato plants bearing fruit. *Sci. Hortic.* **2015**, *185*, 43–47. [CrossRef]

47. Pardossi, A.; Romani, M.; Carmassi, G.; Guidi, L.; Landi, M.; Incrocci, L.; Maggini, R.; Puccinelli, M.; Vacca, W.; Ziliani, M. Boron accumulation and tolerance in sweet basil (*Ocimum basilicum* L.) with green or purple leaves. *Plant Soil* **2015**, *395*, 375–389. [CrossRef]

48. Foyer, C.H.; Noctor, G. Ascorbate and glutathione: The heart of the redox hub. *Plant Physiol.* **2011**, *155*, 2–18. [CrossRef]

49. Qian, H.F.; Peng, X.F.; Han, X.; Ren, J.; Zhan, K.Y.; Zhu, M. The stress factor, exogenous ascorbic acid, affects plant growth and the antioxidant system in *Arabidopsis thaliana*. *Russ. J. Plant Physiol.* **2014**, *61*, 467–475. [CrossRef]

50. Karabal, E.; Yucel, M.; Huseyin, A.O. Antioxidant responses of tolerant and sensitive barley cultivars to boron toxicity. *Plant Sci.* **2003**, *164*, 925–933. [CrossRef]

51. Gunes, A.; Soylemezoglu, G.; Inal, A.; Bagci, E.G.; Coban, S.; Sahin, O. Antioxidant and stomatal responses of grapevine (*Vitis vinifera* L.) to boron toxicity. *Sci. Hortic.* **2006**, *110*, 279–284. [CrossRef]

52. Cervilla, L.M.; Blasco, B.; Rıos, R.; Romero, L.; Ruiz, J. Oxidative stress and antioxidants in tomato (*Solanum lycopericum*) plants subjected to boron toxicity. *Ann. Bot.* **2007**, *100*, 747–756. [CrossRef]

53. Ardic, M.; Sekmen, A.H.; Tokur, S.; Ozdemir, F.; Turkan, I. Antioxidant responses of chickpea plants subjected to boron toxicity. *Plant Biol.* **2009**, *11*, 328–338. [CrossRef] [PubMed]

54. Khan, N.; Bano, A.; Curá, J.A. Role of Beneficial Microorganisms and Salicylic Acid in Improving Rainfed Agriculture and Future Food Safety. *Microorganisms* **2020**, *8*, 1018. [CrossRef] [PubMed]

55. Shaki, F.; Maboud, H.E.; Niknam, V. Effects of salicylic acid on hormonal cross talk, fatty acids profile, and ions homeostasis from salt-stressed safflower. *J. Plant Interact.* **2019**, *14*, 340–346. [CrossRef]

56. Ahmad, P.; Abdel Latef, A.A.; Hashem, A.; Abd-Allah, E.F.; Gucel, S.; Tran, L.S.P. Nitric oxide mitigates salt stress by regulating levels of osmolytes and antioxidant enzymes in chickpea. *Front. Plant Sci.* **2016**, *7*, 347. [CrossRef]

57. Hossain, M.A.; Hoque, M.A.; Burritt, D.J.; Fujita, M. Proline Protects Plants against Abiotic Oxidative Stress: Biochemical and Molecular Mechanisms. In *Oxidative Damage to Plants*; Academic Press: Cambridge, MA, USA, 2014; pp. 477–522.

58. El-Esawi, M.A.; Elkelish, A.; Soliman, M.; Elansary, H.O.; Zaid, A.; Wani, S.H. *Serratia marcescens* BM1 Enhances Cadmium Stress Tolerance and Phytoremediation Potential of Soybean Through Modulation of Osmolytes, Leaf Gas Exchange, Antioxidant Machinery, and Stress-Responsive Genes Expression. *Antioxidants* **2020**, *9*, 43. [CrossRef]

59. Abdelaal, K.A.; EL-Maghraby, L.M.; Elansary, H.; Hafez, Y.M.; Ibrahim, E.I.; El-Banna, M.; El-Esawi, M.; Elkelish, A. Treatment of Sweet Pepper with Stress Tolerance-Inducing Compounds Alleviates Salinity Stress Oxidative Damage by Mediating the Physio-Biochemical Activities and Antioxidant Systems. *Agronomy* **2020**, *10*, 26. [CrossRef]

60. Alhaithloul, H.A.; Soliman, M.H.; Ameta, K.L.; El-Esawi, M.A.; Elkelish, A. Changes in Ecophysiology, Osmolytes, and Secondary Metabolites of the Medicinal Plants of *Mentha piperita* and *Catharanthus roseus* Subjected to Drought and Heat Stress. *Biomolecules* **2020** *10*, 43. [CrossRef]

61. El-Esawi, M.A.; Alaraidh, I.A.; Alsahli, A.A.; Alzahŕani, S.M.; Ali, H.M.; Alayafi, A.A.; Ahmad, M. *Serratia liquefaciens* KM4 Improves Salt Stress Tolerance in Maize by Regulating Redox Potential, Ion Homeostasis, Leaf Gas Exchange and Stress-Related Gene Expression. *Int. J. Mol. Sci.* **2018**, *19*, 3310. [CrossRef]

62. Zafar-ul-Hye, M.; Naeem, M.; Danish, S.; Khan, M.J.; Fahad, S.; Datta, R.; Brtnicky, M.; Kintl, A.; Hussain, G.S.; El-Esawi, M.A. Effect of Cadmium-Tolerant Rhizobacteria on Growth Attributes and Chlorophyll Contents of Bitter Gourd under Cadmium Toxicity. *Plants* **2020**, *9*, 1386. [CrossRef]

63. Naveed, M.; Bukhari, S.S.; Mustafa, A.; Ditta, A.; Alamri, S.; El-Esawi, M.A.; Rafique, M.; Ashraf, S.; Siddiqui, M.H. Mitigation of Nickel Toxicity and Growth Promotion in Sesame through the Application of a Bacterial Endophyte and Zeolite in Nickel Contaminated Soil. *Int. J. Environ. Res. Public Health* **2020**, *17*, 8859. [CrossRef]

64. Imran, M.; Hussain, S.; El-Esawi, M.A.; Rana, M.S.; Saleem, M.H.; Riaz, M.; Ashraf, U.; Potcho, M.P.; Duan, M.; Rajput, I.A.; et al. Molybdenum Supply Alleviates the Cadmium Toxicity in Fragrant Rice by Modulating Oxidative Stress and Antioxidant Gene Expression. *Biomolecules* **2020**, *10*, 1582. [CrossRef] [PubMed]

65. Ali, Q.; Shahid, S.; Ali, S.; El-Esawi, M.A.; Hussain, A.I.; Perveen, R.; Iqbal, N.; Rizwan, M.; Nasser Alyemeni, M.; El-Serehy, H.A.; et al. Fertigation of Ajwain (*Trachyspermum ammi* L.) with Fe-Glutamate Confers Better Plant Performance and Drought Tolerance in Comparison with $FeSO_4$. *Sustainability* **2020**, *12*, 7119. [CrossRef]

66. Soliman, M.; Alhaithloul, H.A.; Hakeem, K.R.; Alharbi, B.M.; El-Esawi, M.; Elkelish, A. Exogenous Nitric Oxide Mitigates Nickel-Induced Oxidative Damage in Eggplant by Upregulating Antioxidants, Osmolyte Metabolism, and Glyoxalase Systems. *Plants* **2019**, *8*, 562. [CrossRef] [PubMed]

67. Arnon, D.I. Copper enzymes in isolated chloroplasts. Polyphenoloxidase in *Beta vulgaris*. *Plant Physiol.* **1949**, *24*, 1–15. [CrossRef] [PubMed]

68. Mukherjee, S.P.; Choudhuri, M.A. Implications of water stress-induced changes in the levels of endogenous ascorbic acid and hydrogen peroxide in Vigna seedlings. *Physiol. Plant* **1983**, *58*, 166–170. [CrossRef]

69. Stark, D.; Wray, V. Anthocyanins. In *Methods in Plant Biology*; Volume 1: Plant Phenolics; Harborne, J.B., Ed.; Academic Press: London, UK; Harcourt Brace Jovanovich: London, UK, 1989; pp. 325–356.

8

α-Tocopherol Foliar Spray and Translocation Mediates Growth, Photosynthetic Pigments, Nutrient Uptake and Oxidative Defense in Maize (*Zea mays* L.) under Drought Stress

Qasim Ali [1,*], Muhammad Tariq Javed [1], Muhammad Zulqurnain Haider [1]⬤, Noman Habib [1], Muhammad Rizwan [2]⬤, Rashida Perveen [3], Shafaqat Ali [2,4,*]⬤, Mohammed Nasser Alyemeni [5], Hamed A. El-Serehy [6] and Fahad A. Al-Misned [6]

[1]   Department of Botany, Government College University, New Campus, Jhang Road,
      Faisalabad 38000, Pakistan; mtariqjaved@gcuf.edu.pk (M.T.J.); dr_mzhaider@yahoo.com (M.Z.H.);
      nomi4442003@yahoo.com (N.H.)
[2]   Department of Environmental Sciences and Engineering, Government College University,
      Allama Iqbal Road, Faisalabad 38000, Pakistan; mrazi1532@yahoo.com
[3]   Department of Physics, University of Agriculture, Faisalabad 38040, Pakistan; 2007ag942@uaf.edu.pk
[4]   Department of Biological Sciences and Technology, China Medical University, Taichung 40402, Taiwan
[5]   Botany and Microbiology Department, College of Science, King Saud University, Riyadh 11451, Saudi Arabia;
      mnyemeni@ksu.edu.sa
[6]   Department of Zoology, College of Science, King Saud University, Riyadh 11451, Saudi Arabia;
      helserehy@ksu.edu.sa (H.A.E.-S.); almisned@ksu.edu.sa (F.A.A.-M.)
*   Correspondence: drqasimali@gcuf.edu.pk (Q.A.); shafaqataligill@gcuf.edu.pk or
     shafaqataligill@yahoo.com (S.A.)

Abstract: A pot experiment was conducted to assess the induction of drought tolerance in maize by foliar-applied α-tocopherol at early growth stage. Experiment was comprised two maize cultivars (Agaiti-2002 and EV-1098), two water stress levels (70% and 100% field capacity), and two α-tocopherol levels (0 mmol and 50 mmol) as foliar spray. Experiment was arranged in a completely randomized design in factorial arrangement with three replications of each treatment. α-tocopherol was applied foliary at the early vegetative stage. Water stress reduced the growth of maize plants with an increase in lipid peroxidation in both maize cultivars. Contents of non-enzymatic antioxidants and activities of antioxidant enzymes increased in studied plant parts under drought, while the nutrient uptake was decreased. Foliary-applied α-tocopherol improved the growth of both maize cultivars, associated with improvements in photosynthetic pigment, water relations, antioxidative mechanism, and better nutrient acquisition in root and shoot along with tocopherol contents and a decrease in lipid peroxidation. Furthermore, the increase of tocopherol levels in roots after α-Toc foliar application confers its basipetal translocation. In conclusion, the findings confer the role of foliar-applied α-tocopherol in the induction of drought tolerance of maize associated with tissue specific improvements in antioxidative defense mechanism through its translocation.

Keywords: α-Tocopherol; antioxidants; drought; nutrient dynamics; tissue specific response

1. Introduction

Among different environmental adversities, water shortage is of major focus, which has hampered the production of global agricultural systems [1,2]. At a global level, about 45% of all land is prevailed by drought [3]. On the other hand, an estimated increase in the world population will be about

2.5 billion in the next 25 years, which will exert huge pressure on agriculture to fulfill world food demand and on the available freshwater resources. From the last two decades, Pakistan has also faced the problem of agricultural productivity to fulfill the food demand of the sixth largest population in the world. With an agriculture-based economy, Pakistan is predominantly categorized as arid country lying within the geographic coordinates of 23.38°–30.25° N latitude and 61.78°–74.30° E longitude, with a total land area of 796,096 km^2 [4]. The interannual rainfall variability makes the arid region (covering 75% land area of Pakistan) more susceptible to drought risks. Approximately 34.15 Mha of land area is in agriculture use, and uncultivated land is 23.60 Mha. About 25% of the cultivated land is rainfed, which plays a vital role in the country's economy [5]. Due to the major contribution of the agriculture sector in Pakistan's economy, Pakistan is more susceptible to drought risks [6]. In recent decades, unexpected and rapid changes in climate have severely affected socioeconomic and environmental conditions in Pakistan [6]. The major cause of drought stress is a decrease in soil water contents in combination with evaporation due to over-changing atmospheric conditions [7]. Shortage of water induces drastic changes in plants' physio-biochemical and molecular properties that ultimately affects all growth stages of a plant's life cycle, including the final yield [8,9]. At present (and in the near future), the maintenance of crop productivity for a large population under limited water supply is a challenge for the researchers working in the agriculture sector.

To survive under water deficit conditions, plants have manipulated metabolic defensive systems/mechanisms, which are species- and genotype-specific [10–12]. Disturbance in plant water status is the important effect of water shortage that triggers various other metabolic processes to survive under water stress [10,11,13,14]. It results in reduced growth and final grain yield due to perturbations in photosynthesis by disturbances in the biosynthesis of photosynthetic pigments and impaired nutrient uptake [15,16]. Water deficit conditions cause sub-optimal plant photosynthetic efficiency due to limited CO_2 diffusion into the leaves due to less stomatal opening or reduced Rubisco activity [17,18]. To cope with a stressful environment, the plant mineral uptake mechanism plays a significant role in improving resistance [19,20]. Generally, under water deficit conditions, mineral uptake and transport reduces due to a decrease in the nutrient diffusion rate [16,21]. Among different nutrients, potassium (K$^+$), nitrogen (N), calcium (Ca^{2+}), phosphorus (P), and magnesium (Mg^{2+}) have prime importance due to their vital functions in plant physio-biochemical processes [14,20,22].

The stress tolerance in crop plants that results in better yield is growth-stage and species-specific [23,24]. The seedling stage is of prime importance in potentially contributing to better seed yield. Uniform crop stand leads to better yield, which depends on better seedling growth [25,26]. Furthermore, it was found that at early seedling stages, crop cultivars with better antioxidative potential are more drought tolerant than cultivars with less antioxidative activity [27] because the disturbances in different physiological mechanisms results in another secondary stress (oxidative stress) by excessive production of reactive oxygen species (ROS).

Stress-induced oxidative stress due to production of ROS (O$_2^-$, H$_2$O$_2$, OH$^-$, and O*) is a common phenomenon in all organisms [28]. Over-production of ROS damages membrane lipids [28], thereby increasing malondialdehyde (MDA) accumulation due to limited activity of antioxidative defense mechanisms [29]. Under stressful environments, the levels of MDA are parallel with antioxidant enzyme activities, which are the indices to assess the status of the extent of damage due to the overproduction of ROS [30]. Other than the levels of antioxidant enzyme to counteract ROS damage, plants also have non-enzymatic antioxidative defense mechanisms such as the production of ascorbic acid, phenolic acid, carotenoids, tocopherols, etc. [31]. Furthermore, it is well known that the antioxidative defensive phenomenon is inter-species, cultivar, and growth-stage-specific. However, most of the higher yielding genotypes are not drought tolerant when considering stress tolerance mechanisms [32].

Furthermore, some high-yield crop cultivars are deficit with regard to such anti-stress mechanisms [14,28,33]. For the induction of drought tolerance, different approaches have been adopted, including the exogenous application of secondary growth metabolic compounds [34–37]. Exogenous application such as the foliar spray of different secondary metabolites of which the plant is in deficit is considered as an effective means among others for stress tolerance induction [38,39]. It is well known that foliar application of such compounds is translocatable to different plant parts. Furthermore, after their translocation to different plant parts, they play a potential role in the induction of drought tolerance. Along with modulating metabolic activities, plants also control their own metabolisms [34,40]. Among different secondary metabolic compounds, the tocopherols are lipophilic in nature and scavenge ROS, with the ability to recycle themselves and, as a result, reduce lipid peroxidation. Tocopherols belong to a family of eight members including α, β, γ, and δ tocopherols, along with their respective precursors (tocotrienols) that have high antioxidative activity and protect plants from stress through different metabolic processes [41]. Among these, α-Toc is largely known as vitamin E, with large antioxidant potential in comparison with other family members, but the production of α-Toc to reduce oxidative damage is cultivar-specific [42]. However, α-Toc exogenous application was found to be helpful for stress tolerance induction. For example, in wheat, exogenous application of α-Toc improved salt-stress tolerance [43]. In flax, genotypes foliar-applied tocopherol significantly improved salt stress tolerance [44]. Most of the studies presented are regarding salt tolerance induction and the application of α-Toc on adult-stage plants, and there is a lack of knowledge regarding its exogenous use at other growth stages. However, the discovery of the proper plant stage for better drought-stress induction through exogenous use of this compound is of prime importance [45].

Furthermore, there are missing gaps in understanding the proper physiological mechanism for the induction of stress tolerance at different growth stages by the exogenous use of organic compounds like that of α-Toc, also considering its translocation to specific plant parts. Therefore, the current work was aimed to quantify to which extent the foliar applied α-Toc could modulate growth in water-stressed maize plants and when it should be applied in the early growth stage. The goal of the study was to draw parallels among tissue-specific alterations in endogenous tocopherol levels, antioxidative defense mechanisms, and nutrient mobility patterns after α-Toc foliar application in maize plants grown in a drought-stressed rhizosphere. The research outcomes are helpful for optimizing strategies for growing maize with limited irrigation and in semi-arid and arid regions for better growth and production.

Maize (*Zea mays* L.) is the third most commonly produced cereal, after wheat and rice. It has a potential to grow in a wide range of environmental conditions and has gained great economic priority due to its potential nutritional quality all over the world, including in Pakistan [46]. In Pakistan, 1.016 million hectares are under maize cultivation, and 35% of the total cultivated area is rainfed, which is now facing problems in getting better production under dry environmental spells; this situation has further become more severe due to the present change in environmental conditions. Maize kernels are not only good and cheap source of carbohydrates but are also a rich source of carotenoids, proteins, and edible oil. However, due to changes in rainfall patterns along with the shortage of fresh water for irrigation, its production is under threat, along with that of other crops.

2. Materials and Methods

The present experiment was arranged in the research area of the Department of Botany, Government College University Faisalabad, Pakistan, (latitude 30°30 N, longitude 73°10 E, and altitude 213 m) under natural environmental conditions during August–September 2018. To avoid disturbances due to rain, the experimental area was covered with a polyethylene sheet. The design of the experiment was completely randomized in factorial arrangement, with three replications of each treatment. The experiment consisted of two drought levels (control and 70% field capacity), two highly yielding maize genotypes (EV-1098 and Agaiti-2002), and two levels of α-Toc (0 mmol and 50 mmol) in solution form applied as foliar spray with three replications of each treatment. The 70% field capacity used in the present study was selected following some earlier studies [47,48]. These two maize cultivars selected for study are used frequently in breeding programs to produce high-yielding hybrid genotypes. The experimental unit was comprised a total 24 equal-size plastic pots (28 cm × 30 cm), each filled with 10 kg soil. The soil was fully irrigated with canal water before seed sowing. When the soil was at field capacity, seeds of both maize genotypes were hand sown. Before sowing, the soil was prepared well by hand digging. The seeds of both maize genotypes were purchased from Maize and Millet Research Institute, Yousafwala Sahiwal, Pakistan. Ten healthy seeds were sown in each pot. After five days of the completion of seed germination, five seedlings per pot were maintained by thinning. The water stress treatment was started just after the thinning of the seedlings by controlling the irrigation of half of the pots at 70% field capacity, and the other half of the pots were treated as control plants and irrigated to maintain 100% field capacity. Average mean daily length was 13/11 h, mean minimum and maximum day/night temperatures were $38 \pm 3/30 \pm 3$ °C and $25 \pm 2.5/20 \pm 2.5$ °C, respectively, the mean relative humidity during whole experiment (at daytime) was 50%. During the whole experimental period the averaged photosynthetically available radiation (PAR) measured at noon was varied from 794 $\mu molm^{-2} s^{-1}$ to 1154 $\mu molm^{-2} s^{-1}$. Soil moisture content was maintained on daily basis and using a tensiometer, (Irrometer, Model RT-12 inch Riverside, CA, USA). Ten days after thinning, the seedlings were supplied exogenously as foliar spray with 0 mmol and 50 mmol solution of α-toc. Foliar spray of α-Toc solution was done in evening before sunset for the maximum absorption of the solution in leaf. The spray of α-Toc solution was made only once during the whole experimental period. An aliquot of 50 mL solution of each of α-Toc level was applied manually per replicate as foliar spray that costs only $0.015 USD for six plants and $65 USD per acre. The solution was prepared by dissolving the required measured quantity in minimal amount of ethanol, and then the final volume was maintained with distilled water. The 0 mmol treatment without α-Toc was considered as control treatment. Before foliar spray, 0.1% of Tween-20 was added as the surfactant to the finally prepared solution for the maximum absorption of the solution. The data for varying attributes was calculated after 15 days of α-Toc foliar spray. Fresh leaf material was taken in liquid nitrogen and stored at −80 °C for different biochemical studies.

2.1. Soil Analysis

The soil used was sandy loam with a saturation percentage of 47.5, average pH, and the ECe of the soil solution was 7.63 ds.m^{-1} and 0.045 ds.m^{-1}, respectively, organic matter (1.21%), with the available P (0.051 mg kg^{-1}), K (30 mg kg^{-1}), and total N (6.1 mg kg^{-1}). The soil solution had soluble CO_3^{2-} (traces), HCO_3^- (5.01 meq L^{-1}), Cl^- (8.49 meq L^{-1}), SO_4^{-2} (2.01 meq L^{-1}), Na (3.01 meq L^{-1}), $Ca^{2+}+Mg^{2+}$ (13.91 meq L^{-1}), and SAR (0.079 meq L^{-1}).

2.2. Estimation of Different Growth Parameters

Two plants per replicate were uprooted and washed with distilled water for the estimation of different growth attributes. After calculating root and shoot lengths, number of leaves, leaf area, and fresh masses of roots and shoots, the same plants was then oven-dried using an electric oven at 70 °C for 48 h, and their dry masses were calculated.

2.3. Estimation of Leaf Photosynthetic Pigments

For the estimation of leaf chlorophyll (Chl.) *a*, *b*, total Chl, and Chl *a/b*, we followed the method described by Arnon [49]. The content of carotenoids (Car) was estimated following Kirk and Allen [50]. The extraction of the pigments was done using 80% acetone. Briefly, fresh leaf material (0.1 g) was chopped and put in 10 mL acetone for overnight at 4 °C and the absorbance of the extract was read at 663, 645, and 480 nm using a spectrophotometer (Hitachi U-2001, Tokyo, Japan). The quantities were computed using the specific formulae:

$$\text{Chl. } a = [12.7 \text{ (OD 663)} - 2.69 \text{ (OD 645)}] \times V/1000 \times W \tag{1}$$

$$\text{Chl. } b = [22.9 \text{ (OD 645)} - 4.68 \text{ (OD 663)}] \times V/1000 \times W \tag{2}$$

$$\text{Total Chl.} = [20.2 \text{ } (\Delta A645) - 8.02(\Delta A663)] \times v/w \times 1/1000 \tag{3}$$

$$A \text{ carotenoid } (\mu g/g \text{ FW}) = \Delta A480 + (0.114 \times \Delta A663) - (0.638 \times \Delta A645) \tag{4}$$

$$Car = A\ Car./Em\ 100\% \times 100 \tag{5}$$

$$Emission = Em\ 100\% = 2500 \tag{6}$$

$$\Delta A = \text{absorbance at respective wavelength} \tag{7}$$

$$V = \text{volume of the extract (mL)} \tag{8}$$

$$W = \text{weight of the fresh leaf tissue (g)} \tag{9}$$

2.4. Leaf Relative Water Content (LRWC)

For the estimation of LRWC, the second one from top was used. In first step, after excising the leaf, the fresh weight was measured and tagged with a specific mark. Then, the leaf was soaked in dH_2O for 4 h. Then, the leaf was taken out of the water, it absorbed the extra surface water, and we measured its weight again and termed the result the turgid weight. The same leaf was then oven-dried at 75 °C for 48 h and again weighed and termed this the dry weight of leaf. Then LRWC was estimated using the formula from the obtained data

$$LRWC\ (\%) = \frac{\text{Fresh weight of leaf } - \text{ dry weight of leaf}}{\text{Turgid weight of leaf } - \text{ dry weight of leaf}} \times 100 \tag{10}$$

2.5. Leaf Relative Membrane Permeability

We followed the method described by Yang et al. [51] to find out the leaf relative membrane permeability (LRMP). The known amount (0.5 g) of excised leaf was cut into small pieces (approximately 1 cm) and put in test tubes having 20 mL of deionized dH_2O. After vortexing well for 5 s, the EC of the assayed material was measured and termed as EC0. The test tubes containing leaf were then kept at 4 °C for 24 h, and the EC1 was measured. These test tubes containing leaf material were then autoclaved for 30 min at 120 °C and assayed the EC2. The LRMP was measured using the following equation:

$$RMP\ (\%) = \frac{EC1 - EC0}{EC2 - EC0} \times 100 \tag{11}$$

2.6. Estimation of Leaf Malondialdehyde Content

Content of malondialdehyde (MDA) was measured using the method given by Cakmak and Horst [52] as the measure of lipid peroxidation. The trichloroacetic acid (TCA) method was used for the estimation of MDA content. One gram of freshly taken leaf material was ground in TCA (10% solution). The supernatant (0.5 mL) was obtained from the homogenized material and mixed with 3 mL of thiobarbituric acid (TBA), prepared in 20% TCA. Test tubes having the triturate were kept at 95 °C for 50 min and then cooled immediately in chilled water. After centrifugation (10,000× g) of mixture for 10 min, the absorbance of colored part was read at 600 nm and 532 nm. The content of MDA was calculated using the following formula:

$$MDA\ (nmol) = \Delta\ (A532\ nm - A\ 600\ nm)/1.56 \times 105 \tag{12}$$

Absorption coefficient for the calculation of MDA is 156 $mmol^{-1}\ cm^{-1}$.

2.7. Extraction of Antioxidant Enzymes and Total Soluble Proteins from Different Plant Parts

For the extraction of antioxidant enzymes and total soluble proteins (TSP) from each plant part (root, stem, leaf), fresh material was ground (0.5 g) in chilled (10 mL) 50-mM phosphate buffer (pH 7.8). The mixture was then centrifuged at $10,000 \times g$ for 20 min at 4 °C. The supernatant so obtained was then used for the estimation of total soluble proteins (TSP) and estimation of antioxidative enzymes activities.

2.7.1. Estimation of Total Soluble Proteins in Different Plant Parts

TSP in the buffer extracts was estimated following the method of Bradford [53]. The absorbance of the triturate was measured at 595 nm, and the quantities of the TSP in samples were computed using a series of protein standards (200–1400 mg/kg) prepared from analytical-grade bovine serum albumin (BSA).

2.7.2. Estimation of the Activities of Superoxide Dismutase, Peroxidase, and Catalase in Different Plant Parts

Activity of superoxide dismutase (SOD) was estimated using the method of Giannopolitis and Ries [54]. The method works based on the principle of photochemical reduction inhibition of nitroblue tetrazolium (NBT), which was used, and absorbance was read at 560 nm using an UV-visible spectrophotometer. However, the method of Chance and Maehly [55] was followed to measure the peroxidase (POD) and catalase (CAT) activities.

2.8. Determination of Non-Enzymatic Antioxidants in Different Plant Parts

Ascorbic acid (ASA) content in different plant parts was determined following Mukherjee and Choudhuri [56] after extraction in TCA. The flavonoid contents in different plant parts were determined following the methods ascribed by Karadeniz et al. [57]. However, the total tocopherol content in different plant parts was assayed following the method of Backer et al. [58]. The contents of ASA, flavonoids, and tocopherol were measured quantitively using the standard curves prepared with known concentration of analytical grade ASA, rutin, and α-toc, respectively, obtained from Sigma-Aldrich Chemie GmbH - Schnelldorf, Germany.

2.9. Determination Mineral Nutrients

2.9.1. Estimation of K^+, Ca^{2+}, and Mg^{2+} in Different Plan Parts

For the estimation of mineral elements in different plant parts, 0.1 g dry material was digested using a 2 mL digestion mixture (prepared from H_2O_2, H_2SO_4, $LiSO_4$, and Se metal). The final volume was maintained 50 mL using a volumetric flask. Flame photometer was used for determination of the contents of K^+ and Ca^{2+}, while of Mg^{2+}, contents were estimated using an Atomic Absorption Spectrophotometer (Hitachi, Model 7JO-8024, Tokyo, Japan).

2.9.2. Determination of N and P

The nitrogen (N) content from the digested material was determined following the method described by Bremner and Keeney [59]. The phosphorus (P) content from the digested material was estimated using Barton's reagent by spectrophotometrically, and quantity was estimated spectroscopically.

2.10. Statistical Analysis

Microsoft Excel software 2010, US was used for the estimation of means and standard errors from the collected. To find the significant differences among treatments, analysis of variance (ANOVA) was performed using Co-Stat window version 6.3, Cohorts, Berkeley, California, USA. To compare means for significant differences among treatments at 5% levels, Tukey's test (HSD-test) was performed. Correlations and PCA analysis were performed of the studied parameters using the XLSTAT software, version 2014.5, New York, USA and the significance among the generated values of each attribute was found using the Spearman's correlation table.

3. Results

3.1. Different Growth Attributes and Content of Leaf Photosynthetic Pigments of Water-Stressed Maize Plants Foliar-Applied Alpha Tocopherol

Data for different morphological and growth attributes as presented in Table 1, which shows that water shortage imposed significant adverse impacts on the lengths of shoots and roots, the number of leaves, and the total leaf area of both maize cultivars (Table 2). Foliar application of α-Toc significantly reduced the adverse impacts of water shortage on these growth attributes for both cultivars, and both wheat genotypes showed similar increasing response in this regard. However, root length and root fresh weights remained unaffected due to foliar spray of alpha tocopherols.

Reduced water supply significantly decreased the roots and shoots fresh and dry masses of both maize genotypes (Tables 1 and 2). Foliar spray of α-Toc significantly reduced the adverse effects of water stress on these growth attributes. A similar increase in the root and shoot fresh and dry biomasses was found in both genotypes due to foliary-supplied α-Toc, both under stressed and non-stressed conditions.

Leaf Chl. *a*, Chl. *b*, and total Chl. contents decreased significantly of both maize cultivars when grown under limited water supply. Both maize genotypes showed similar decreasing trend in leaf Chl. *a*, Chl. *b*, and total Chl. contents under drought stress. Significant increasing the effect of foliary-supplied α-Toc was recorded on the contents of leaf Chl. *a*, Chl. *b*, and total Chl. of both maize cultivars both under non-stressed and stressed conditions (Tables 2 and 3).

Chl. *a/b* ratio was also significantly affected due to drought stress in both maize genotypes. An improvement in Chl. *a/b* was recorded in cv. EV-1098, but the opposite was true for cv. Agaiti-2002. α-Toc foliar spray significantly improved the leaf Chl. *a/b* only in cv. Agaiti-2002 under conditions of limited water supply. However, the carotenoids content in different plant parts increased significantly due to water shortage in both maize genotypes (Tables 2 and 3), but this increase was cultivar and plant-part-specific. A significantly higher increase in carotenoids was found in leaf and root of cv. Agaiti-2002 in comparison to cv. EV-1098, but in relation with stem carotenoids content, this cultivar-specific difference was not found under drought stress. Foliar spray of α-Toc further enhanced the content of carotenoids in all studied plant parts. Significantly more increase was recorded in the leaf and root of cv. Agaiti-2002 in comparison to cv. EV-1098. However, this improvement in stem carotenoids due to α-Toc foliar application was same in both genotypes. Similar increasing trend in carotenoids under normal irrigation in all studied plant parts was also found in both genotypes due to α-Toc foliar application (Tables 2 and 3).

Table 1. Influence of foliar-applied alpha tocopherols on different growth and morphological attributes of maize cultivars grown under different water regimes (mean ± SE; n = 3). SL = shoot length; RL = root length; NOL = number of leaves; PLA = plant leaf area; SFW = shoot fresh weight; SDW = shoot dry weight; RFW = root fresh weight; RDW = root dry weight; α-toc = alpha tocopherol.

	α-Toc	Cultivars	SL (cm)	RL (cm)	NOL/plant	PLA (cm²)	SFW (g)	SDW (g)	RFW (g)	RDW (g)
Control	0 mmol	Agaiti-2002	12.23 ± 1.10 [a]	7.77 ± 0.48 [a]	3.75 ± 0.20 [b]	69.10 ± 5.73 [bc]	2.92 ± 0.215 [b]	0.37 ± 0.04 [a]	1.18 ± 0.15 [a]	0.13 ± 0.019 [b]
		EV-1098	12.03 ± 0.67 [a]	6.70 ± 0.50 [ab]	3.85 ± 0.20 [b]	62.40 ± 4.04 [cd]	2.76 ± 0.224 [bc]	0.32 ± 0.03 [abc]	1.18 ± 0.11 [a]	0.13 ± 0.022 [b]
	50 mmol	Agaiti-2002	13.33 ± 0.82 [a]	6.20 ± 0.35 [bc]	4.25 ± 0.30 [a]	79.98 ± 5.08 [a]	3.47 ± 0.402 [a]	0.40 ± 0.03 [a]	1.39 ± 0.14 [a]	0.16 ± 0.014 [a]
		EV-1098	12.85 ± 1.11 [a]	7.50 ± 0.79 [a]	4.50 ± 0.35 [a]	73.60 ± 0.63 [ab]	3.74 ± 0.140 [a]	0.38 ± 0.03 [a]	1.35 ± 0.15 [a]	0.17 ± 0.024 [a]
Drought	0 mmol	Agaiti-2002	8.27 ± 1.11 [d]	4.67 ± 0.49 [cd]	3.25 ± 0.13 [c]	45.50 ± 1.06 [f]	2.28 ± 0.145 [c]	0.23 ± 0.03 [bd]	0.74 ± 0.13 [c]	0.09 ± 0.021 [d]
		EV-1098	8.45 ± 0.66 [d]	4.00 ± 0.39 [d]	3.00 ± 0.35 [c]	47.50 ± 1.13 [ef]	2.37 ± 0.135 [c]	0.24 ± 0.01 [cd]	0.78 ± 0.12 [c]	0.10 ± 0.011 [cd]
	50 mmol	Agaiti-2002	10.30 ± 1.51 [b]	5.15 ± 0.59 [e]	3.67 ± 0.20 [b]	55.00 ± 1.74 [de]	2.56 ± 0.136 [b]	0.27 ± 0.01 [cd]	0.93 ± 0.16 [b]	0.11 ± 0.007 [bcd]
		EV-1098	9.77 ± 1.16 [c]	4.73 ± 0.37 [cd]	3.75 ± 0.25 [b]	54.50 ± 1.07 [def]	2.72 ± 0.139 [b]	0.28 ± 0.03 [bcd]	0.99 ± 0.12 [b]	0.12 ± 0.018 [bc]
	LSD 5%		1.311	1.11	0.414	4.26	0.50	0.081	0.15	0.025

Values in column with same alphabets in superscript do not differ significantly.

Table 2. Mean squares from analysis of variance of the data for the studied attributes of water stressed maize plants foliar-applied with α-Toc at seedling stage.

SOV	d.f	SL	RL	NOL	LA	SFW	SDW	RFW	RDW
WS	1	66.7***	34.08**	2.37**	2440***	2.85***	0.045**	1.06***	0.011***
Toc	1	8.13**	0.02 ns	2.01**	41.5***	1.67**	0.003*	0.23 ns	0.004***
CV	1	0.53 ns	0.42 ns	0.01 ns	37.5*	0.04 ns	0.007 ns	0.002 ns	5.04×10^{-4} ns
WS*Toc	1	0.54 ns	0.88 ns	0.02 ns	8.16 ns	0.34 ns	0.004 ns	6×10^{-4} ns	3.37×10^{-4} ns
WS*CV	1	0.16 ns	0.66 ns	0.17 ns	121***	0.001 ns	0.014 ns	0.010 ns	1.04×10^{-4} ns
Toc*CV	1	0.67 ns	1.92 ns	0.09 ns	0.17 ns	0.08 ns	0.002 ns	2.7×10^{-4} ns	1.04×10^{-4} ns
WS*Toc*CV	1	0.10 ns	2.66 ns	0.04 ns	0.17 ns	0.06 ns	0.002 ns	0.002 ns	1.04×10^{-4} ns
Error	16	0.78	3.31	0.17	7.08	0.13	0.003	0.065	2.17×10^{-4}

SOV	d.f	Protein L	Protein R	Potein S	LRWC	LRMP	Chl. a	Chl. b	Chl a/b
WS	1	46728***	6834***	2281***	3314***	517***	0.15***	0.043***	1.270***
Toc	1	11051***	3978***	253**	172**	50.0*	0.10***	0.015***	0.130 ns
CV	1	651.04**	1395**	13.5 ns	118*	0.18 ns	0.01 ns	2.8×10^{-4} ns	0.004 ns
WS*Toc	1	35.04 ns	234***	181.5***	157***	76.5*	2.4×10^{-6} ns	0.004**	0.290*
WS*CV	1	1717***	513***	37.5 ns	22.6 ns	0.80 ns	4.8×10^{-4} ns	2.2×10^{-5} ns	0.003 ns
Toc*CV	1	1785***	18.4 ns	37.5 ns	1.25 ns	16.1 ns	0.002 ns	5.8×10^{-5} ns	0.010 ns
WS*Toc*CV	1	3.37 ns	408***	1.5 ns	6.56 ns	0.19 ns	1.14×10^{-54} ns	0.002*	0.199 ns
Error	16	62.83	18	19.5	17.06	10.8	0.002	3.1×10^{-4}	0.051

Table 2. *Cont.*

SOV	d.f	Total Chl.	POD R	POD S	POD L	CAT R	CAT S	CAT L	SOD S
WS	1	0.351 ***	2223 ***	260 ***	1932 ***	1220 ***	329.8 ***	2011 ***	270 ***
Toc	1	0.190 ***	198 ***	37.6 ***	146 ***	359 ***	12.01 ns	268 ***	4.74 ns
CV	1	0.008 ns	135 **	149 *	157 **	45.8 **	126.1 **	414 ***	71.4 *
WS * Toc	1	0.004 ns	30.4 ns	1.7×10^{-4} ns	0.89 ns	18.6 ns	1.054 ns	0.219 *	1.1 ns
WS * CV	1	7×10^{-4} ns	9.4 ns	1.50 ns	1.21 ns	3.2 ns	12.07 ns	19.65 *	27.4 ns
Toc * CV	1	0.001 ns	9.3 ns	1.53 ns	1.21 ns	3.2 ns	1.054 ns	0.22 ns	0.39 ns
WS * Toc * CV	1	0.003 ns	3.4 ns	5.93 ns	0.08 ns	3.5 ns	1.032 ns	2.85 ns	0.018 ns
Error	16	0.002	9.5	7.10	12.1	6.9	8.286	8.91	8.485

SOV	d.f	SOD L	SOD R	MDA S	MDA L	MDA R	TOC S	TOC L	TOC R
WS	1	1229 ***	7222 ***	3174 ***	1890 ***	3313 ***	1493 ***	1666 ***	1741 ***
Toc	1	86.98 *	176 **	73.5 **	828 ***	793 ***	83.7 **	1148 ***	489 ***
CV	1	81.40 *	218 **	4.0 *	84.3 ***	384 ***	9.0 ns	60.11 *	55.2 *
WS * Toc	1	107 **	24.7 ns	1.5 ns	108 ***	726 ***	3.24 ns	170 **	1.30 ns
WS * CV	1	0.06 ns	16.1 ns	6.0 ns	18.3 ns	337 ***	0.30 ns	2.66 ns	1.71 ns
Toc * CV	1	62.3 *	0.05 ns	1.5 ns	30.3 *	253 ***	0.42 ns	0.16 ns	1.71 ns
WS * Toc * CV	1	28.8 ns	0.80 ns	1.5 ns	18.3 ns	216 ***	3.24 ns	0.66 ns	0.01 ns
Error	16	10.9	15.65	4.9	5.12	8.62	5.52	12.0	11.3

SOV	d.f	AsA S	AsA L	AsA R	Car S	Car L	Car R	Flav S	Flav L
WS	1	14259 ***	23814 ***	26498 ***	5953 ***	10688 ***	5730 ***	45.37 **	84.38 ***
Toc	1	1053. ***	14113 ***	4858 ***	1722 ***	4776 ***	273 ns	0.37 ns	18.38 *
CV	1	630 ***	4056 ***	8720 ***	67.3 ns	412.7 *	894 *	9.37 ns	9.375 ns
WS * Toc	1	108 *	937 ***	881 ***	146 ns	265	8.77 ns	0.38 ns	3.35 ns
WS * CV	1	63.3 ns	0.01 ns	930 ***	2.53 ns	39.9	6.20 ns	0.37 ns	0.37 ns
Toc * CV	1	30.4 ns	1.5 ns	12.7 ns	0.26 ns	27.9	8.19 ns	0.38 ns	0.38 ns
WS * Toc * CV	1	30.3 ns	253 *	304 **	0.25 ns	9.9	3.04 ns	0.37 ns	0.37 ns
Error	16	8.62	31.75	29.7	59.4	64.9	160.6	4.5	3.5

SOV	d.f	Flav R	K S	K L	K R	Ca S	Ca L	Ca R	Mg S
WS	1	135.4 ***	261 **	403 ***	396 ***	56.8 ***	172 ***	32.08 ***	0.014 ns
Toc	1	30.3 *	53.2 ns	68.2 *	45.7 ns	1.67 ns	15.6 ***	14.7 **	0.144 ns
CV	1	9.37 ns	2.83 ns	1.82 ns	0.55 ns	2.66 *	4.61 *	0.26 ns	0.768 ns

Table 2. Cont.

SOV	d.f	MgL	MgR	PS	PL	PR	NS	NL	NR
WS*Toc	1	3.375 ns	0.15 ns	2.95 ns	0.40 ns	6.36 **	0.90 ns	0.011 ns	1.264 *
WS*CV	1	3.375 ns	7.67 ns	0.12 ns	0.22 ns	2.11 ns	1.14 ns	0.206 ns	9×10^{-4} ns
Toc*CV	1	0.375 ns	8.89 ns	11.5 ns	2.01 ns	0.77 ns	0.98 ns	0.061 ns	1.92 **
WS*Toc*CV	1	0.375 ns	0.81 ns	0.26 ns	1.02 ns	0.85 ns	0.14 ns	0.160 ns	0.069 ns
Error	16	4.125	20.1 ns	11.9	14.2	0.57	0.79	1.035	0.221
WS	1	0.29 ns	3.93 ***	10.5 ***	11.7 ***	5.15 *	260 **	321.84 ***	368.8 ***
Toc	1	0.28 ns	0.02 ns	2.45 *	1.15 *	1.56 ns	52.3 ns	45.49 ns	109.4 *
CV	1	2.86 ***	0.01 ns	0.40 ns	0.02 ns	0.36 ns	1.53 ns	26.44 ns	36.03 ns
WS*Toc	1	0.056 ns	0.20 ns	0.34 ns	0.295 ns	0.43 ns	0.58 ns	3.37 ns	1.89 ns
WS*CV	1	2.95 ***	0.16 ns	0.32 ns	0.039 ns	0.001 ns	1.89 ns	3.26 ns	3.07 ns
Toc*CV	1	0.43 ns	0.04 ns	0.30 ns	0.005 ns	0.36 ns	1.73 ns	2.72 ns	3.50 ns
WS*Toc*CV	1	1.23 **	0.12 ns	1.11 ns	0.324 ns	0.35 ns	1.81 ns	1.52 ns	2.14 ns
Error	16	0.11	0.13	0.51	0.203	0.69	21.4ns	11.65	19.3

*, ** and *** = Significant at 0.5, 0.1 and 0.01 levels respectively; ns = non-significant.

Table 3. Influence of foliar-applied alpha tocopherol on photosynthetic pigments of maize cultivars grown under different water regimes (mean ± SE; n = 3). Chl. a = leaf chlorophyll a; Chl. b = leaf chlorophyll b; Chl a/b = chlorophyll a/b ratio; Total Chl. = total chlorophyll; Leaf Car = leaf carotenoids; Root Car = root carotenoids; Stem Car = stem carotenoids.

Stress	α-Toc	Cultivars	Chl. a (mg/g FW)	Chl. b (mg/g FW)	Chl a/b	Total Chl. (mg/g FW)	Leaf Car (μg/g FW)	Root Car (μg/g FW)	Stem Car (μg/g FW)
Control	0 mmol	Agaiti-2002	1.52 ± 0.11 ab	0.36 ± 0.01 bc	4.53 ± 0.16 a	1.88 ± 0.12 ab	102.16 ± 1.78 e	99.26 ± 9.60 de	82.70 ± 5.97 e
		EV-1098	1.47 ± 0.18 bc	0.38 ± 0.02 bc	3.87 ± 0.25 d	1.85 ± 0.17 b	93.52 ± 4.41 f	84.16 ± 12.26 f	80.41 ± 6.24 e
	50 mmol	Agaiti-2002	1.63 ± 0.06 a	0.47 ± 0.01 a	4.40 ± 0.14 ab	2.00 ± 0.16 ab	120.27 ± 13.12 d	105.34 ± 10.99 d	105.00 ± 2.83 d
		EV-1098	1.60 ± 0.14 ab	0.44 ± 0.01 ab	3.64 ± 0.33 ed	2.04 ± 0.18 a	118.23 ± 4.89 d	94.79 ± 6.55 e	101.89 ± 4.40 d
Drought	0 mmol	Agaiti-2002	1.34 ± 0.04 cd	0.33 ± 0.02 c	3.62 ± 0.09 e	1.67 ± 0.05 b	139.98 ± 10.40 b	129.65 ± 9.31 ab	120.00 ± 5.83 bc
		EV-1098	1.31 ± 0.12 c	0.31 ± 0.02 c	4.28 ± 0.05 bc	1.62 ± 0.10 c	127.25 ± 5.61 c	118.75 ± 2.98 c	116.00 ± 5.75 c
	50 mmol	Agaiti-2002	1.46 ± 0.12 bcd	0.34 ± 0.02 c	4.29 ± 0.16 bc	1.80 ± 0.10 b	176.08 ± 2.74 a	134.73 ± 11.52 a	132.00 ± 6.54 a
		EV-1098	1.46 ± 0.09 bcd	0.35 ± 0.02 c	4.17 ± 0.16 c	1.81 ± 0.12 b	163.33 ± 5.61 b	124.26 ± 9.60b c	128.00 ± 4.89 ab
LSD 5%			0.15	0.12	0.64	0.52	8.74	11.81	8.15

Values in column with same alphabets in superscript do not differ significantly.

3.2. Leaf Relative Water Content, Leaf Relative Membrane Permeability, Total Soluble Proteins, and H_2O_2 Contents of Leaf Photosynthetic Pigments of Maize Plants Foliar-Applied with Alpha Tocopherol

Data presented in Table 3 reveals that the imposition of water stress decreased the LRWC of both genotypes, and a slightly higher decrease in LRWC was found in cv. EV-1098 in comparison to cv. Agaiti-2002. The foliar application of α-Toc significantly increased the LRWC of both genotypes, and this increase was found only under drought-stressed conditions; both cultivars showed a similar increasing trend in this regard (Tables 2 and 4).

Leaf relative membrane permeability (LRMP) increased significantly under water deficit conditions, and this increase was similar in both maize cultivars. Exogenous application of α-Toc as foliar spray was found to be effective in decreasing the LRMP in both maize cultivars under water-stressed conditions, and both maize cultivars showed similar responses in this regard (Tables 2 and 4).

Drought stress exerted a tissue-specific increment in leaf, root, and stem TSP contents of both genotypes when grown without foliar application of α-Toc. In leaf and root, this improvement in TSP was higher in cv. Agaiti-2002 in comparison to cv. EV-1098, but in relation to stem TSP, both cultivars showed the same increasing trend. Exogenous application of α-Toc further improved TSP accumulation in all studied plant parts in both maize cultivars under stressed and non-stressed conditions. Alpha-toc-induced this improvement in TSP contents was significantly more prominent in leaves of cv. Agaiti-2002 in comparison to cv. EV-1098 under limited water supply, but a similar increasing trend was recorded in root and stem (Tables 2 and 4).

Under stressful conditions, the extent of oxidative damage is measured in terms of MDA contents. The data presented shows that MDA contents in all studied plant parts of both the cultivars increased significantly under limited water supply. α-Toc foliar-application significantly reduced the MDA accumulation in all studied plant parts, and a more prominent reduction was found in leaves in comparison to other plant parts in both maize genotypes (Tables 2 and 4).

Table 4. Influence of foliar application of α-Toc on leaf relative water content, leaf relative membrane permeability, tissue specific total soluble proteins and malondialdehyde content of maize cultivars grown under different water regimes (mean ± SE; $n = 3$). LRWC = leaf relative water content; RMP = leaf relative membrane permeability; Leaf TSP = leaf total soluble proteins; Root TSP = root total soluble proteins; Stem TSP = stem total soluble proteins; Leaf MDA = leaf malondialdehyde; Root MDA = root malondialdehyde; Stem MDA = stem malondialdehyde.

Stress	α-Toc	Cultivars	LRWC (%)	RMP (%)	Leaf TSP (mg/kg FW)	Root TSP (mg/kg FW)	Stem TSP (mg/kg FW)	Leaf MDA (nmol/g FW)	Root MDA (nmol/g FW)	Stem MDA (nmol/g FW)
Control	0 mmol	Agaiti-2002	87.54 ± 1.71 [a]	32.33 ± 3.92 [e]	210.00 ± 29.89 [f]	144.33 ± 7.50 [f]	73.00 ± 15.15 [c]	68.10 ± 1.19 [bc]	66.18 ± 6.40 [c]	62.02 ± 4.48 [cd]
		EV-1098	81.73 ± 5.00 [b]	33.96 ± 2.33 [de]	233.33 ± 23.34 [e]	160.33 ± 4.58 [e]	72.00 ± 9.36 [c]	62.35 ± 2.94 [c]	66.10 ± 8.17 [c]	60.31 ± 5.18 [cd]
	50 mmol	Agaiti-2002	88.36 ± 1.62 [a]	34.83 ± 2.76 [d]	267.33 ± 24.22 [d]	186.67 ± 2.57 [cd]	83.67 ± 9.01 [b]	60.18 ± 8.75 [c]	65.23 ± 7.33 [c]	60.35 ± 2.12 [c]
		EV-1098	81.38 ± 5.36 [b]	33.55 ± 3.19 [de]	257.00 ± 18.26 [e]	182.00 ± 3.45 [d]	86.67 ± 14.81 [b]	55.49 ± 3.26 [bc]	66.53 ± 4.37 [c]	56.42 ± 3.29 [c]
Drought	0 mmol	Agaiti-2002	58.23 ± 3.52 [d]	45.32 ± 0.50 [b]	312.33 ± 31.67 [c]	183.67 ± 10.05 [d]	101.67 ± 16.96 [a]	90.65 ± 6.94 [a]	99.83 ± 6.21 [a]	85.59 ± 4.37 [ab]
		EV-1098	51.34 ± 3.09 [e]	47.88 ± 3.78 [a]	304.67 ± 25.01 [c]	201.33 ± 5.64 [b]	94.33 ± 13.58 [ab]	84.84 ± 3.74 [a]	102.84 ± 1.98 [a]	84.04 ± 4.32 [b]
	50 mmol	Agaiti-2002	67.87 ± 0.62 [c]	40.29 ± 2.08 [c]	376.67 ± 22.30 [a]	196.00 ± 9.98 [bc]	99.00 ± 9.93 [a]	70.72 ± 1.83 [b]	64.3 ± 7.68 [b]	81.92 ± 4.91 [ab]
		EV-1098	66.17 ± 2.02 [c]	39.08 ± 2.44 [c]	330.67 ± 33.06 [b]	227.67 ± 5.81 [a]	98.00 ± 10.98 [a]	72.22 ± 3.74 [b]	92.50 ± 6.40 [b]	80.60 ± 3.67 [a]
LSD 5%			4.06	3.49	24.44	12.64	10.26	5.82	7.88	6.12

Values in column with same alphabets in superscript do not differ significantly.

3.3. Root, Stem, and Leaf Total Tocopherol (Figure 1A–C); Ascorbic Acid (Figure 1D–F); and Total Flavonoid Contents (Figure 1G–I) of Maize Plants Foliar-Applied with α-Toc

Imposition of water stress significantly increased the accumulation of total Toc contents in the studied plant parts of both maize cultivars. This accumulation in total-Toc content in all studied plant parts was increased further due to the foliar application of α-Toc. This increased accumulation in internal total-Toc in all studied plant parts due to its foliar application was more in root and leaf in comparison to stem in both genotypes under both non-stressed and stressed conditions. α-Toc applied this increase in all studied plant parts and was similar in both maize cultivars (Figure 1A–C).

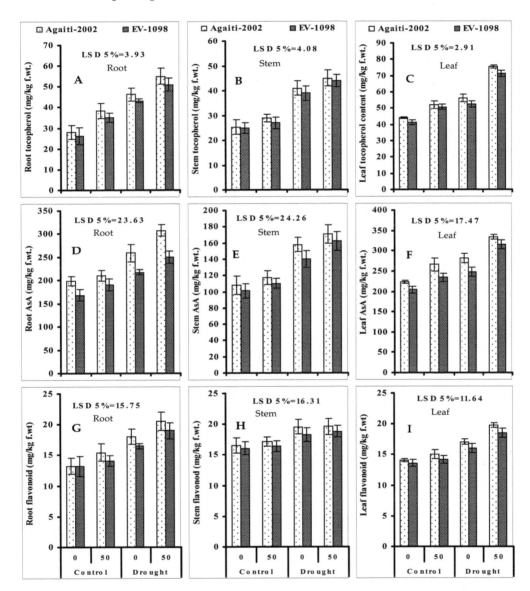

Figure 1. Root, stem, and leaf total-Toc (**A–C**), AsA (**D–F**), and total flavonoids (**G–I**) of maize plants foliar-applied with α-Toc when grown under water deficit conditions (mean ± SE; *n* = 4); AsA = ascorbic acid; 0 and 50 = mmol solution of α-Tocopherol for foliar spray.

AsA and flavonoid contents in different studied plant parts also increased significantly in both genotypes under water deficit conditions, and this improvement in AsA and flavonoid accumulation was more in root and leaf in cv. Agaiti-2002 in comparison to cv. EV-1098 (Figure 1; Table 2). Exogenous application of α-Toc as foliar spray further enhanced the AsA accumulation in all studied plant parts of both maize genotypes; accumulation was higher in cv. Agaiti-2002, both under stressed and non-stressed conditions. However, improvement in flavonoids was found only in the leaf and root of both maize genotypes when grown under water deficit conditions; this improved accumulation in flavonoids was not found in stem flavonoids (Figure 1D–I).

3.4. Activities of CAT (Figure 2A–C), SOD (Figure 2D–F) and POD (2G–I) in Root, Stem, and Leaf of Maize Plants Foliar-Applied with α-Toc

Activities of CAT and SOD in all studied plant parts increased significantly in both genotypes when grown under limited water supply, and comparatively more improvement was found in root and leaf of cv. Agaiti-2002 in comparison with stem. Alpha-toc application further enhanced the CAT and SOD activities in root and leaf in both genotypes, but such improvement in CAT and SOD activities was not found in the stem of both genotypes. In leaf, significantly more improvement in CAT activity due to α-Toc application was recorded in cv. Agaiti-2002 as compared with cv. EV-1098; however, in relation with SOD activity in root and leaf, cv. Agaiti-2002 was superior in comparison to cv. EV-1098 due to α-Toc application (Figure 2A–F; Table 2).

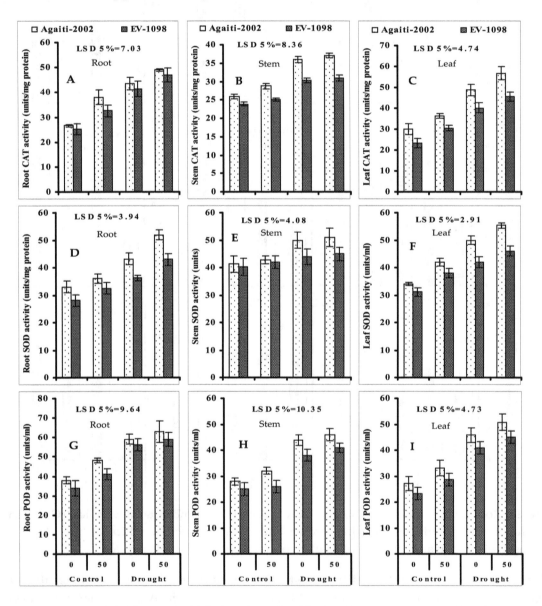

Figure 2. Activities of CAT (**A–C**), SOD (**D–F**), and POD (**G–I**) in root, stem and leaf, respectively, of drought-stressed maize plants applied with α-Toc as foliar spray when grown under water deficit conditions (mean ± SE; *n* = 4). CAT = catalase; SOD = superoxide dismutase; POD = peroxidase 0 and 50 = mmol solution of α-Tocopherol for foliar spray.

Like CAT and SOD activities, POD activity was also improved significantly under limited water supply in both genotypes in all studied plant parts. Foliar spray of α-Toc further enhanced the POD activity in root and leaf of both genotypes under non-stressed and stressed conditions, but this improvement in POD activity was not found in stem of both genotypes. A non-significant difference between genotypes was found in this regard (Figure 2G–I; Table 2).

3.5. Contents of K, Ca, Mg, N, and P in Different Parts of Maize Plants Foliary-Applied with α-Toc When Grown under Different Water Regimes

Drought stress significantly altered the tissue-specific acquisition patterns of macro-nutrients of both the studied cultivars when grown without α-Toc application (Tables 2 and 5). Potassium contents of leaf, root, and stem were reduced significantly grown under water stress without foliar spray of α-Toc. Foliar application of α-Toc increased the potassium content in specific organs under non-stressed and stressed conditions, and the impact was significant for leaf and root K of cv. Agaiti-2002 under water stress. The leaf, root and stem Ca and Mg uptake was also significantly improved after foliar application of α-Toc in both maize genotypes under non-stressed and stressed conditions, which was impaired due to limited water supply. This prominent difference in the uptake of K^+, Ca^{2+}, and Mg^{2+} due to α-Toc foliar application was similar in both maize genotypes under non-stressed and stressed conditions.

Like other nutrients, drought stress also negatively affected the P and N uptake in leaf, root, and stem of both the cultivars and this impact was more prominent on leaf and stem N. Exogenous application of α-Toc helped both the cultivars to maintain their N and P nutrition of root, leaf and stem under non-stressed and stressed conditions. Regarding the N contents in studied plant parts, comparatively more improvement in N uptake due to α-Toc foliar spray was found in leaf and root than stem (Tables 2 and 5).

Table 5. Influence of foliar-applied α-Toc on tissue specific organic and inorganic minerals of two maize (*Zea mays* L.) cultivars under non-stress and water-stressed conditions (mean ± SE; $n = 3$).

Stress	α-Toc	Cultivars	K⁺ Leaf (mg g⁻¹ DW)	K⁺ Root (mg g⁻¹ DW)	K⁺ stem (mg g⁻¹ DW)	Ca²⁺ leaf (mg g⁻¹ DW)	Ca²⁺ Root (mg g⁻¹ DW)	Ca²⁺ stem (mg g⁻¹ DW)	Mg²⁺ leaf (mg g⁻¹ DW)	Mg²⁺ root (mg g⁻¹ DW)	Mg²⁺ stem (mg g⁻¹ DW)
Control	0 mmol	Agaiti-2002	39.30 ± 3.18 ^{a}	41.35 ± 2.99 ^{ab}	33.09 ± 3.20 ^{bc}	11.35 ± 0.20 ^{d}	10.34 ± 0.75 ^{c}	10.03 ± 0.61 ^{b}	3.41 ± 0.06 ^{b}	4.13 ± 0.30 ^{a}	3.31 ± 0.32 ^{ab}
		EV-1098	41.17 ± 1.47 ^{ab}	40.21 ± 2.12 ^{b}	31.99 ± 2.38 ^{c}	12.92 ± 0.61 ^{c}	10.05 ± 0.03 ^{c}	9.02 ± 0.43 ^{c}	2.73 ± 0.15 ^{d}	4.02 ± 0.12 ^{a}	3.14 ± 0.31 ^{b}
	50 mmol	Agaiti-2002	43.15 ± 2.19 ^{a}	42.90 ± 1.41 ^{a}	35.12 ± 1.67 ^{ab}	14.36 ± 1.46 ^{b}	10.99 ± 0.18 ^{b}	11.71 ± 1.22 ^{a}	4.01 ± 0.44 ^{a}	4.29 ± 0.14 ^{a}	3.51 ± 0.33 ^{a}
		EV-1098	43.67 ± 2.45 ^{a}	42.93 ± 1.77 ^{a}	36.00 ± 1.21 ^{a}	15.33 ± 0.40 ^{a}	12.23 ± 0.31 ^{a}	10.40 ± 0.36 ^{b}	3.27 ± 0.16 ^{bc}	4.09 ± 0.22 ^{a}	3.45 ± 0.07 ^{a}
Drought	0 mmol	Agaiti-2002	30.33 ± 3.47 ^{e}	35.67 ± 2.82 ^{cd}	25.00 ± 1.85 ^{ef}	6.48 ± 0.59 ^{g}	7.33 ± 0.40 ^{e}	5.74 ± 0.71 ^{g}	2.03 ± 0.35 ^{e}	2.64 ± 0.29 ^{b}	2.38 ± 0.22 ^{d}
		EV-1098	32.33 ± 1.76 ^{de}	33.00 ± 2.79 ^{e}	24.33 ± 1.76 ^{f}	7.67 ± 0.40 ^{f}	7.67 ± 0.81 ^{f}	6.87 ± 0.72 ^{f}	2.24 ± 0.19 ^{e}	2.34 ± 0.29 ^{b}	2.82 ± 0.26 ^{c}
	50 mmol	Agaiti-2002	36.00 ± 1.85 ^{c}	37.64 ± 2.29 ^{c}	27.33 ± 1.76 ^{d}	8.45 ± 0.30 ^{e}	9.15 ± 0.82 ^{d}	8.33 ± 0.40 ^{e}	2.54 ± 0.09 ^{d}	2.66 ± 0.33 ^{b}	2.49 ± 0.31 ^{d}
		EV-1098	34.81 ± 2.19 ^{cd}	34.67 ± 2.45 ^{de}	27.00 ± 3.89 ^{de}	8.33 ± 0.40 ^{e}	9.96 ± 0.66 ^{c}	8.88 ± 0.13 ^{d}	3.11 ± 0.19 ^{c}	2.71 ± 0.25 ^{b}	2.47 ± 0.20 ^{d}
LSD 5%			2.99	2.20	2.26	0.50	0.47	0.40	0.29	0.41	0.22

Stress	α-Toc	Cultivars	N Leaf (mg g⁻¹ DW)	N Root (mg g⁻¹ DW)	N Stem (mg g⁻¹ DW)	P Leaf (mg g⁻¹ DW)	P Root (mg g⁻¹ DW)	P Stem (mg g⁻¹ DW)
Control	0 mmol	Agaiti-2002	38.63 ± 2.28 ^{b}	41.35 ± 2.99 ^{ab}	33.09 ± 3.20 ^{b}	5.20 ± 0.33 ^{b}	5.91 ± 0.43 ^{b}	4.50 ± 0.46 ^{c}
		EV-1098	41.17 ± 1.47 ^{b}	40.21 ± 2.12 ^{ab}	32.72 ± 2.93 ^{b}	5.55 ± 0.25 ^{b}	5.74 ± 0.27 ^{b}	4.01 ± 0.28 ^{de}
	50 mmol	Agaiti-2002	45.82 ± 2.77 ^{a}	42.90 ± 1.41 ^{a}	38.16 ± 1.90 ^{a}	6.06 ± 0.34 ^{a}	6.13 ± 0.20 ^{ab}	5.02 ± 0.22 ^{b}
		EV-1098	47.00 ± 2.52 ^{a}	43.93 ± 2.72 ^{a}	39.06 ± 2.73 ^{a}	6.01 ± 0.23 ^{a}	6.28 ± 0.31 ^{a}	5.28 ± 0.49 ^{a}
Drought	0 mmol	Agaiti-2002	25.33 ± 3.46 ^{d}	33.33 ± 2.45 ^{c}	26.00 ± 2.27 ^{c}	4.33 ± 0.50 ^{cd}	4.62 ± 0.42 ^{d}	3.84 ± 0.22 ^{e}
		EV-1098	23.33 ± 2.64 ^{d}	34.41 ± 3.00 ^{e}	23.00 ± 3.20 ^{d}	4.06 ± 0.27 ^{d}	4.86 ± 0.50 ^{cd}	3.58 ± 0.25 ^{f}
	50 mmol	Agaiti-2002	30.00 ± 2.10 ^{c}	36.61 ± 3.27 ^{b}	31.00 ± 1.85 ^{b}	4.29 ± 0.27 ^{cd}	5.02 ± 0.47 ^{c}	4.08 ± 0.23 ^{d}
		EV-1098	32.67 ± 0.81 ^{c}	37.67 ± 1.45 ^{b}	27.67 ± 2.02 ^{c}	4.54 ± 0.22 ^{c}	5.05 ± 0.53 ^{c}	3.82 ± 0.11 ^{e}
LSD 5%			2.95	4.00	2.80	0.39	0.32	0.22

Values in column with same alphabets in superscript do not differ significantly.

3.6. PCA Analysis and Spearman's Correlation Coefficient (r²) Values Extracted from XLSTAT Software of All the Studied Attributes of Maize Plants Foliar-Applied with α-Toc

PCA and correlations coefficients among studied attributes revealed a significant positive correlation of total-Toc contents in leaf, root, and stem with morphological and growth attributes, levels of antioxidants, and uptake of mineral nutrients (K, Ca, Mg, N, and P) in all studied tissues of maize. A positive correlation of leaf and stem Toc was found with leaf area (0.768 *** and 0.664 **) and fresh weights (0.921 *** and 0.661 ***), respectively, that depicts the role of Toc in the improved growth under drought stress. Positive correlation was also recorded of shoot dry weight with Toc levels in studied plant tissues such as in leaf (0.578 **) and root (0.643 ***), respectively. Significantly positive correlation was found of Toc levels in the root with LRWC (0.721 ***). CAT, POD, and SOD activities in different plant parts like leaf (0.966 ***, 0.961 *** and 0.936 ***) and stem (0.863 ***, 0.872 *** and 0.859 ***), respectively, were also positively correlated with plant Toc levels. Tocopherol contents were also positively correlated with potassium and calcium contents in leaf (0.553 ** and 0.606 **, 0.569 ** and 0.633 ***), root (0.555 ** and 0.675 ***, 0.674 ** and 0.461 *), and stem (0.470 * and 0.673 ***, 0.749 *** and 0.437 *), respectively. Furthermore, a positive correlation was also recorded between nitrogen and phosphorus contents with Toc levels in studied plant tissues such as in leaf (0.610 ** and 0.613 **, 0.539 **, and 0.683 ***), root (0.669 ***, 0.494 * and 0.488 * and 0.729 ***, 0.430 * and 0.620 **), and stem (0.626 *** and 0.601 **, 0.536 **, and 0.688 ***), respectively. Figure 3 shows the PCA analysis of varying studied attributes that confirmed correlation studies. Of the extracted components, F1 has a major contribution (67.43%) that has divided the studied attributes in different groups. Of them, the major group encircled has parameters that are positively correlated include Pr L, RFW, RDW, N R, S L, SDW, K L, Ca L, P S, P L, K S, and LRWC, and L A, Ca R, Ca S, Mg L, Pr R, P R, N R, and N L contributed maximally in determining the variance. The F2 component has less variance (17.70%). Both components have a total variance of 80.13% (Figure 3; Table 6).

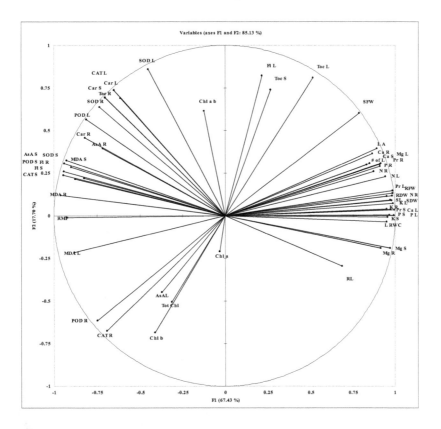

Figure 3. Principle component analysis of tocopherol levels in different plant tissues of maize with studied growth and physio-biochemical attributes, and nutrient accumulation.

Table 6. Spearman correlation coefficient values (r^2) of Toc levels in different plant parts of maize with growth, biochemical attributes, and nutrient uptake.

	Toc L	Toc R	Toc S
Toc L	1.000		
Toc R	0.204 ns	1.00	
Toc S	0.847 ***	0.302 ns	1.00
SL	0.534 **	−0.644 ***	0.287ns
RL	0.192 ns	−0.661 ***	0.122 ns
NL	0.797 ***	−0.353 ns	0.557 **
L A	0.768 ***	−0.371 ns	0.664 ***
SFW	0.921 ***	−0.124 ns	0.661 ***
SDW	0.578 **	0.643 ***	0.368 ns
RFW	0.627 ***	−0.597 **	0.381 ns
RDW	0.780 ***	−0.420*	0.531 ns
L RWC	0.391 ns	0.721 ***	0.164 ns
Chl. a	−0.088 ns	−0.059 ns	0.266 ns
Chl. b	−0.787 ***	−0.181 ns	−0.423*
Chl. a/b	0.553 **	0.538 **	0.740 ***
Tot Chl.	−0.624 ***	−0.133 ns	−0.239 ns
RMP	−0.416*	0.664 ***	−0.084 ns
MDA L	−0.534 **	−0.463 *	−0.202 ns
MDA R	−0.342 ns	−0.751 ***	−0.052 ns
MDA S	−0.240 ns	−0.845 ***	−0.037 ns
Protien L	0.561 **	0.591 **	0.392 ns
Protein R	0.695 ***	0.324 ns	0.314 ns
Protein S	0.468 *	0.666 ***	0.159 ns
AsA L	−0.434 *	−0.115 ns	−0.149 ns
AsA R	−0.180 ns	0.709 ***	−0.023 ns
AsA S	−0.196 ns	0.797 ***	0.145 ns
Car L	0.191 ns	0.925 ***	0.339 ns
Car R	−0.078 ns	0.870 ***	0.190 ns
Car S	0.218 ns	0.974 ***	0.411 ns
Flav L	0.669 ***	0.443*	0.426*
Flav R	−0.348 ns	0.679 ***	−0.269 ns
Flav S	−0.211 ns	0.733 ***	0.047 ns
CAT L	0.275 ns	0.966 ***	0.434 *
CAT R	−0.884 ***	0.041 ns	−0.757 ***
CAT S	−0.283 ns	0.863 ***	−0.114 ns
POD L	0.049 ns	0.961 ***	0.255 ns
POD R	−0.831 ***	0.116 ns	−0.633 ***
POD S	−0.263 ns	0.872 ***	−0.107 ns
SOD L	0.438 *	0.936 ***	0.430 *
SOD R	0.173 ns	0.951 ***	0.273 ns
SOD S	−0.220 ns	0.859 ***	0.047 ns
K L	0.553 **	0.606 **	0.282 ns
K R	0.555 **	0.675 ***	0.333 ns
K S	0.470 *	0.673 ***	0.319 ns
Ca L	0.569 **	0.633 ***	0.339 ns
Ca R	0.674 ***	0.461 *	0.416 *
Ca S	0.749 ***	0.437 *	0.469 *
Mg L	0.656 ***	0.439 *	0.470 *
Mg R	0.372 ns	0.761 ***	0.201 ns
Mg S	0.335 ns	0.833 ***	0.151 ns
N L	0.610 **	0.613 **	0.368 ns
N R	0.669 ***	0.494 *	0.488 *
N R	0.626 ***	0.601 **	0.337 ns
P L	0.539 **	0.683 ***	0.383 ns
P R	0.729 ***	0.430*	0.620 **
P S	0.536 **	0.688 ***	0.281 ns

4. Discussion

The exogenous application of water-soluble antioxidants have been widely investigated to improve stress tolerance, but plant growth modulations by foliar application of lipophilic antioxidants like α-Toc has been little studied, probably due to limited information regarding their application, absorption, and translocation within the plant. Kumar et al. [60] and Ali et al. [61] reported that the exogenously applied α-Toc can partly alleviate the deleterious impacts of heat and water stress in wheat. In another study, it was found that the exogenous application of α-Toc effectively decreased the adverse effects of salt stress in flax cultivars [44]. In most of the earlier studies, the α-Toc was applied at adult growth stages. However, the seedling stage (among other growth stages) is considered important due to its involvement in better seed yield by establishing better crop stand [14]. In view of the available information in literature, the present experiment was planned with the objective to study the involvement of α-Toc in the improvement of water stress tolerance in relation to the growth modulations of maize depending upon tissue specific partitioning of macro-nutrients and antioxidants in relation with its own translocation/synthesis in specific terms. For this purpose, the response of selected maize genotypes (Agaiti-2002 and EV-1098) was examined under water stress at an early growth stage with and without foliar spray of α-Toc.

4.1. Tocopherol Content in Different Plant Parts

Foliar spray of α-Toc significantly increased the leaf tocopherol levels under non-stressed and stressed conditions, which pointed out the existence of an appropriate mechanism for the uptake of α-Toc in the leaves of maize. The increments in root tocopherol contents exhibited a similar pattern, as did the leaves, after foliar application, which suggests an efficient basipetal translocation of α-Toc in maize. Our findings are in agreement with Kumar et al. [60], who reported an elevation in the endogenous levels of α-Toc in heat-stressed wheat plants after its exogenous application. Furthermore, it has been reported that the exogenous application of these organic compounds, along with altering the cellular metabolic activities, also controls the plant's own metabolism. In the present study, the improvement in the internal levels of α-Toc by its exogenous application might also be due to its involvement in regulating plant metabolism [34,36,40].

4.2. Growth, Water Relations, and Photosynthetic Pigments

Seedling growth of maize plants was adversely affected in plants grown without foliar application of α-Toc under water stress, which is in line with the findings that drought-caused growth reduction is a clear phenomenon in crop plants [8,14]. Similarly, in the present study, a drought-induced decrease was recorded in root and shoot lengths, root and shoot fresh and dry weights, leaf area, and number of leaves of both maize genotypes. Growth is dependent on physiological factors, including the content of plant photosynthetic pigments and water relations that directly influences the leaf photosynthetic rate by affecting the capacity of light capturing and assimilation process [61,62]. Different plant species and even cultivars in the same species have different potentials to tolerate the adverse conditions regarding these attributes [63].

In the present study, water-stress-induced reduction in biomass is associated with reduced photosynthetic pigment along with disturbed plant water relations, and this reduction was less in cv. Agaiti-2002, showing its better tolerance to drought [34]. The foliar spray of α-Toc substantially elevated the plant's endogenous levels and resulted in significant growth improvement under stressed and non-stressed conditions. Increments in plant biomass production is positively associated with the improvement in plant water relations and biosynthesis of biosynthetic pigments such as chlorophyll and carotenoids under the influence of α-Toc foliar application. The increment in plant water status might probably be due to impact of α-Toc on H-ATPase system showing its role in cellular osmotic adjustment, due to a necessary part of cellular membranes. This involvement of alpha tocopherol in cellular osmotic adjustment confers its role in maintaining the cellular water relations under stressful conditions. Similar might be in present study where foliar application of alpha tocopherol improved the leaf relative water content of water stressed maize plants. This improvement in plant water relations further confers its role in improving the leaf net photosynthetic efficiency because plant better water content is necessary to regulate stomatal regulation for better photosynthesis [62]. Furthermore, it is found that α-Toc, being a part of cellular membranes, plays a significant role in decreasing the degradation of photosynthetic pigments in a stressful environment [64]. Tocopherols also protect D_1 protein [65] and chloroplastic membranes from damaging effects when grown under stressful conditions.

In the present study, foliar-applied α-Toc under drought stress further enhanced its internal levels in parallel with the improvement in leaf photosynthetic pigments, which might be due to the significant role of alpha tocopherol in reducing the adverse effects on leaf photosynthetic pigments, resulting in improved photosynthetic efficiency along with better plant water relations that resulted in better plant biomass production. In an earlier study, it was found by Sakr and El-Metwally [43] and El-Quesni [66] in wheat and *Hibiscus rosa sinensis*, respectively, that exogenous application of α-Toc enhanced plant biomass production, which might be due to the role of α-Toc in the accumulation of total carbohydrates and protein biosynthesis, confirming its role in photosynthesis and assimilation [67]; this can be correlated with present findings, where higher biomass production was associated with α-Toc levels in different parts that improved plant water relations and net photosynthesis as a result of better net assimilation with improved biomass production. Furthermore, this study reveals the increased plant dry weights due to foliar application of α-Toc, which points toward the improved photosynthetic activity and assimilation with the establishment of new binding sites [68] after its exogenous application.

Furthermore, in the present study, both maize cultivars maintained an optimum level of their carotenoid contents even under drought and α-Toc supplementation, which further enhanced the plant carotenoid contents, especially in leaf and root. These observations point out that α-Toc-induced improvement in the growth of maize plants might be due to an improvement in the contents of accessory pigments as additional support to different photosynthetic attributes. In an earlier study, it was found that, in different wheat cultivars [43] and *Vicia faba* [69], foliar-applied α-Toc improved the leaf carotenoid concentration in association with its enhanced growth. Without α-Toc application, a decrease in leaf water contents was found in maize plants, which is a well-known phenomenon in all plants. α-Toc foliar application significantly increased the leaf water content of water-stressed maize plants, showing its protective role in drought-stressed plants, which might be due to its role in the management of cellular turgor potential through imparting its role in cellular osmotic adjustment by enhancing biosynthesis of osmolytes [7], resulting in better growth by providing an environment for increased cell division and provide an environment for better photosynthesis.

4.3. Lipid Peroxidation and Antioxidative Defence Mechanism

An increase in the levels of ROS under stressful environment is a general phenomenon due to O_2 excitation to form singlet oxygen or its conversion to hydroxyl radicals (OH^-), hydrogen peroxide (H_2O_2), or superoxide (O^{-2}) due to the transfer of excited electrons, respectively [34,70], with restricted e^- transfer at different steps in photosynthesis and respiration under reduced metabolic activities. These overly produced ROS directly affect different cellular membranes through lipid peroxidation. As a defense for the protection of the cellular membranes and other components from the deleterious and damaging effects of overproduced ROS, plants have evolved well-developed mechanisms for the antioxidation of ROS, i.e., comprised of non-enzymatic (AsA, phenolics, carotenoids, flavonoids, tocopherol, etc.) and enzymatic (SOD, POD, CAT, APX) components [14,34,71]. This antioxidative system works well in combination. In the present study, the α-Toc-treated plants suffered significantly lower oxidative damage, especially in root and leaf, as depicted by the lower MDA contents in these plant parts relative to untreated ones (as reported earlier for wheat) [60]. Drought stress significantly increased oxidative stress in maize plants; this is obvious from increased levels of MDA, a product of lipid peroxidation. Damage to biological membranes due to oxidative stress is a general phenomenon that generally increases in specific environments [14,45]. In an earlier study, significantly lower oxidative stress was recorded in α-Toc applied plants as obvious from lower membrane permeability which is in line with its role in quenching lipid peroxyl radicals, responsible for propagating lipid peroxidation [69,72,73]. It was reported that during early growth stages, α-Toc played a significant role in counteracting the adverse effects of membrane lipid peroxidation. Furthermore, being lipophilic, α-Toc has a significant role in membrane stabilization [74] and also protects them from ROS [75]. Furthermore, α-Toc directly scavenges singlet oxygen [76], giving rise an intermediate tocopherol quinone, which again yields α-Toc in chloroplasts, thereby conferring the recycling for oxidized tocopherols [77]. Reports exist that α-Tocopherol is also an excellent quencher and scavenger of singlet oxygen by controlling the lifetime of ROS. By resonance energy transfer, one α-Toc molecule can neutralize up to 120 molecules of singlet oxygen [78]. The activities of antioxidants such as SOD, POD, and CAT were found to be higher in leaves and roots of maize plant after α-Toc treatment, which suggested their antioxidative role to be stimulated in the presence of α-Toc.

Higher activity rates of these enzymes were found in leaves and roots where more accumulation of α-Toc was found in comparison with stem, showing the supportive role of α-Toc in the activities of antioxidative enzymes. Furthermore, the higher levels of non-enzymatic antioxidant in root and leaf as compared to stem (such as AsA, phenolics, and flavonoids) are also associated with high content of α-Toc in these plant parts. These findings show that α-Toc application after its translocation to the studied plant parts played a significant role in increasing the activities of antioxidative enzymes and the levels of non-enzymatic antioxidant compounds and thus played an imperative role in protecting cellular membranes by boosting the plant's own mechanism. It was found by Fahrenholtz et al. [79] that α-Toc acts as an antioxidative defense mechanism in plants. It was also found that α-Toc minimizes the oxidative changes in the cellular membrane in a significant way with other antioxidants [80–82].

4.4. Uptake of Mineral Nutrients

Drought-induced growth reduction can also be attributed to disturbances in the uptake of mineral nutrients along with other physiological attributes. It is well known that disturbance or reductions in the leaf uptake of mineral nutrient in plants is probably due to nutrient availability, partitioning, and transport, which is negatively affected under drought conditions. Plant mineral nutrients status played a major role in determining drought tolerance [83]. In the present study, the PCA analysis and the correlations studied suggest that an improvement in the levels of α-Toc contents in different plant parts induced by its foliar application increased the uptake of mineral nutrients (K, Ca, N, and P). Mineral nutrients effectively decrease the harsh effects of water stress by various mechanisms [22]. For example, it has been found that better uptake of mineral nutrients like Ca^{2+}, N, and K^+ reduces the deleterious effects of over produced ROS by increasing the concentration of antioxidants like CAT, POD, and SOD [22]. It has been reported that P, K, and Mg improve root growth, which results in improved water intake conferring the drought tolerance. It can be interpreted that optimum nutrient levels maintained after α-Toc application confer drought tolerance induction in maize plants in parallel with improved growth. This is more likely because leaf water contents were significantly improved by foliar spray of α-Toc. The supportive role of α-Toc after its application in the absorption of nutrients from the soil in stressful environment has been found extensively [44,78], and it is reported that α-Toc induced increase uptake of nutrients due to α-Toc being an antioxidant, along with membrane permeability. Furthermore, previous studies found that α-Toc induced an increase in growth, water relation, and nutrient uptake associated with improved stem and leaf anatomy, which further improved translocation to different plant parts. Therefore, the studies confirm that, as in the present study, α-Toc application might improve the uptake and translocation of different nutrients from the soil solution to the roots and then to different plant parts, resulting in better assimilation and growth.

5. Conclusions

It can be concluded that endogenous levels of α-Toc have an important role in enhancing water stress tolerance of maize cultivars, and its foliar application is found to be effective in reducing water-stress-induced adversative effects on growth by modulating different metabolic activities. Our results confirmed that α-Tocopherol application resulted in membrane protection through increased activities of antioxidative enzymes (CAT, POD, and SOD) and the content of non-enzymatic antioxidants with improved water relations. The correlations and PCA analysis revealed that the increase in α-Tocopherol contents in different plant parts after its foliar application increased the uptake of mineral nutrients (K^+, Ca^{2+}, N, and P). Optimum water content and nutrients, along with better antioxidant potential, ultimately resulted in drought tolerance in both maize cultivars that increased growth. In relation to translocation-dependent effects, it was found that α-Toc followed basipetal translocation, concentrating mainly in the roots rather than the shoot after its foliar application. Therefore, analysis of the impact of foliar application of α-Toc on seed yield and nutritional quality of arable crops under stressful environment should be the subject of future studies.

Author Contributions: Conceptualization, Q.A., M.R., S.A., M.N.A., H.A.E.-S., and F.A.A.-M.; Data curation, Q.A., M.T.J., M.Z.H., and R.P.; Formal analysis, Q.A., M.T.J., N.H., and R.P.; Funding acquisition, Q.A., M.N.A., H.A.E.-S., and F.A.A.-M.; Investigation, N.H. and R.P.; Methodology, M.T.J., M.Z.H., and R.P.; Resources, M.R., S.A., M.N.A., H.A.E.-S., and F.A.A.-M.; Software, M.R., S.A., M.N.A., H.A.E.-S., and F.A.A.-M.; Supervision, Q.A. and M.T.J.; Validation, M.T.J., M.Z.H., and N.H.; Visualization, M.Z.H. and N.H.; Writing—original draft, Q.A. and F.A.A.-M.; Writing—review and editing, M.R., S.A., M.N.A., and H.A.E.-S. All authors have read and agreed to the published version of the manuscript.

Acknowledgments: The authors are grateful to the Higher Education Commission (HEC) Islamabad, Pakistan for its support.

Abbreviations

Toc L	leaf tocopherol
Toc R	root tocopherol
Toc S	stem tocopherol
SL	shoot length
RL	root length
NL	number of leaves
LA	leaf area
SFW	shoot fresh weight
SDW	shoot dry weight
RFW	root fresh weight
RDW	root dry weight
L RWC	leaf relative water content
Chl. a	chlorophyll a
Chl. b	chlorophyll b
Chl. a/b	chlorophyll a/b ratio
Tot Chl.	total chlorophyll
RMP	relative membrane permeability
MDA L	MDA leaf
MDA R	MDA root
MDA S	MDA stem
Protien L	protein leaf
Protein R	protein root
Protein S	protein stem

AsAL	ascorbic acid leaf
AsA R	ascorbic acid root
AsA S	ascorbic acid stem
Car L	carotenoids leaf
Car R	carotenoids root
Car S	carotenoids stem
Flav L	flavonoids leaf
Flav R	flavonoids root
Flav S	flavonoids stem
CAT L	catalase leaf
CAT R	catalase root
CAT S	catalase stem
POD L	peroxidase leaf
POD R	peroxidase root
POD S	peroxidase stem
SOD L	superoxide dismutase leaf
SOD R	superoxide dismutase root
SOD S	superoxide dismutase stem
K L	potassium leaf
K R	potassium root
K S	potassium stem
Ca L	calcium leaf
Ca R	calcium root
Ca S	calcium stem
Mg L	magnesium leaf
Mg R	magnesium root
Mg S	magnesium stem
N L	nitrogen leaf
N R	nitrogen root
N S	nitrogen stem
P L	phosphorus leaf
P R	phosphorus root
P S	phosphorus stem

References

1. Gobin, A. Impact of Heat and Drought Stress on Arable Crop Production in Belgium. *Nat. Hazards Earth Syst. Sci.* **2012**, *12*, 1911–1922. [CrossRef]
2. Khan, N.; Bano, A. Rhizobacteria and Abiotic Stress Management. In *Plant Growth Promoting Rhizobacteria for Sustainable Stress Management*; Springer: Singapore, 2019; pp. 65–80.
3. Cairns, J.E.; Sanchez, C.; Vargas, M.; Ordoñez, R.; Araus, J.L. Dissecting Maize Productivity: Ideotypes Associated with Grain Yield Under Drought Stress and Well-Watered Conditions. *J. Integr. Plant Biol.* **2012**, *54*, 1007–1020. [CrossRef] [PubMed]
4. Adnan, S.; Ullah, K.; Gao, S.; Khosa, A.H.; Wang, Z. Shifting of Agro Climatic Zones, their Drought Vulnerability, and Precipitation and Temperature Trends in Pakistan. *Int. J. Climatol.* **2017**, *37*, 529–543. [CrossRef]
5. Kazmi, D.H.; Li, J.; Rasul, G.; Tong, J.; Ali, G.; Cheema, S.B.; Liu, L.; Gemmer, M.; Fisherr, T. Statistical Downscaling and Future Scenario Generation of Temperatures for Pakistan Region. *Theor. Appl. Climatol.* **2015**, *120*, 341–350. [CrossRef]
6. Hina, S.; Saleem, F. Historical Analysis (1981–2017) of Drought Severity and Magnitude over a Predominantly Arid Region of Pakistan. *Clim. Res.* **2019**, *78*, 189–204. [CrossRef]
7. Um, M.; Kim, Y.; Park, D.; Jung, K.; Wang, Z.; Kim, M.M.; Shin, H. Impacts of Potential Evapotranspiration on Drought Phenomena in Different Regions and Climate Zones. *Sci. Total Environ.* **2020**, *703*, 135590. [CrossRef]

8.   Ali, Q.; Anwar, F.; Ashraf, M.; Saari, N.; Perveen, R. Ameliorating Effects of Exogenously Applied Proline on Seed Composition, Seed Oil Quality and Oil Antioxidant Activity of Maize (*Zea mays* L.) Under Drought Stress. *Int. J. Mol. Sci.* **2013**, *14*, 818–835. [CrossRef]

9.   Khan, N.; Bano, A.; Rahman, M.A.; Guo, J.; Kang, Z.; Babar, M.A. Comparative physiological and metabolic analysis reveals a complex mechanism involved in drought tolerance in chickpea (*Cicer arietinum* L.) induced by PGPR and PGRs. *Sci. Rep.* **2019**, *9*, 1–19. [CrossRef]

10.   Rana, V.; Singh, D.; Dhiman, R.; Chaudhary, H.K. Evaluation of Drought Tolerance Among Elite Indian Bread Wheat Cultivars. *Cereal. Res. Commun.* **2014**, *42*, 91–101. [CrossRef]

11.   Chaves, M.M.; Maroco, J.P.; Pereira, J.S. Understanding Plant Responses to Drought From Genes to the Whole Plant. *Physiol. Mol. Biol. Plants* **2003**, *30*, 239–264. [CrossRef]

12.   Khan, N.; Bano, A.; Rahman, M.A.; Rathinasabapathi, B.; Babar, M.A. UPLC-HRMS-based untargeted metabolic profiling reveals changes in chickpea (Cicer arietinum) metabolome following long-term drought stress. *Plant Cell Environ.* **2019**, *42*, 115–132. [CrossRef] [PubMed]

13.   Demirevska, K.; Simova-Stoilova, L.; Fedina, I.; Georgieva, K.; Kunert, K. Response of *Oryza cystatin* L. Transformed Tobacco Plants to Drought, Heat and Light Stress. *J. Agron. Crop Sci.* **2010**, *196*, 90–99. [CrossRef]

14.   Ali, Q.; Javed, M.T.; Noman, A.; Haider, M.Z.; Waseem, M.; Iqbal, N.; Perveen, R. Assessment of Drought Tolerance in Mung Bean Cultivars/Lines as Depicted by the Activities of Germination Enzymes, Seedling's Antioxidative Potential and Nutrient Acquisition. *Arch. Agron. Soil. Sci.* **2018**, *64*, 84–102. [CrossRef]

15.   Chen, J.; Junying, D. Effect of Drought on Photosynthesis and Grain Yield of Corn Hybrids With Different Drought Tolerance. *Acta-Agron Sin.* **1996**, *22*, 757–762.

16.   Khan, N.; Bano, A.; Curá, J.A. Role of Beneficial Microorganisms and Salicylic Acid in Improving Rainfed Agriculture and Future Food Safety. *Microorganisms* **2020**, *8*, 1018. [CrossRef]

17.   Lawlor, D.W.; Cornic, G. Photosynthetic Carbon Assimilation and Associated Metabolism in Relation to Water Deficits in Higher Plants. *Plant Cell Environ.* **2002**, *25*, 275–294. [CrossRef]

18.   Rapacz, M.; Kościelniak, J.; Jurczyk, B.; Adamska, A.; Wójcik, M. Different Patterns of Physiological and Molecular Response to Drought in Seedlings of Malt- and Feed-Type Barleys (*Hordeum vulgare*). *J. Agron. Crop Sci.* **2010**, *196*, 9–19. [CrossRef]

19.   Marschner, P. *Mineral Nutrition of Higher Plants*; Academic Press: London, UK, 2011.

20.   Mukhtar, I.; Shahid, M.A.; Khan, M.W.; Balal, R.M.; Iqbal, M.M.; Naz, T.; Zubair, M.; Ali, H.H. Improving salinity tolerance in chili by exogenous application of calcium and sulphur. *Soil Environ.* **2016**, *1*, 35.

21.   Heidari, M.; Karami, V. Effects of Different Mycorrhiza Species on Grain Yield, Nutrient Uptake and Oil Content of Sunflower Under Water Stress. *J. Saudi Soc. Agric. Sci.* **2014**, *13*, 9–13. [CrossRef]

22.   Waraich, E.A.; Ahmad, R.; Ashraf, M.Y.; Ahmad, S.M. Improving Agricultural Water Use Efficiency by Nutrient Management in Crop Plants. *Acta Agric. Scand. B Soil Plant Sci.* **2011**, *61*, 291–304. [CrossRef]

23.   Fahad, S.; Bajwa, A.A.; Nazir, U.; Anjum, S.A.; Farooq, A.; Zohaib, A.; Ihsan, M.Z. Crop Production Under Drought and Heat Stress: Plant Responses and Management Options. *Front. Plant Sci.* **2017**, *8*, 1147. [CrossRef] [PubMed]

24.   Hussain, H.A.; Hussain, S.; Khaliq, A.; Ashraf, U.; Anjum, S.A.; Men, S.; Wang, L. Chilling and Drought Stresses in Crop Plants: Implications, Cross Talk, and Potential Management Opportunities. *Front. Plant Sci.* **2018**, *9*, 393. [CrossRef]

25.   Rajala, A.; Niskanen, M.; Isolahti, M.; Peltonen-Sainio, P. Seed Quality Effects on Seedling Emergence, Plant Stand Establishment and Grain Yield in Two-Row Barley. *Agric. Food Sci.* **2011**, *20*, 228–234. [CrossRef]

26.   Singla, S.; Grover, K.; Angadi, S.V.; Schutte, B.; VanLeeuwen, D. Guar Stand Establishment, Physiology, and Yield Responses to Planting Date in Southern New Mexico. *Agron. J.* **2016**, *8*, 2289–2300. [CrossRef]

27.   Srivalli, B.; Sharma, G.; Khanna-Chopra, R. Antioxidative Defense System in an Upland Rice Cultivar Subjected to Increasing Intensity of Water Stress Followed by Recovery. *Physiol. Plant.* **2003**, *119*, 503–512. [CrossRef]

28.   Ashraf, M. Inducing Drought Tolerance in Plants: Recent Advances. *Biotechnol. Adv.* **2010**, *28*, 169–183. [CrossRef]

29.   Zhang, L.X.; Li, S.X.; Zhang, H.; Liang, Z.S. Nitrogen Rates and Water Stress Effects on Production, Lipid Peroxidation and Antioxidative Enzyme Activities in Two Maize (*Zea mays* L.) genotypes. *J. Agron. Crop Sci.* **2007**, *193*, 387–397. [CrossRef]

30. Farooq, M.; Wahid, A.; Lee, D.J.; Cheema, S.A.; Aziz, T. Comparative Time Course Action of the Foliar Applied Glycinebetaine, Salicylic Acid, Nitrous Oxide, Brassinosteroids and Spermine in Improving Drought Resistance of Rice. *J. Agron. Crop Sci.* **2010**, *196*, 336–345. [CrossRef]

31. Posmyk, M.M.; Kontek, R.; Janas, K.M. Antioxidant Enzymes Activity and Phenolic Compounds Content in Red Cabbage Seedlings Exposed to Copper Stress. *Ecotoxicol. Environ. Saf.* **2009**, *72*, 596–602. [CrossRef]

32. Farooq, M.; Basra, S.M.A.; Wahid, A.; Cheema, Z.A.; Cheema, M.A.; Khaliq, A. Physiological Role of Exogenously Applied Glycinebetaine to Improve Drought Tolerance in Fine Grain Aromatic Rice (*Oryza sativa* L.). *J. Agron. Crop. Sci.* **2008**, *194*, 325–333. [CrossRef]

33. Khan, S.; Hasan, M.; Bari, A.; Khan, F. Climate Classification of Pakistan. In Proceedings of the Balwois Conference, Ohrid, Macedonia, 25–29 May 2010; pp. 1–47.

34. Ali, Q.; Ashraf, M. Induction of Drought Tolerance in Maize (*Zea mays* L.) Due to Exogenous Application of Trehalose: Growth, Photosynthesis, Water Relations and Oxidative Defence Mechanism. *J. Agron. Crop Sci.* **2011**, *197*, 258–271. [CrossRef]

35. Tayebi-Meigooni, A.; Awang, Y.; Biggs, A.R.; Mohamad, R.; Madani, B.; Ghasemzadeh, A. Mitigation of Salt-Induced Oxidative Damage in Chinese Kale (*Brassica alboglabra* L.) Using Ascorbic Acid. *Acta Agric. Scand. Sect. B-Soil Plant Sci.* **2014**, *64*, 13–23.

36. Noman, A.; Ali, Q.; Maqsood, J.; Iqbal, N.; Javed, M.T.; Rasool, N.; Naseem, J. Deciphering Physio-Biochemical, Yield, and Nutritional Quality Attributes of Water-Stressed Radish (*Raphanus sativus* L.) Plants Grown From Zn-Lys Primed Seeds. *Chemosphere* **2018**, *195*, 175–189. [CrossRef] [PubMed]

37. Khan, N.; Bano, A.; Zandi, P. Effects of exogenously applied plant growth regulators in combination with PGPR on the physiology and root growth of chickpea (Cicer arietinum) and their role in drought tolerance. *J. Plant Interact.* **2018**, *1*, 239–247. [CrossRef]

38. Khan, N.; Zandi, P.; Ali, S.; Mehmood, A.; Adnan Shahid, M.; Yang, J. Impact of salicylic acid and PGPR on the drought tolerance and phytoremediation potential of Helianthus annus. *Front. Microbiol.* **2018**, *9*, 2507. [CrossRef]

39. Khan, N.; Bano, A.; Ali, S.; Babar, M.A. Crosstalk amongst phytohormones from planta and PGPR under biotic and abiotic stresses. *Plant Growth Regul.* **2020**, *90*, 189–203. [CrossRef]

40. Jamil, S.; Ali, Q.; Iqbal, M.; Javed, M.T.; Iftikhar, W.; Shahzad, F.; Perveen, R. Modulations in Plant Water Relations and Tissue-Specific Osmoregulation by Foliar-Applied Ascorbic Acid and the Induction of Salt Tolerance in Maize Plants. *Braz. J. Bot.* **2015**, *38*, 527–538. [CrossRef]

41. Hasegawa, P.M.; Bressan, R.A.; Zhu, J.K.; Bohnert, H.J. Plant Cellular and Molecular Responses to High Salinity. *Annu. Rev. Plant Biol.* **2000**, *51*, 463–499. [CrossRef]

42. Fritsche, S.; Wang, X.; Jung, C. Recent Advances in Our Understanding of Tocopherol Biosynthesis in Plants: An Overview of Key Genes, Functions, and Breeding of Vitamin E Improved Crops. *Antioxidants* **2017**, *6*, 99. [CrossRef]

43. Sakr, M.T.; El-Metwally, M.A. Alleviation of the Harmful Effects of Soil Salt Stress on Growth, Yield and Endogenous Antioxidant Content of Wheat Plant by Application of Antioxidants. *Pak. J. Biol. Sci.* **2009**, *12*, 624–630. [CrossRef]

44. Sadak, S.M.; Dawood, S. Role of Ascorbic Acid and α Tocopherol in Alleviating Salinity Stress on Flax Plant (*Linum usitatissimum* L.). *J. Stress Physiol. Biochem.* **2014**, *10*, 93–111.

45. Gill, S.S.; Tuteja, N. Reactive Oxygen Species and Antioxidant Machinery in Abiotic Stress Tolerance in Crop Plants. *Plant Physiol. Biochem.* **2010**, *48*, 909–930. [CrossRef]

46. Saleem, A.; Saleem, U.; Subhani, G.M. Correlation and Path Coefficient Analysis in Maize (*Zea mays* L.). *J. Agric. Sci.* **2007**, *45*, 177–183.

47. Aslam, M.; Zamir, M.S.I.; Anjum, S.A.; Khan, I.; Tanveer, M. An Investigation into Morphological and Physiological Approaches to Screen Maize (*Zea mays* L.) Hybrids for Drought Tolerance. *Cereal Res. Commun.* **2015**, *43*, 41–51. [CrossRef]

48. Anjum, S.A.; Ashraf, U.; Tanveer, M.; Khan, I.; Hussain, S.; Shahzad, B.; Zohaib, A.; Abbas, F.; Saleem, M.F.; Ali, I.; et al. Drought Induced Changes in Growth, Osmolyte Accumulation and Antioxidant Metabolism of Three Maize Hybrids. Front. *Plant Sci.* **2017**, *8*, 69. [CrossRef]

49. Arnon, D.I. Copper Enzyme in Isolated Chloroplast Polyphenol Oxidase in *Beta vulgaris*. *Plant Physiol.* **1949**, *24*, 1–15.

50. Kirk, J.T.O.; Allen, R.L. Dependence of Chloroplast Pigment Synthesis on Protein Synthesis: Effect of Actidione. *Biochem. Biophys. Res. Commun.* **1965**, *21*, 523–530. [CrossRef]

51. Yang, G.; Rhodes, D.; Joly, R.J. Effects of High Temperature on Membrane Stability and Chlorophyll Fluorescence in Glycinebetaine-Deficient and Glycinebetaine-Containing Maize Lines. *Physiol. Mol. Biol. Plants* **1996**, *23*, 437–443. [CrossRef]

52. Cakmak, I.; Horst, W.J. Effect of Aluminium on Lipid Peroxidation, Superoxide Dismutase, Catalase, and Peroxidase Activities in Root Tips of Soybean (*Glycine max*). *Physiol. Plant.* **1991**, *83*, 463–468. [CrossRef]

53. Bradford, M.M.A. Rapid and Sensitive Method for the Quantitation of Microgram Quantities of Protein Utilizing the Principle of Protein-Dye Binding. *Anal. Biochem.* **1976**, *72*, 248–254. [CrossRef]

54. Giannopolitis, C.N.; Ries, S.K. Superoxide Occurrence in Higher Plants. *Plant Physiol.* **1977**, *59*, 309–314. [CrossRef] [PubMed]

55. Chance, B.; Maehly, A.C. Assay of Catalase and Peroxidase. *Methods Enzymol.* **1955**, *2*, 764–775.

56. Mukherjee, S.P.; Choudhuri, M.A. Implications of Water Stress-Induced Changes in the Levels of Endogenous Ascorbic Acid and Hydrogen Peroxide in Vigna Seedlings. *Physiol. Plant.* **1983**, *58*, 166–170. [CrossRef]

57. Karadeniz, F.; Burdurlu, H.S.; Koca, N.; Soyer, Y. Antioxidant Activity of Selected Fruits and Vegetables Grown in Turkey. *Turk. J. Agric. For.* **2005**, *29*, 297–303.

58. Backer, H.; Frank, O.; De Angelis, B.; Feingold, S. Plasma Tocopherol in Man at Various Times After Ingesting Free or Acetylated Tocopherol. *Nutr. Rep. Int.* **1980**, *21*, 531–536.

59. Bremner, J.M.; Keeney, D.R. Steam Distillation Methods for Determination of Ammonium, Nitrate and Nitrite. *Anal. Chim. Acta* **1965**, *32*, 485–495. [CrossRef]

60. Kumar, S.; Singh, R.; Nayyar, H. α-Tocopherol Application Modulates the Response of Wheat (*Triticum aestivum* L.) Seedlings to Elevated Temperatures by Mitigation of Stress Injury and Enhancement of Antioxidants. *J. Plant Growth Regul.* **2012**, *32*, 307–314. [CrossRef]

61. Ali, Q.; Ali, S.; Iqbal, N.; Javed, M.T.; Rizwan, M.; Khaliq, R.; Wijaya, L. Alpha-Tocopherol Fertigation Confers Growth Physio-Biochemical and Qualitative Yield Enhancement in Field Grown Water Deficit Wheat (*Triticum aestivum* L.). *Sci. Rep.* **2019**, *9*, 1–15. [CrossRef]

62. Taize, L.; Zeiger, E.; Moller, I.M.; Murphy, A. *Plant Physiology and Development*, 6th ed.; Sinauer Associates, Inc.: Sunderland, MA, USA, 2015.

63. Foryer, C.; Noctor, G. Oxygen Processing in Photosynthesis: Regulation and Signaling. *New Phytol.* **2000**, *146*, 359–388. [CrossRef]

64. Munné-Bosch, S.; Alegre, L. The Function of Tocopherols and Tocotrienols in Plants. *Crit. Rev. Plant Sci.* **2002**, *21*, 31–57. [CrossRef]

65. Trebst, A.; Depka, B.; Holländer-Czytko, H. A Specific Role for Tocopherol and of Chemical Singlet Oxygen Quenchers in the Maintenance of Photosystem II Structure and Function in *Chlamydomonas reinhardtii*. *FEBS Lett.* **2002**, *516*, 156–160. [CrossRef]

66. El-Quesni, F.E.M.; Abd El-Aziz, N.G.; Kandil, M.M. Some Studies on the Effect of Ascorbic Acid and a-Tocopherol on the Growth and Some Chemical Composition of *Hibiscus rosa sinensis* L. at Nubaria. *Ozean J. Appl. Sci.* **2009**, *2*, 159–167.

67. Sadak, M.S.; Rady, M.M.; Badr, N.M.; Gaballah, M.S. Increasing Sun Flower Salt Tolerance Using Nicotinamide and a-Tocopherol. *Int. J. Acad. Res.* **2010**, *2*, 263–270.

68. Javed, M.T.; Greger, M. Cadmium Triggers *Elodea canadensis* to Change the Surrounding Water pH and Thereby Cd uptake. *Int. J. Phytoremediation* **2010**, *13*, 95–106. [CrossRef] [PubMed]

69. Orabi, S.A.; Abdelhamid, M.T. Protective Role of α-Tocopherol on Two *Vicia faba* Cultivars Against Seawater-Induced Lipid Peroxidation by Enhancing Capacity of Anti-Oxidative System. *J. Saudi Soc. Agric. Sci.* **2016**, *15*, 145–154. [CrossRef]

70. Shigeoka, S.; Ishikawa, T.; Tamoi, M.; Miyagawa, Y.; Takeda, T.; Yabuta, Y.; Yoshimura, K. Regulation and Function of Ascorbate Peroxidase Isoenzymes. *J. Exp. Bot.* **2002**, *53*, 1305–1319. [CrossRef]

71. Ali, Q.; Haider, M.Z.; Iftikhar, W.; Jamil, S.; Javed, M.T.; Noman, A.; Perveen, R. Drought Tolerance Potential of *Vigna mungo* L. Lines as Deciphered by Modulated Growth, Antioxidant Defense, and Nutrient Acquisition Patterns. *Braz. J. Bot.* **2016**, *39*, 801–812. [CrossRef]

72. Maeda, H.; Sakuragi, Y.; Bryant, D.A.; DellaPenna, D. Tocopherols Protect *Synechocystis* sp. Strain PCC 6803 From Lipid Peroxidation. *Plant Physiol.* **2005**, *138*, 1422–1435. [CrossRef]

73. Sattler, S.E.; Cahoon, E.B.; Coughlan, S.J.; DellaPenna, D. Characterization of Tocopherol Cyclases From Higher Plants and Cyanobacteria. Evolutionary Implications for Tocopherol Synthesis and Function. *Plant Physiol.* **2003**, *132*, 2184–2195. [CrossRef]

74. Wang, X.; Quinn, P.J. Preferential Interaction of α-Tocopherol With Phosphatidylcholines in Mixed Aqueous Dispersions of Phosphatidylcholine and Phosphatidylethanolamine. *Eur. J. Biochem.* **2000**, *267*, 6362–6368. [CrossRef]

75. Maeda, H.; DellaPenna, D. Tocopherol Functions in Photosynthetic Organisms. *Curr. Opin. Plant Biol.* **2007**, *10*, 260–265. [CrossRef]

76. Fukuzawa, K.; Matsuura, K.; Tokumura, A.; Suzuki, A.; Terao, J. Kinetics and Dynamics of Singlet Oxygen Scavenging by α-Tocopherol in Phospholipid Model Membranes. *Free Radic. Biol. Med.* **1997**, *22*, 923–930. [CrossRef]

77. Kobayashi, N.; DellaPenna, D. Tocopherol Metabolism, Oxidation and Recycling Under High Light Stress in Arabidopsis. *Plant J.* **2008**, *55*, 607–618. [CrossRef]

78. Fahrenholtz, S.R.; Doleiden, F.H.; Trozzolo, A.M.; Lamola, A.A. On the Quenching of Singlet Oxygen by α-Tocopherol. *J. Photochem. Photobiol. Biol.* **1974**, *20*, 505–509. [CrossRef]

79. Semida, W.M.; Taha, R.S.; Abdelhamid, M.T.; Rady, M.M. Foliar-Applied α-Tocopherol Enhances Salt-Tolerance in *Vicia faba* L. Plants Grown Under Saline Conditions. *S. Afr. J. Bot.* **2014**, *95*, 24–31. [CrossRef]

80. Khattab, H. Role of Glutathione and Polyadenylic Acid on the Oxidative Defense Systems of Two Different Cultivars of Canola Seedlings Grown Under Saline Conditions. *Aust. J. Basic Appl. Sci.* **2007**, *1*, 323–334.

81. Pourcel, L.; Routaboul, J.M.; Cheynier, V.; Lepiniec, L.; Debeaujon, I. Flavonoid Oxidation in Plants: From Biochemical Properties to Physiological Functions. *Trends Plant Sci.* **2007**, *12*, 29–36. [CrossRef]

82. Shao, H.B.; Chu, L.Y.; Shao, M.A.; Jaleel, C.A.; Hong-mei, M. Higher Plant Antioxidants and Redox Signaling Under Environmental Stresses. *C. R. Biol.* **2008**, *331*, 433–441. [CrossRef]

83. Hu, Y.; Schmidhalter, U. Drought and Salinity: A Comparison of Their Effects on Mineral Nutrition of Plants. *J. Plant Nutr. Soil Sci.* **2005**, *168*, 541–549. [CrossRef]

Rhizobacteria Isolated from Saline Soil Induce Systemic Tolerance in Wheat (*Triticum aestivum* L.) against Salinity Stress

Noshin Ilyas [1,*] , **Roomina Mazhar** [1] , **Humaira Yasmin** [2] , **Wajiha Khan** [3] , **Sumera Iqbal** [4] , **Hesham El Enshasy** [5,6,7,*] and **Daniel Joe Dailin** [5,6]

[1]   Department of Botany, PMAS-Arid Agriculture University, Rawalpindi 46300, Pakistan; roominamazhar83@gmail.com
[2]   Department of Bio-Sciences, COMSATS University, Islamabad 45550, Pakistan; humaira.yasmin@comsat.edu.pk
[3]   Department of Biotechnology, COMSATS University Islamabad, Abbottabad Campus, Abbottabad 22010, Pakistan; wajihak@cuiatd.edu.pk
[4]   Department of Botany, Lahore College for Women University, Lahore 54000, Pakistan; sumeraiqbal2@yahoo.com
[5]   Institute of Bioproduct Development (IBD), Universiti Teknologi Malaysia (UTM), Skudai, Johor 81310, Malaysia; jddaniel@utm.my
[6]   School of Chemical and Energy Engineering, Faculty of Engineering, Universiti Teknologi Malaysia (UTM), Skudai, Johor 81310, Malaysia
[7]   City of Scientific Research and Technology Applications (SRTA), New Burg Al Arab, Alexandria 21934, Egypt
*   Correspondence: noshinilyas@yahoo.com (N.I.); henshasy@ibd.utm.my (H.E.E.)

Abstract: Halo-tolerant plant growth-promoting rhizobacteria (PGPR) have the inherent potential to cope up with salinity. Thus, they can be used as an effective strategy in enhancing the productivity of saline agro-systems. In this study, a total of 50 isolates were screened from the rhizospheric soil of plants growing in the salt range of Pakistan. Out of these, four isolates were selected based on their salinity tolerance and plant growth promotion characters. These isolates (SR$_1$. SR$_2$, SR$_3$, and SR$_4$) were identified as *Bacillus* sp. (KF719179), *Azospirillum brasilense* (KJ194586), *Azospirillum lipoferum* (KJ434039), and *Pseudomonas stutzeri* (KJ685889) by 16S rDNA gene sequence analysis. In vitro, these strains, in alone and in a consortium, showed better production of compatible solute and phytohormones, including indole acetic acid (IAA), gibberellic acid (GA), cytokinin (CK), and abscisic acid (ABA), in culture conditions under salt stress. When tested for inoculation, the consortium of all four strains showed the best results in terms of improved plant biomass and relative water content. Consortium-inoculated wheat plants showed tolerance by reduced electrolyte leakage and increased production of chlorophyll a, b, and total chlorophyll, and osmolytes, including soluble sugar, proline, amino acids, and antioxidant enzymes (superoxide dismutase, catalase, peroxidase), upon exposure to salinity stress (150 mM NaCl). In conclusion, plant growth-promoting bacteria, isolated from salt-affected regions, have strong potential to mitigate the deleterious effects of salt stress in wheat crop, when inoculated. Therefore, this consortium can be used as potent inoculants for wheat crop under prevailing stress conditions.

Keywords: salinity; PGPR; wheat; compatible solutes; antioxidant enzymes

1. Introduction

Globally, the production rate of agriculture is far less than the estimated food requirement of the ever-increasing population and the gap will be widened over time [1] (GAP Report, 2018).

Agro-ecosystems are influenced by environmental and climatic conditions, farming techniques, and management practices. It is estimated that internationally, salinity affects 22% of the total cultivated and 33% of the total irrigated agricultural area, which is increasing at an alarming rate of 10% annually. Pakistan is also facing severe salinity issues and a total area of 6.30 million hectares is salt affected, out of which 1.89 million hectares is marked as saline [2].

Due to a higher concentration of sodium chloride (NaCl), plants growing in salt-affected soils suffer from both hyperosmotic and hyperionic effects. These stresses result in reduced water uptake; altered ion and mineral absorption rates; increased production of reactive oxygen species, causing disorganization of the cell membrane; and reduction of metabolic activities [3]. Halophytes adapt themselves to saline conditions by adjusting their physiological activities, maintaining their water balance by osmotic adjustments, producing compatible solutes, and modifying the antioxidant system [4]. Some plants overcome salinity stress through the production of osmolytes, particularly glycine betaine, proline, soluble sugars, and proteins [5].

Improvement in the crop yield of saline soils requires a multidimensional approach consisting of salt-tolerant varieties or amelioration by chemical neutralizers, but there is a dire need for eco-friendly sustainable approaches. Rhizobacteria, showing potential to improve plant growth, are termed as plant growth-promoting rhizobacteria (PGPR) [6]. PGPR have the potential to improve plant growth through various mechanisms, including better plant growth, the production of phytohormones, and amelioration of stresses [7]. Due to the natural coping mechanisms of PGPR, their inoculation can help the amelioration of various abiotic stresses in plants. PGPR inoculation can help to improve the growth and yield of crops, particularly in regions prone to drought and salt stress [8,9]. Natural halotolerant PGPR strains have better potential for the amelioration of salt stress in regional crops for sustainable yields. These native PGPR strains are well acclimated to indigenous conditions and the plant–microbe interactions can help the plants to tolerate stress [10].

In this study, native halotolerant PGPR strains were isolated from local saline soils, and their ability to promote plant growth when inoculated under salt stress was investigated. The objective of the present research was to focus on the evaluation of isolated bacterial strains to stimulate salinity tolerance and the promotion of wheat growth, as well as the identification and characterization of the candidate strain both bio-physiochemically and genetically. This study provides a basis to identify and characterize PGPR from natural saline conditions and testing their potential for improving salinity tolerance in wheat, the major staple crop across the world.

2. Materials and Methods

2.1. Soil Sampling and Physicochemical Analysis

The rhizospheric soil of four halophytes namely, *Abutilon bidentatum*, *Maytenus royleanus*, *Leptochloa fusca* (Kallar grass), and *Dedonia viscose*, was collected from a salt range of Pakistan (313–360 m.a.s.l; 32°23–33°00 north latitude and 71°30–73°30 east longitude). The rhizospheric soil was sieved and stored at 4 °C for future analysis. Rhizospheric soil was analyzed for pH and electrical conductivity (EC) [11], soil texture, macro and micronutrients [12], and available nutrients [13].

2.2. Strain Isolation and HaloTolerance Assay

Rhizobacteria were isolated from rhizospheric soil of *Abutilon bidentatum*, *Maytenus royleanus*, *Leptochloa fusca*, and *Dedonia viscose* by using the serial dilution and spread plate techniques [14]. The soil suspension was made by adding 1 g of soil in 9 mL of Milli-Q distilled water. An aliquot of soil suspension was inoculated on Luria-Bertani (LB) agar plates and incubated at 28 ± 2 °C for 48 h. The obtained colonies were purified by sub-culturing. The colony-forming unit (CFU) was calculated according to the formula given by [15]:

$$CFU/g = (\text{colonies number} \times \text{dilution factor/volume of inoculum}).$$

Distinct bacterial colonies were examined for colony characteristics (shape, size, margin, elevation, appearance, texture, pigmentation, and optical properties) as well as for cellular characteristics (cell shape, gram testing) [16]. QTS-24 kits were used to determine the carbon/nitrogen (C/N) source utilization pattern of bacterial isolates. Isolated bacterial strains were tested for their halotolerance abilities by growing them in LB media supplemented with NaCl (2%, 4%, 6%, 8%, 10%, 15%) [16].

2.3. Plant Growth-Promoting (PGP) Traits

All the bacterial isolates were evaluated for their PGP characteristics. Phosphorous (P) solubilization was done by spot inoculating overnight grown cultures onto pikovaskaya's agar (Sigma) containing tri-calcium phosphate as an insoluble P source [17]. The colonies, which produced clearing zones in the pikovaskaya's agar plates, were considered positive for phosphorous solubilization. Total solubilized phosphate was measured by using the phosphomolybdate blue color method [18]. Modified pikovaskaya's broth medium was inoculated with each strain and incubated at 30 °C for 5 days. The cultures were centrifuged at 6000 rpm for 15 min. The supernatant (500 µL) was mixed with 40 µL of 2,4-dinitrophenol, after which 20 µL of dilute sulfuric acid were added, followed by 5 mL of chromogenic reagent, and the volume was diluted to 50 mL using sterilized water and absorbance was recorded at 680 nm. Siderophore production was done by spot inoculation on chrome azurol S (CAS) media as described by Schwyn and Neilands [19]. Bacterial strains were spot inoculated on petri plates containing CAS media. An uninoculated plate was taken as the control. After inoculation, plates were incubated at 28 °C for 5–7 days and observed for the formation of an orange zone around the bacterial colonies. Bacterial isolates were tested for hydrogen cyanide production through the method of Lorck [20]. Bacterial strains were streaked on nutrient agar medium (pre-soaked in 0.5% picric acid and 2% sodium carbonate w/v), supplemented with glycine (4.4 g/L). Plates were sealed with parafilm paper and incubated at 30 °C for 4 days. The appearance of an orange or red color indicates the production of hydrogen cyanide.

2.4. Germination Experiment

This experiment was carried in the Plant Physiology Laboratory of PMAS-Arid Agriculture University. Seeds of the wheat variety (Galaxy 2013) obtained from the National Agricultural Research Centre, Islamabad were surface sterilized by treatment with sodium hypochlorite (1%) solution for 5 min. After, seeds were successively washed with distilled water. All the isolated strains were tested for germination attributes. Sterilized seeds of wheat were placed in pre-soaked filter paper in Petri dishes. NaCl solution (50 mM, 100 mM, 150 mM, 200 mM) was given instead of normal water. The germination experiment was carried out under laboratory conditions with an average photoperiod of 10 h day/14 h night at 24 °C. The germination percentage, seedling vigor index, and promptness index were measured for each treatment [21]. Four strains were selected for further analysis, based upon their efficacy in the germination experiment and were labeled as SR_1, SR_2, SR_3, and SR_4.

2.5. Production of Osmolytes

To analyze proline and total soluble sugars, the supernatant of PGPR grown in LB broth supplemented with NaCl concentrations (0%, 2%, 4%, 6%, 8%, and 10%) were analyzed as described by Upadhyay et al. [22]. For the estimation of the proline contents, centrifugation of the culture broth was done at 1000× g for 10 min and the supernatant was used for estimation. Total soluble sugar (TSS) was estimated by mixing 1 mL of supernatant with 4 mL of anthrone reagent, the mixture was later boiled in a water bath for 8 min. After rapid cooling, the optical density was measured at 630 nm, and the amount of TSS was calculated from a standard curve.

2.6. Phytohormone Production

The ability of four selected halotolerant strains to produce phytohormones (IAA, GA, CK, ABA) in the culture media was measured by the method of Tien et al. [23]. The extraction of hormones was done by centrifugation of bacterial cultures at 10,000 rpm for 15 min. For adjustment of the pH (2.8), 1 N HCl was used. In the next step, an equal volume of ethyl acetate was used for hormone extraction. The resulting solution was evaporated at 35 °C and the end residue was mixed in 1500 μL of methanol. Finally, the samples were run on High Performance Liquid Chromatography (HPLC) (Agilent 1100), which had a C18 column (39 × 300 mm) and a UV detector. For standardization of HPLC, pure grade chemicals of the hormones IAA, CK, GA, and ABA (Sigma Chemical Co., St. Louis, MO, USA) were dissolved in HPLC-grade methanol and were used. The wavelength used for the detection was as follows: IAA at 280 nm; and GA, CK, and ABA at 254 nm. The phytohormone content of LB media, without inoculum, was used to normalize the data.

2.7. 16S rRNA Gene Sequence and Phylogenetic Analysis

DNA was extracted from pure LB broth cultures as described by Chen and Kuo [24]. Amplification of genomic DNA of isolated strains was done as described by Weisburg et al. [25]. The PCR was carried out for amplification of the 16S rRNA gene with universal nucleotide sequence forward primer (fd1) AGAGTTTGATCCTGGCTCAG, and reverse primer (rd1) (AAGGAGGTGATCCAGCC). DNA was purified and sequenced on an automated sequencer by gel purification kits (JET quick, Gel Extraction Spin Kit, GENOMED). The strains were identified by using a nearly complete sequence of the 16s rRNA gene on (BLAST) NCBI by comparing sequence homology with other strains. The maximum parsimony method was used for the analysis of evolutionary linkages [26].

2.8. Plant Inoculation

A pot experiment was conducted in the greenhouse of the Botany Department, PMAS-AAUR, Rawalpindi. A complete randomized design was applied with three replications. Each selected halotolerant strain was grown overnight in LB media. To obtain a cell pellet, the supernatant was discarded after centrifugation at 3000 rpm for 3 min. The cell pellet was washed three times with autoclaved water and the absorbance was recorded with a spectrophotometer at 600 nm to obtain the desired concentration, i.e., 10^7 CFU. Ten sterilized seeds were sown in each pot (containing 10 kg of soil) in the greenhouse with the day 10 h/14 h night at a temperature of 21/15 °C. Soil moisture was maintained at 15 ± 1%. Four strains and their consortium were evaluated under two treatment controls and 150 mM NaCl stress. The salt level was maintained with EC of 4.0 dS m^{-1} (first irrigation) or 8.5 dS m^{-1} (second irrigation). Plants were harvested after 45 days of sowing. Fresh and dry biomass was recorded. Leaf area was measured with the help of a leaf area meter. All the samples were collected in zipper bags and stored at −20 °C freezer for further biochemical assays. The percent of water content was determined by measuring the ratio between the fresh and dry weight of the upper fully developed leaf by using the following formula [27]:

$$RWC = [FW − DW]/[TW − DW] × 100 \qquad (1)$$

2.9. Electrolyte Leakage (%)

Electrolyte leakage was determined by the method of Srairam [28]. Leaf discs weighing 0.1 g were heated in 10 mL of distilled water for 30 min at 40 °C and the electrical conductivity (C1) was recorded. The same discs were then heated at 100 °C and again electrical conductivity (C2) was recorded. Whereas, calculations were done by the following formula:

$$MSI = [1 − (C1/C2)] × 100 \qquad (2)$$

2.10. Chlorophyll and Carotenoid Content

Leaf chlorophyll a, b, total chlorophyll, and carotenoid contents were estimated by the method of Arnon [29]. Fresh leaves (0.5 g) were ground in 10 mL of 80% acetone. The readings of the filtrate were measured at 470 nm, 663 nm, and 645 nm. Calculations were done by the following equations:

$$Chla\ (mg/g) = [12.7A_{663} - 2.69A_{645}]\ (v/w) \tag{3}$$

$$Chlb\ (mg/g) = [22.9A_{645} - 4.68A_{663}]\ (v/w) \tag{4}$$

$$Total\ chlorophyll\ (mg/g) = [(20.2A_{645} + 8.02A_{663})\ v/w] \tag{5}$$

$$Carotenoids\ content(mg/g) = (1000\ A_{470} - 1.8\ Chl_a - 85.02\ Chl_b)/198 \tag{6}$$

where A is the optical density at a specific wavelength.

2.11. Proline Content

Proline contents were determined by following the protocols of Bates [30]. Fresh leaves (0.5 g) were homogenized with 10 mL of sulfosalicylic acid (3.0%). The solution was filtered, and the filtrate was mixed with equal amounts of glacial acetic acid and ninhydrin reagent. The mixture was heated for 1 h in a water bath at 90 °C and the reaction was stopped by transferring the mixture to ice. Toluene (1 mL) was added to the mixture and the solution was mixed and the solution separated into two layers. The upper layer was isolated in separate test tubes and the reading was measured at 520 nm. Proline was determined as follows:

$$Proline = (Reading\ of\ sample \times Diluted\ concentration \times K\ value)/material\ weight \tag{7}$$

2.12. Total Soluble Sugar and Amino Acid

Soluble sugars were estimated after the method of Dubois et al. [31]. Ground plant tissue (0.1 g) was mixed with 3 mL of 80% methanol. The solution was heated in a water bath for 30 min at 70 °C. An equal volume of extract (0.5 mL) and 5% phenol was mixed with concentrated sulphuric acid (1.5 mL) and was again incubated in the dark for 30 min. The absorbance of the sample was checked at 490 nm and the calculations were done by applying the following formula:

$$Sugar\ (\mu g/mL) = Absorbance\ of\ sample \times Dilution\ factor \times K\ value \tag{8}$$

Fresh tissue in grams.

The standard curve was prepared for glucose solution, which was used for the determination of the amount of sugar, expressed in $mg\ g^{-1}\ fw^{-1}$.

The Ninhydrin method was used for the determination of free amino acids [32]. Leaf extract (1 mL) was mixed with the same volume of 0.2 M citrate buffer (pH-5) and 80% ethanol, and 2 mL of the ninhydrin reagent. The absorbance of the reaction mixture was taken to 570 nm. Amino acids were computed with the equation:

$$Amino\ acids = Absorption \times volume \times Diluted\ concentration/Sample\ weight \times 1000.$$

The amino acid, leucine, was used for preparing the standard curve, and results were expressed in mg of amino acid per g of dry tissue.

2.13. Total Protein Content

The concentration of protein was quantified by the Bradford assay [33]. Bovine serum albumin was used as a standard. Proteins were extracted by dissolving 0.2 g of leaf samples in 4 mL of sodium phosphate buffer (pH 7), and 0.5 mL of the extract was mixed with 3 mL of Comassive bio red dye. The optical density of the solution was measured at 595 nm. Protein was determined by:

$$\text{Protein} = \text{Reading of extract} \times \text{Diluted concentration} \times \text{value of K/sample weight} \qquad (9)$$

2.14. Antioxidant Enzyme Assay

Enzyme extract was prepared by grinding one gram of leaf in liquid nitrogen. The obtained powder was added in 10 mL of 50 mM phosphate buffer (pH 7.0) and was mixed with 1 mM Ethylene Diamine Tetra Acetic acid (EDTA) and 1% polyvinylpyrrolidone (PVP). The whole mixture was centrifuged at $13,000 \times g$ for 20 min at 4 °C. The supernatant was used for the enzyme assay.

The catalase (CAT) content was estimated by observing the degradation of H_2O_2 at 240 nm [34]. Catalase activity (U mg protein^{-1}) was calculated from the molar absorption coefficient of $40 \text{ mm}^{-1}\text{cm}^{-1}$ for H_2O_2. Peroxidase dismutase (POD) was determined by following the procedure of Rao [35]. The reaction mixture consisted of 10 μL of crude enzyme extract, 20 μL of 100 mM guaiacol, 10 μL of 100 mM H_2O_2, and 160 μL of 50 mM sodium acetate (pH 5.0). Absorbance was recorded at 450 nm.

Superoxide dismutase (SOD) activity was done by using the procedure of Giannopolitis and Ries [36]. The composition of the reaction mixture was 50 mM sodium phosphate buffer (pH 7.8), 0.1 M tris-HCL, 14 mM methionine, 1.05 mM riboflavin, 0.03% TritonX-100, 50 mM nitroblue tetrazolium chloride (NBT), 100 mM EDTA, and 20 μL enzyme extracts. After adding riboflavin, the glass tubes were illuminated for 5 min, and reactions were stopped by turning off lights. The absorbance was recorded at 560 nm.

2.15. Statistical Analysis

Three replicates were used for the mean and standard deviation values of the data. The obtained data were further analyzed by Duncan's multiple range tests using MSTAT-C version 1.4.2. The correlation coefficient of the data was done using the software Statistix version 8.1. Mean values were compared by the least significant difference (LSD) at $p \leq 0.05$ [37]. The heatmap for the correlation coefficient was prepared by using web tool clustvis (https://biit.cs.ut.ee/clustvis/).

3. Results

3.1. Soil Analysis

Analysis of the rhizospheric soil samples of all four plants showed the soil was sandy clay loam with an EC range of 0.76–0.85 dSm^{-1}, pH in the range of 7.99–8.12, high Na/K ratio, and a low concentration of nutrients (Table 1).

Table 1. Physiochemical properties of the rhizosphere soil and rhizobacterial population.

Host Plant Species	pH	EC (dSm⁻¹)	Soil Texture	SAR (mmol/L)	OC (%)	Macronutrient (meq/L)						Available Nutrients (kg/ha)			PGPR Population (cfu × 10⁵ g⁻¹ of Soil)
						CO_3	HCO_3	Cl	Ca^+	Na^+	K^+	N	P	K	
Abutilom bidiantum	7.99	0.82	Sandy clay loam	39.6	0.72	3.9	15	50	3	82.95	0.3	240	200	320	69
Maytenus royleanus	8.12	0.85	Sandy clay loam	40.3	0.54	4	14	48	2.5	80.6	0.4	238	196	325	64
Kallar grass	7.80	0.76	Sandy clay loam	42.5	0.62	4.2	17	47	2.8	78.9	0.5	243	190	315	62
Dedonia viscoca	8.01	0.79	Sandy clay loam	38.1	0.64	4.5	17.5	48	3.1	81.2	0.4	237	205	330	65

3.2. Isolation and Screening of Salt-Tolerant PGPR Strains

A total of 50 isolates were obtained from the rhizospheric soil of four halophytic plants. Among all isolates, 90% of colonies were round, creamy, and had entire margins (Supplementary Materials Table S1). Further, 78% of isolates were Gram-negative and rod-shaped (Supplementary Materials Table S2).

In the halotolerant assay, 70% of strains were able to grow up to 6%, 20% strains showed tolerance at 10%, while four strains SR_1, SR_2, SR_3, and SR_4 were able to grow at 15% NaCl (Supplementary Materials Table S2). These four strains also showed positive results for phosphorous solubilization, hydrogen cyanide, and siderophore production (Supplementary Materials Table S3).

3.3. Effect of Bacterial Isolates on Germination of Wheat

Salt stress resulted in a considerable reduction in the germination parameters of the wheat seeds. Under salt-stressed conditions, the seedling vigor index and germination index showed a 12.5% and 31% decrease compared to the control. Though most of the strains showed a significant increase in seed germination, four strains SR_1, SR_2, SR_3, and SR_4 showed prominent results (14.28%, 35%, 42%, and 55%), respectively, as compared to the non-inoculated control under the salt stress condition (Supplementary Materials Table S4).

3.4. Identification of Isolates

Initially, the four strains were identified based on the C/N source utilization pattern (Supplementary Materials Table S5). Molecular identification of the screened halotolerant strains was done based on 16S rRNA sequences and on the comparison of the 1500-bp sequence of 16S rRNA gene subjected to BLAST to confirm the relatedness with other bacterial strains. The isolate SR_1 (1485 base pair) was closely related (98% nucleotide identity) to sequences of bacteria annotated as *Bacillus* strain JQ 926435 in the GenBank database. The sequence of SR_2 (1480 base pairs) was 99% identical to *Azospirillum brasilense* DQ 288686.1, SR_3 (1482 base pairs), and 96% identical to strain *Azospirillum lipoferum* accession no. M. 5906.1. Furthermore, the isolated strain SR_4 showed a 99% homology with *Pseudomonas stutzeri* JQ 926435. The accession numbers of the identified strains were obtained from NCBI and are given in Table 2.

Table 2. Molecular identification of the isolates based on partial 16S rDNA analysis.

No	Isolates	Base Pair Length	Similarity (%)	Strain Identification	Accession No.
1	SR_1	1485	98%	*Bacillus* sp.	KF719179
2	SR_2	1480	99%	*Azospirillum brasilense*	KJ194586
3	SR_3	1482	96%	*Azospirillum lipoferum*	KJ434039
4	SR_4	1263	99%	*Pseudomonas stutzeri*	KJ685889

Further phylogenetic analysis of the identified bacteria was conducted in MEGA4 software to determine their affiliation [38]. The evolutionary history was inferred using the maximum parsimony method [26]. The results are shown in Supplementary Materials Figures S1–S4.

3.5. Production of Phytohormones

Based on the halotolerance assays, PGP traits, and germination assay results, four isolates were selected for further analysis. All the halotolerant PGPR strains showed the production of phytohormones in liquid culture (Figure 1). Halotolerant PGPR strains were able to produce IAA (0.5–2.1 μg mL^{-1}), gibberellic acid (1.5–2.5 μg mL^{-1}), CK (0.39–0.64 μg mL^{-1}), and ABA (1.9–3.4 μg mL^{-1}). The PGPR strains SR_2 and SR_3 produced higher concentrations of phytohormones than those of SR_1 and SR_4; however, the bacterial consortium produced maximum concentrations of IAA (2.1 μg mL^{-1}), gibberellic acid (2.5 μg mL^{-1}), CK (0.64 μg mL^{-1}), and ABA (3.4 μg mL^{-1}).

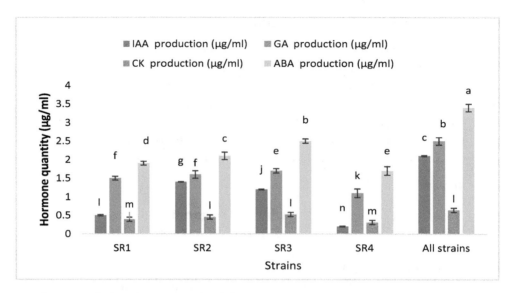

Figure 1. Production of phytohormones (Indole Acetic Acid (IAA), Gibberellic Acid (GA), Cytokinin (CK), and Abscisic Acid (ABA) by PGPR strains and their consortium in culture media. (SR$_1$: Inocualted with *Bacillus* sp; SR$_2$: Inocualted with *Azospirillum brasilense*; SR$_3$: Inocualted with *Azospirillum lipoferum*; SR$_4$: Inocualted with *Pseudomonas stutzeri*; Consortium is a combination of all four strains *Bacillus* sp, *Azospirillum brasilense, Azospirillum lipoferum, Pseudomonas stutzeri*). This data displays the means and standard deviation ($n = 3$). Different letters show significant differences between treatments ($p < 0.05$).

3.6. Production of Compatible Solutes

A considerable amount of proline was produced by all the screened halotolerant strains when subjected to different salinity levels. Production of proline by SR$_2$ and SR$_3$ was the highest in the 10% saline condition than the control. The maximum amount of proline (12.1 µg mg^{-1}) was produced by the bacterial consortium, which was 23% greater than SR$_2$ and SR$_3$. For the carbohydrate contents, a significant amount of soluble sugars was recorded by all the strains (Figure 2). The production of soluble sugars was more pronounced at different salinity levels than the control. The bacterial strains SR$_2$ and SR$_3$ produced a greater amount of (89–111 µg mg^{-1}) soluble sugar as compared to the control, but the consortium of bacterial isolates recorded the maximum values at 10% NaCl (222 µg mg^{-1}) (Figure 3).

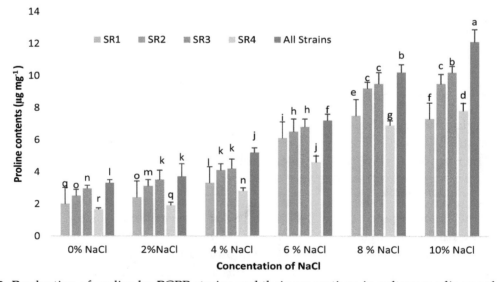

Figure 2. Production of proline by PGPR strains and their consortium in culture media supplemented with different concentrations of NaCl (2%, 4%, 6%, 8%, and 10%). The treatment details are the same as in Table 3. This data displays the means and standard deviation ($n = 3$). Different letters show significant differences between treatments ($p < 0.05$).

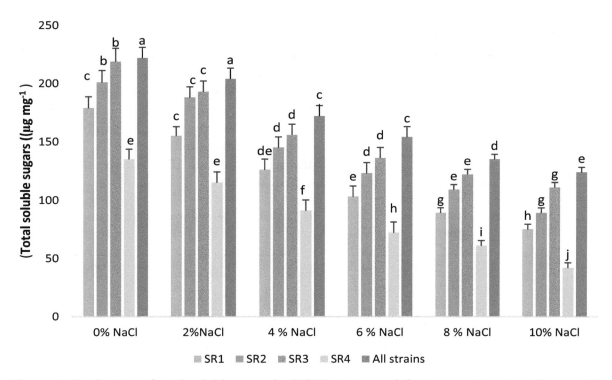

Figure 3. Production of total soluble sugar by PGPR strains and their consortium in culture media supplemented with different concentrations of NaCl (2%, 4%, 6%, 8%, and 10%). The treatment details are the same as in Table 3. This data displays the means and standard deviation ($n = 3$). Different letters show significant differences between treatments ($p < 0.05$).

3.7. Effect of PGPR Inoculation on the Biomass of Wheat (Triticum aestivum L.) Plants Grown under Salinity Stress

The overall decrease of 30% in the plant biomass of wheat plants was observed due to salt stress. However, the bacterial isolates exerted a significant positive influence on wheat growth and resulted in an increase in the biomass of plants in the control and stressed conditions, respectively. The relative increase in the fresh and dry biomass due to bacterial isolates ranged between 39% and 67% as compared to the uninoculated plants under saline conditions.

The best results were obtained when plants were inoculated with a consortium of all four isolated strains, which caused an increase of 93% in stress and 60% in controlled conditions. Moreover, pronounced results were also encountered for dry biomass, when plants were inoculated with a consortium, which resulted in an increase of 65.4% in salt stress and 78.7% in control conditions (Table 3).

Table 3. Effect of inoculation of halotolerant PGPR on the fresh and dry biomass and leaf area of wheat plants grown under salinity stress.

Treatments	Fresh Biomass (g)		Dry Biomass (g)		Leaf Area (cm^2)	
	0 mM	150 mM	0 mM	150 mM	0 mM	150 mM
Control	10 ± 1g	7.3 ± 0.4i	3.3 ± 0.1c	2.2 ± 0.04d	140 ± 12e	120 ± 17f
SR$_1$	11 ± 0.5f	8.2 ± 0.9h	3.5 ± 0.4c	2.8 ± 0.06d	150 ± 15d	130 ± 14f
SR$_2$	13.2 ± 0.9e	10.3 ± 0.7g	4.1 ± 0.1b	3.5 ± 0.09c	167 ± 12c	140 ± 24e
SR$_3$	16.9 ± 1.2 b	13.9 ± 1.4d	4.7 ± 0.5b	3.8 ± 0.03 c	177 ± 17b	147 ± 17e
SR$_4$	11.50.98f	7.8 ± 0.54i	3.6 ± 0.5c	3.0 ± 0.08d	160 ± 14c	135 ± 12f
Consortium	20.3 ± 1.8a	14.1 ± 1.9c	5.9 ± 0.2a	4.3 ± 0.03b	186 ± 19a	152 ± 13d

This data displays the means and standard deviation ($n = 3$). Different letters show significant differences ($p < 0.05$). (SR$_1$: Inocualted with *Bacillus* sp; SR$_2$: Inocualted with *Azospirillum brasilense*; SR$_3$: Inocualted with *Azospirillum lipoferum*; SR$_4$: Inocualted with *Pseudomonas stutzeri*; Consortium is a combination of all four strains *Bacillus* sp, *Azospirillum brasilense, Azospirillum lipoferum, Pseudomonas stutzeri*).

3.8. Effect on the Membrane Stability Index and Water Content

Results of the percent electrolytic leakage showed that the inoculation remains significant under stress as well as normal conditions However, co-inoculation with bacterial consortium successfully decreased (34%) the ionic discharge at the 150 mM NaCl level compared to the control (Table 4). Furthermore, the percent of water content showed a significant reduction of 33% in wheat plants under salt stress as compared to the uninoculated control plants. More pronounced results were obtained with SR_2 and SR_3, causing an increase of 10.5% and 17.54% in the stress condition. The consortium-inoculated plants recorded the maximum amount of water of 21% and 17.64% in the stress and control conditions. A similar trend was observed by SR_1 and SR_4 (Table 4).

Table 4. Effect of inoculation of halotolerant PGPR strains on the leaf water content and electrolyte leakage of wheat plants grown under salinity stress.

Treatments	Percent Water Content		Electrolyte Leakage (%)	
	0 mM	150 mM	0 mM	150 mM
Control	85 ± 1.5b	57 ± 0.9e	33 ± 0.4d	55 ± 0.5a
SR_1	86.3 ± 1.6b	60 ± 1d	30 ± 0.3d	50 ± 0.45a
SR_2	89 ± 1.9b	63 ± 1.4d	26.2 ± 0.25e	41.3 ± 0.33c
SR_3	95 ± 2.1a	67 ± 1.7d	25.7 ± 0.4e	42.08 ± 0.11c
SR_4	88.7 ± 1.9b	61 ± 1.15d	31.2 ± 0.22e	47.8 ± 0.44b
Consortium	97 ± 2.0a	70 ± 1.75c	22.1 ± 0.22f	35.2 ± 0.23d

This data displays the means and standard deviation ($n = 3$). Different letters show significant differences ($p < 0.05$). Treatment details are the same as in Table 3.

3.9. Chlorophyll Contents

Salinity stress negatively affected the photosynthetic pigments of wheat plants. A considerable decrease of 30.4%, 22%, and 25% was observed in chlorophyll a, b, and total chlorophyll. The response to the consortium was effective ($p \leq 0.05$) and resulted in a 13.23%, 12.49%, 12.9%, and 11.76% increase as compared to the control under salt-stress conditions (Table 5).

Table 5. Effect of halotolerant PGPR on the chlorophyll a, chlorophyll b, total chlorophyll, and carotenoid contents of wheat plants grown under salinity stress.

	Chlorophyll a (mg/g Fresh Weight)		Chlorophyll b (mg/g Fresh Weight)		Total Chlorophyll (mg/g Fresh Weight)		Carotenoid (mg/g Fresh Weight)	
Treatments	0 mM	150 mM	0 mM	150 mM	0 mM	150 mM	0 mM	150 mM
Control	1.06 ± 0.01d	0.59 ± 0.01h	0.27 ± 0.02d	0.12 ± 0.01h	1.18 ± 0.10e	0.86 ± 0.05f	46.9 ± 0.1f	65.8 ± 0.15k
SR_1	1.13 ± 0.03b	0.75 ± 0.03g	0.29 ± 0.04b	0.13 ± 0.02g	1.26 ± 0.09d	1.01 ± 0.03k	47.3 ± 0.3e	67.6 ± 0.5i
SR_2	1.18 ± 0.04c	0.81 ± 0.02f	0.32 ± 0.03c	0.15 ± 0.02f	1.33 ± 0.7c	1.13 ± 0.02j	50.5 ± 0.4c	69.8 ± 0.4h
SR_3	1.2 ± 0.05c	0.85 ± 0.04g	0.33 ± 0.05c	0.17 ± 0.03f	1.37 ± 0.8b	1.18 ± 0.04k	51.1 ± 0.2b	69.2 ± 0.5i
SR_4	1.12 ± 0.02b	0.77 ± 0.03b	0.28 ± 0.01b	0.13 ± 0.02g	1.13 ± 0.6f	1.05 ± 0.01d	48.2 ± 0.4d	68.1 ± 0.6j
Consortium	1.4 ± 0.04a	0.9 ± 0.02e	0.35 ± 0.05a	0.19 ± 0.04e	1.59 ± 0.5a	1.25 ± 0.03h	52.8 ± 0.6a	70.4 ± 0.8g

This data displays the means and standard deviation ($n = 3$). Different letters show significant differences ($p < 0.05$). Treatment details are the same as in Table 3.

3.10. Proline Contents

Salinity stress increased proline accumulation in wheat plants. A considerable increase of 50% in the proline content of wheat plants was recorded in saline stress conditions as compared to their respective control. Inoculation with halotolerant PGPR increased the levels of proline in the leaves. All four inoculants increased the proline contents in the range of 18–36%, respectively. The accumulation of proline was maximum in consortium-treated plants, with an increase of 46.67% under stress conditions (Table 6).

Table 6. Effects of halotolerant PGPR on the total soluble sugar, amino acid, protein, and proline contents of wheat plants grown under salinity stress.

Treatments	Total Soluble Sugar ($\mu g\ g^{-1}$ FW)		Total Amino Acid ($\mu g\ g^{-1}$ FW)		Proline ($\mu g\ g^{-1}$ FW)	
	0 mM	150 mM	0 mM	150 mM	0 mM	150 mM
Control	27 ± 2d	33 ± 5i	330 ± 10g	368 ± 20e	40 ± 03d	120 ± 5j
SR$_1$	29 ± 3d	35 ± 7h	345 ± 12 f	379 ± 17e	44 ± 05d	128 ± 6i
SR$_2$	31 ± 5c	39 ± 6g	360 ± 15e	401 ± 27c	51 ± 3c	130 ± 7g
SR$_3$	33 ± 3c	33 ± 1.0f	370 ± 24e	420 ± 25b	54 ± 3b	135 ± 9h
SR$_4$	29 ± 4 d	36 ± 8h	350 ± 12f	387 ± 24.4d	43 + 3 c	125 ± 5i
Consortium	39 ± 5 c	43 ± 5e	381 ± 10d	439 ± 15a	57 ± 4b	145 ± 7f

This data displays the means and standard deviation ($n = 3$). Different letters show significant differences ($p < 0.05$). Treatment details are the same as in Table 3.

3.11. Amino Acid Content

The amino acid content was highest in the consortium of halotolerant PGPR strains, with an increase of 19.29% and 15.54% under salt stress and control conditions. Moreover, plants inoculated with SR$_2$ and SR$_3$ contained 10% and 14.1% greater concentrations of amino acids as compared to the uninoculated stressed plants (Table 6).

3.12. Total Soluble Sugar

Salinity stress produced a significant increase of 12.5% for the soluble sugar contents of wheat plants as compared to the control. The best outcomes were obtained when plants were inoculated with SR$_2$ and SR$_3$, which resulted in an increase of 9.52% and 15.87%, respectively, under stress conditions. However, a more prominent effect was revealed with the inoculation of a consortium of strains, with an increase of 28.57% and 23.2%, respectively, under the stress and control condition (Table 6).

3.13. Antioxidants Enzyme Assay

The antioxidant enzymes of the wheat plants showed a significant increase under salinity stress. Inoculation with all four halotolerant PGPR improved the production of antioxidant enzymes in plants. However, the best results were shown by the consortium of all strains. The consortium increased the superoxide dismutase activity by 21.4% as compared to stressed plants. Similarly, a significant increase of 16% in the catalase activity was recorded by the inoculation with the consortium. A significant increase of 34.4% in the peroxidase content of plants was recorded as compared to the control (Table 7).

Table 7. Effects of halotolerant PGPR on the antioxidant enzymes activity of wheat plants grown under salinity stress.

Treatments	Superoxide Dismutase (EU mg^{-1} Protein)		Catalase (EU mg^{-1} Protein)		Peroxidase (EU mg^{-1} Protein)	
	0 mM	150 mM	0 mM	150 mM	0 mM	150 mM
Control	0.74 ± 0.06k	1.83 ± 0.02f	2.5 ± 0.03h	4.13 ± 0.02f	144 ± 3f	255 ± 5f
SR$_1$	0.76 ± 0.04j	1.85 ± 0.01e	2.7 ± 0.02k	4.3 ± 0.04e	148 ± 7j	260 ± 4d
SR$_2$	0.8 ± 0.03i	1.9 ± 0.04c	3.01 ± 0.04i	4.7 ± 0.09c	153 ± 4i	263 ± 6c
SR$_3$	0.82 ± 0.05h	1.91 ± 0.03b	3.12 ± 0.05h	4.8 ± 0.10b	155 ± 6h	267 ± 7b
SR$_4$	0.78 ± 0.3j	1.85 ± 0.4d	2.6 ± 0.04j	4.5 ± 0.05d	150 ± 4j	257 ± 5d
Consortium	0.86 ± 0.07g	1.96 ± 0.05a	3.25 ± 0.05g	5.05 ± 0.04a	162 ± 3g	270 ± 6a

This data displays the means and standard deviation ($n = 3$). Different letters show significant differences ($p < 0.05$). Treatment details are the same as in Table 3.

3.14. Heatmap Responses of Pearson's Correlation Coefficient (r)

From the heat map analysis, the data of the osmolyte production, electrolyte leakage, chlorophyll contents, antioxidant enzymes, and halotolerant PGPR showed positive correlations (Figure 4). A comparative analysis of the parameters related to salinity tolerance (presented by green boxes) showed that salinity tolerance had a positive correlation with amino acid, osmotic potential, soluble sugars, proline, SOD, POD, and CAT activities (Figure 5).

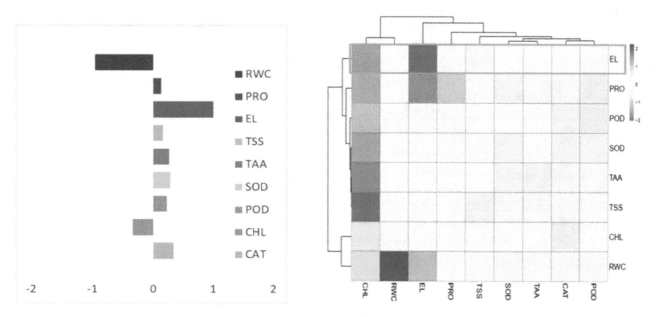

Figure 4. Heatmap of the correlation coefficient (r) for the antioxidant enzymes, stress determinants, and relative water content of wheat leaves treated with bacterial isolates and their consortium. Whereas, EL = Electrolyte leakage, Pro = Proline, POD = Peroxidase, SOD = Superoxide dismutase, CHL = Total chlorophyll, TAA = Total amino acids, TSS = Total soluble sugars, RWC = relative water content.

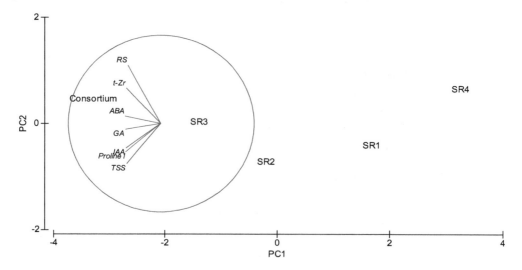

Figure 5. Principle component analysis (PCA) of phytohormones, proline, total soluble sugars, and reducing sugars of halotolerant bacterial isolates and their consortium grown under salt stress in culture conditions. Whereas, IAA = Indole acetic acid, GA = gibberellic acid, CK = Trans zeatin riboside, RS = Reducing sugars, TSS = Total soluble sugars.

4. Discussion

Soil bacteria associated with rhizosphere have been known as growth promotors as well as biotic and abiotic stress alleviators [8]. Bacteria associated with the roots of halophytes and saline soil, capable of tolerating higher levels of salts, are termed as halotolerant [39]. In the current study, bacterial isolates SR_1, SR_2, SR_3, and SR_4 showed the best salt tolerance abilities among all 50 bacterial isolates from the roots–soil interface of plants growing in the saline area. Phenotypic and molecular genotyping (16S RNA sequencing) of four potent isolates proved that SR_2 and SR_3 strains belong to the *Azospirillum* genus (*Azospirillum brasilense* and *Azospirillum lipoferum*) and the other two (SR$_I$ and SR$_4$) belong to the genus *Bacillus* (*Bacillus sturtezi)* and *Pseudomonas* (*Paeudomonas stutzeri*) (Table 2). These beneficial PGPR belonged to different genera, which indicate that plant growth promotion has been distributed across different taxons Halotolerant strains from the genera of *Pseudomonas*, *Bacillus*, *Azospirillum*, *Klebsiella*, and *Ochromobacter* have shown remarkable performance in the amelioration of salt stress in a wide range of crops [40].

Halotolerant PGPR has been reported to promote plant growth as well as mitigate salinity stress [41]. In the current study, we attempted to identify the key mechanisms used by halotolerant strains to alleviate the salinity stress in wheat plants by regulating plant defense mechanisms. The ability of halotolerant PGPR to produce phytohormones is associated with improved growth of plants under saline conditions [42]. The halotolerant PGPR produced IAA, GA, CK, and ABA. The results showed that *Azospirillum* strains produced higher amounts of GA, IAA, and CK than those of *Bacillus* and *Pseudomonas* strains in liquid media (Figure 1). The production of hormones by halotolerant PGPR is thoroughly supported by previous literature and many halotolerant strains of *Azotobacter*, *Bacillus*, *Arthrobacter*, *Azospirillum*, and *Pseudomonas* have been shown to produce IAA, GA, CK, and ABA [43]. These phytohormones regulate the stress defense responses in plants. They influence all aspects of plant growth, like cell wall elongation (IAA), cell division (CK), germination (gibberellin), and stress tolerance (ABA) [44–46]. Various reports suggest that these phytohormones produced under salinity stress help plants to survive and impart tolerance in them under abiotic stresses [46].

Here, the results proved that rhizobacteria secrete more compatible solutes (soluble sugars and proline) in culture media supplemented with a higher NaCl (10%) content. Various studies documented that bacterial cells can accumulate a considerable amount of compatible solutes inside their cells, acting as osmolytes and helping them to survive under severe osmotic stress [47].

Salinity is one of the common factors that can limit agricultural productivity due to its effects on seed germination, plant growth, and crop yield. Wheat is an important staple crop, but as it is a moderately salt-tolerant crop, high salt stress strictly limits its growth and development. Salt stress ultimately reduces the crop yield and nutritive value of wheat. The regulation of physiological, enzymatic, and biochemical changes in plants after inoculation with PGPR helps to alleviate salt or drought stress [40,48].

We demonstrated that salinity reduced the growth and development and relative water content of wheat plants. It also caused curling and wilting of leaves, early leaf senescence, and ultimately a reduction in the growth of plants. This is consistent with what was found in a previous study that salinity restricts cell differentiation and the cell cycle due to osmotic and ionic stress, deficiency of nutrients, oxidative damage, and limited water uptake, which affects plant germination, growth development, and physiological processes, ultimately leading to growth inhibition [49].

In this study, a consortium of four strains produced a prominent result for the dry biomass and leaf area than the control and individual inoculants. These results are in line with Walker et al. [50], who reported that inoculation with a consortium of *Azospirillum-Pseudomonas-Glomus* improved the root architecture in maize under salinity. A better adaptability of PGPR to stress conditions is correlated with efficient root colonization, phosphate solubilization, and nitrogen fixation abilities [51]. From the results, it is clear that salinized plants inoculated with halotolerant strains and their consortium exhibited a higher relative water content of leaves. Rakshapal et al. [52] also observed that PGPR-treated plants not only cope with stress but also that these microbes help to maintain higher water levels in comparison to control plants.

Salinity decreases the photosynthetic efficiency of plants and results in the production of reactive oxygen species (ROS), which cause damage to DNA, proteins, and membranes [53]. We described the results of photosynthetic pigments of wheat plants, which showed that treatment with a consortium showed a pronounced effect of reducing the damage caused by salinity on the photosynthetic apparatus. A similar pattern of results was reported by El-Esawi et al. [54], who observed an increase in the photosynthetic efficiency of plants by PGPR inoculation under salinity.

Salt stress can develop more discharge of electrolytes through the misplacement of Ca associated with membranes. As a result, the permeability of the membrane is destroyed and accumulates a higher efflux of electrolytes inside plant cells/tissue [55]. In the current study, the successive increase in the electrolyte leakage of wheat plants was observed at 150 mM salt stress than the control. These results are inconsistent with the Bojórquez-Quintal et al. [56], who found salt stress enhances electrolyte leakage and the generation of reactive oxygen species (ROS), having a detrimental effect on plant growth. Our results showed that inoculation with halotolerant PGPR tends to decrease the injurious effect of saline stress and decrease the potential electrolytic leakage of ions in stress-treated plants. This is consistent with what was found in previous studies [57,58].

In the present study, the concentration of compatible solutes was also increased in inoculated wheat plants under salt stress (Table 7). The accumulation of compatible solutes, particularly proline, free amino acid, and soluble sugar, is correlated with the adaptability of the plant to stress conditions. We reported that halotolerant PGPR produces compatible osmolytes, which help the plants to maintain their ionic balance. PGPR also induce osmolyte accumulation [59] and phytohormone signaling [40], which facilitates plants in overcoming the initial osmotic shock after salinization. In a previous study, it was found that rice inoculation with salt-tolerant *Bacillus amyloliquefaciens* under salinity increased the plant's salt tolerance and affected the expression of genes involved in osmotic and ionic stress response mechanisms [60].

Proline is the most important osmolyte, which is produced in plants by the hydrolysis of proteins under osmotic stress [61]. From the results, it is clear that a consortium of halotolerant PGPR plants improved proline levels under salt stress. These results are in line with Wang et al. [62]. The production of osmolytes helps the plant to maintain a high turgor potential, prevent oxidative damage by scavenging reactive oxygen species, and protect the membrane structure [63].

We also reported a pronounced increase in the production of soluble sugars with a consortium of halotolerant strains in wheat under salinity stress. PGPR can stimulate carbohydrate metabolism and transport, which results in changes in the source–sink relations, photosynthesis, and growth rate. In previous reports, seeds inoculated with *B. aquimaris* strains showed an increased production of total soluble sugars in wheat under salinity conditions, which resulted in higher biomass and plant growth [64].

An increase in the antioxidant enzyme activity of wheat plants grown under salinity stress was observed by a consortium of halotolerant PGPR strains. This indicates that these bacteria can help the plant to combat the deleterious effects of ROS generated during salinity stress. These results tie well with the previous studies, where an increase in antioxidant enzyme activity under salinity stress was proven to be associated with salt tolerance [65]. Moreover, Wang et al. [66] reported that the application of PGPR strains alleviates the oxidative damage induced by abiotic stresses, including salinity, by augmenting the activity of antioxidant enzymes.

5. Conclusions

In summary, crop inoculations with halotolerant PGPR consortium can serve as a potential tool for alleviating salinity stress. Halotolerant PGPR strains have developed several mechanisms to cope with salinity, particularly the potential to produce phytohormones and compatible solutes. Halotolerant PGPR strains can induce salinity tolerance in plants by activating key defense mechanisms like the production of osmoregulators as well as activating ROS scavenging enzymes. Natural microflora adapted to saline conditions can be used for the development of microbial consortia for crop inoculation, ultimately leading to the formulation of biofertilizer for salt-stressed areas. However, further investigation is needed to observe their performance in field conditions.

Supplementary Materials:
Table S1: Morphology of isolates from rhizosphere of plants from saline soil, Table S2: Preliminary screening data of isolated strains (+ indicates groeth, − indicates no growth), Table S3: Growth characters of isolated strains, Table S4: Effect of Isolates on germination attributes of wheat, Table S2: Carbon/Nitrogen source utilization pattern

determined by QTS -24 kits, Figure S1: Phylogenetic analysis of strain SR1, Figure S2: Phylogenetic analysis of strain SR2, Figure S3: Phylogenetic analysis of strain SR3, Figure S4: Phylogenetic analysis of strain SR4.

Author Contributions: Conceptualization, N.I. Writing—Original Draft, N.I. and R.M. Formal Analysis, H.Y. and W.K. Investigation, R.M. Proofreading, H.E.E. and D.J.D. Editing, H.Y., W.K. and S.I. Formatting, D.J.D. Writing—Review, H.Y., S.I. and W.K. Supervision, N.I. Facilitation, H.E.E. Review, H.E.E. and D.J.D. All authors have read and agreed to the published version of the manuscript.

References

1. GAP Report. Global Agricultural Productivity Report (GAP Report) Global Harvest Initiative, Washington. 2018. Available online: https://globalagriculturalproductivity.org/wp-content/uploads/2019/01/GHI_2018-GAP-Report_FINAL-10.03.pdf (accessed on 5 May 2020).
2. Abbas, R.; Rasul, S.; Aslam, K.; Baber, M.; Shahid, M.; Mobeen, F.; Naqqash, T. Halo-tolerant PGPR: A hope for cultivation of saline soils. *J. King Saud Univ. Sci.* **2019**, *31*, 1195–1201. [CrossRef]
3. Mishra, J.; Fatima, T.; Arora, N.K. Role of secondary metabolites from plant growth-promoting rhizobacteria in combating salinity stress. In *Plant Microbiome: Stress Response*; Ahmad, P., Egamberdieva, D., Eds.; Springer: Singapore, 2018; pp. 127–163.
4. Yasin, N.A.; Khan, W.; Ahmad, S.R.; Ali, A.; Ahmad, A.; Akram, W. Imperative roles of halo-tolerant plant growth-promoting rhizobacteria and kinetin in improving salt tolerance and growth of black gram (*Phaseolusmungo*). *Environ. Sci. Pollut. Res.* **2017**, *25*, 4491–4505. [CrossRef]
5. Abdel-Latef, A.A.H.; Chaoxing, H. Does Inoculation with *Glomus mosseae* Improve Salt Tolerance in Pepper Plants? *J. Plant Growth Regul.* **2014**, *33*, 644–653. [CrossRef]
6. Liu, K.; McInroy, J.A.; Hu, C.-H.; Kloepper, J.W. Mixtures of Plant-Growth-Promoting Rhizobacteria Enhance Biological Control of Multiple Plant Diseases and Plant-Growth Promotion in the Presence of Pathogens. *Plant Dis.* **2018**, *102*, 67–72. [CrossRef]
7. Khan, A.; Zhao, X.Q.; Javed, M.T.; Khan, K.S.; Bano, A.; Shen, R.F. *Bacillus pumilus* enhances tolerance in rice (*Oryza sativa* L.) to combined stresses of NaCl and high boron due to limited uptake of Na+. *Environ. Exp. Bot.* **2016**, *124*, 120–129. [CrossRef]
8. Yasmin, H.; Nosheen, A.; Naz, R.; Keyani, R.; Anjum, S. Regulatory role of rhizobacteria to induce drought and salt stress tolerance in plants. In *Field Crops: Sustainable Management by PGPR. Sustainable Development and Biodiversity*; Maheshwari, D., Dheeman, S., Eds.; Springer: Cham, Switzerland, 2019; Volume 23, p. 23.
9. Niu, X.; Song, L.; Xiao, Y.; Ge, W. Drought-tolerant plant growth-promoting rhizobacteria associated with foxtail millet in a semi-arid agroecosystem and their potential in alleviating drought stress. *Front. Microbiol.* **2018**, *8*, 2580. [CrossRef] [PubMed]
10. Chu, T.N.; Tran, B.T.H.; Van-Bui, L.; Hoang, M.T.T. Plant growth-promoting rhizobacterium *Pseudomonas* PS01 induces salt tolerance in *Arabidopsis thaliana*. *BMC Res. Notes* **2019**, *12*, 11. [CrossRef] [PubMed]
11. Radojevic, M.; Bashkin, V.N. *Practical Environmental Analysis*; The Royal Society of Chemistry: Cambridge, UK, 1999; p. 466.
12. Soltanpour, P.N.; Schwab, A.P. A new soil test for simultaneous extraction of macro- and micro-nutrients in alkaline soils. *Commun. Soil Sci. Plant Anal.* **1977**, *8*, 195–207. [CrossRef]
13. Kingston, H.M. *Microwave Assisted Acid Digestion of Sediments, Sludges, Soils and Oils*; Duquesne University: Pittsburgh, PA, USA, 1994; p. 4.
14. Somasegaran, P.; Hoben, H.J. Counting rhizobia by a plant infection method. In *Handbook for Rhizobia*; Springer: New York, NY, USA, 1994; pp. 58–64.
15. James, G.C. *Native Sherman Rockland Community College, State University of New York*; The Benjamin/Coming Publishing Company Inc.: San Francisco, CA, USA, 1987; pp. 75–80.
16. Robinson, R.J.; Fraaije, B.A.; Clark, I.M.; Jackson, R.W.; Hirsch, P.R.; Mauchline, T.H. Endophytic bacterial community composition in wheat (*Triticum aestivum*) is determined by plant tissue type, developmental stage and soil nutrient availability. *Plant Soil* **2016**, *405*, 381–396. [CrossRef]
17. Pikovskaya, R. Mobilization of phosphorus in soil in connection with vital activity of some microbial species. *Mikrobiologiya* **1948**, *17*, 362–370.
18. Murphy, J.A.M.E.S.; Riley, J.P. A modified single solution method for the determination of phosphate in natural waters. *Anal. Chim. Acta* **1962**, *27*, 31–36. [CrossRef]

19. Schwyn, B.; Neilands, J.B. Universal chemical assay for the detection and determination of siderophores. *Anal. Biochem.* **1987**, *160*, 47–56. [CrossRef]

20. Lorck, H. Production of hydrocyanic acid by bacteria. *Physiol. Plant.* **1948**, *1*, 142–146. [CrossRef]

21. Noreen, Z.; Ashraf, M.; Hassan, M.U. Inter-accessional variation for salt tolerance in pea (*Pisum sativum* L.) at germination and screening stage. *Pak. J. Bot.* **2007**, *39*, 2075–2085.

22. Upadhyay, S.K.; Singh, J.S.; Singh, D.P. Exopolysaccharide-producing plant growth-promoting rhizobacteria under salinity condition. *Pedosphere* **2011**, *21*, 214–222. [CrossRef]

23. Tien, T.M.; Gaskins, M.H.; Hubbell, D.H. Plant growth substances produced by *Azospirillum brasilense* and their effect on the growth of pearl millet (*Pennisetum americanum* L.). *Appl. Environ. Microbiol.* **1979**, *37*, 1016–1024. [CrossRef] [PubMed]

24. Chen, W.P.; Kuo, T.T. A simple and rapid method for the preparation of gram-negative bacterial genomic DNA. *Nucleic Acids Res.* **1993**, *21*, 2260. [CrossRef]

25. Weisburg, W.G.; Barns, S.M.; Pelletier, D.A.; Lane, D.J. 16S ribosomal DNA amplification for phylogenetic study. *J. Bacteriol.* **1991**, *173*, 697–703. [CrossRef]

26. Eck, R.V.; Dayhoff, M.O. *Atlas of Protein Sequence and Structure*; National Biomedical Research Foundation: Silverspring, MD, USA, 1966; p. 12.

27. Unyayer, S.; Keles, Y.; Cekic, F.O. The antioxidative response of two tomato species with different tolerances as a result of drought and cadmium stress combination. *Plant Soil Environ.* **2005**, *51*, 57–64. [CrossRef]

28. Sriram, S.; Raguchander, T.; Babu, S.; Nandakumar, R.; Shanmugam, V.; Vidhyasekaran, P. Inactivation of phytotoxin produced by the rice sheath blight pathogen *Rhizoctonia solani*. *Can. J. Microbiol.* **2000**, *46*, 520–524. [CrossRef]

29. Arnon, D.I. Copper enzymes in isolated chloroplasts. Polyphenol oxidase in *Beta vulgaris*. *Plant Physiol.* **1949**, *24*, 1–15. [CrossRef] [PubMed]

30. Bates, L.S.; Waldren, R.P.; Teare, I.D. Rapid determination of free proline for water stress studies. *Plant Soil* **1973**, *39*, 205–207. [CrossRef]

31. Dubois, M.; Gilles, K.A.; Hamilton, J.K.; Rebers, P.A.; Smith, F. A colorimetric method for the determination of sugars. *Nature* **1951**, *168*, 167. [CrossRef] [PubMed]

32. Hamilton, P.B.; VanSlyke, D.D. Amino acid determination with ninhydrin. *J. Biol. Chem.* **1943**, *150*, 231–233.

33. Bradford, M. A rapid and sensitive method for the quantitation of microgram, quantitation of proteins utilizing the principle of protein dye-binding. *Anal. Biochem.* **1976**, *72*, 248–254. [CrossRef]

34. Aebi, H. Catalase in Vitro. *Methods Enzymol.* **1984**, *105*, 121–126. [CrossRef]

35. Rao, M.V.; Paliyath, G.; Ormrod, D.P. Ultraviolet-B- and ozone induced biochemical changes in antioxidant enzymes of *Arabidopsis thaliana*. *Plant Physiol.* **1996**, *110*, 125–136. [CrossRef]

36. Giannopolitis, C.N.; Ries, S.K. Superoxide dismutase in higher plants. *Plant Physiol.* **1977**, *59*, 309–314. [CrossRef]

37. Steel, R.G.D.; Torrie, G.H. *Principles and Procedures of Statistics Singapore*, 2nd ed.; McGraw Hill Book Co Inc.: New York, NY, USA, 1980.

38. Tamura, K.; Dudley, J.; Nei, M.; Kumar, S. MEGA4: Molecular evolutionary genetics analysis (MEGA) software version 4.0. *Mol. Boil. Evol.* **2007**, *24*, 1596–1599. [CrossRef]

39. Ramadoss, D.; Lakkineni, V.K.; Bose, P.; Ali, S.; Annapurna, K. Mitigation of salt stress in wheat seedlings by halotolerant bacteria isolated from saline habitats. *SpringerPlus* **2013**, *2*, 6. [CrossRef]

40. Egamberdieva, D.; Wirth, S.; Bellingrath-Kimura, S.D.; Mishra, J.; Arora, N.K. Salt-Tolerant Plant Growth Promoting Rhizobacteria for Enhancing Crop Productivity of Saline Soils. *Front. Microbiol.* **2019**, *10*, 2791. [CrossRef] [PubMed]

41. Sarkar, A.; Ghosh, P.K.; Pramanik, K.; Mitra, S.; Soren, T.; Pandey, S. A halo-tolerant *Enterobacter* sp. displaying ACC deaminase activity promotes rice seedling growth under salt stress. *Microbiol. Res.* **2018**, *169*, 20–32. [CrossRef] [PubMed]

42. Patel, D.; Saraf, M. Influence of soil ameliorants and microflora on induction of antioxidant enzymes and growth promotion of (*Jatropha curcas* L.) under saline condition. *Eur. J. Soil Biol.* **2013**, *55*, 47–54. [CrossRef]

43. Abd-Allah, E.F.; Alqarawi, A.A.; Hashem, A.; Radhakrishnan, R.; Al-Huqail, A.A.; Al-Otibi, F.A. Endophytic bacterium *Bacillus subtilis* (BERA 71) improves salt tolerance in chickpea plants by regulating the plant defense mechanisms. *J. Plant Interact.* **2017**, *3*, 37–44. [CrossRef]

44. TrParray, A.P.; Jan, S.; Kamili, A.N.; Qadri, R.A.; Egamberdieva, D.; Ahmad, P. Current perspectives on plant growth promoting rhizobacteria. *J. Plant Growth Regul.* **2016**, *35*, 877–902. [CrossRef]

45. Salomon, M.; Bottini, R.; de Souza Filho, G.A.; Cohen, A.C.; Moreno, D.; Gil, M. Bacteria isolated from roots and rhizosphere of *Vitis vinifera* retard water losses, induce abscisic acid accumulation and synthesis of defense-related terpenes in in vitro cultured grapevine. *Physiol. Plant.* **2014**, *151*, 359–374. [CrossRef] [PubMed]

46. Bottini, R.; Cassán, F.; Piccoli, P. Gibberellin production by bacteria and its involvement in plant growth promotion and yield increase. *Appl. Microbiol. Biotechnol.* **2004**, *65*, 497–503. [CrossRef]

47. Parida, A.K.; Das, A.B. Salt tolerance and salinity effect on plants: A review. *Ecotoxicol. Environ. Saf.* **2005**, *60*, 324–349. [CrossRef]

48. Kumar, A.; Verma, J.P. Does plant-microbe interaction confer stress tolerance in plants: A review? *Microbiol. Res.* **2018**, *207*, 41–52. [CrossRef]

49. Isayenkov, S.V.; Maathuis, F.J.M. Plant Salinity Stress: Many Unanswered Questions Remain. *Front. Plant Sci.* **2019**, *10*, 80. [CrossRef]

50. Walker, V.; Couillerot, O.; Felten, A.V.; Bellvert, F.; Jansa, J.; Maurhofer, M.; Bally, R.; Moënne-Loccoz, Y.; Comte, G. Variation of secondary metabolite levels in maize seedling roots induced by inoculation with *Azospirillum*, *Pseudomonas* and *Glomus* consortium under field conditions. *Plant Soil* **2012**, *356*, 151–163. [CrossRef]

51. Nadeem, S.M.; Zahir, Z.A.; Naveed, M.; Arshad, M. Preliminary investigations on inducing salt tolerance in maize through inoculation with rhizobacteria containing ACC deaminase activity. *Can. J. Microbiol.* **2007**, *53*, 1141–1149. [CrossRef] [PubMed]

52. Rakshapal, S.; Sumit, K.S.; Rajendra, P.P.; Alok, K. Technology for improving essential oil yield of *Ocimum basilicum* L. (sweet basil) by application of bioinoculant colonized seeds under organic field conditions. *Ind. Crop. Prod.* **2013**, *45*, 335–342. [CrossRef]

53. Islam, F.; Yasmeen, T.; Arif, M.S.; Ali, S.; Ali, B.; Hameed, S. Plant growth promoting bacteria confer salt tolerance in *Vigna radiata* by up-regulating antioxidant defense and biological soil fertility. *Plant Growth Regul.* **2016**, *80*, 23–36. [CrossRef]

54. El-Esawi, M.A.; Alaraidh, I.A.; Alsahli, A.A.; Alamri, S.A.; Ali, H.M.; Alayafi, A.A. *Bacillus firmus* (SW5) augments salt tolerance in soybean (*Glycine max* L.) by modulating root system architecture, antioxidant defense systems and stress-responsive genes expression. *Plant Physiol. Biochem.* **2018**, *132*, 375–384. [CrossRef]

55. Garg, N.; Manchanda, G. Role of arbuscular mycorrhizae in the alleviation of ionic, osmotic and oxidative stresses induced by salinity in *Cajanus cajan* (L.) Millsp. (pigeonpea). *J. Agron. Crop Sci.* **2009**, *195*, 110–123. [CrossRef]

56. Bojórquez-Quintal, E.; Velarde-Buendía, A.; Ku-González, Á.; Carillo-Pech, M.; Ortega-Camacho, D.; Echevarría-Machado, I.; Pottosin, I.; Martínez-Estévez, M. Mechanisms of salt tolerance in habanero pepper plants (*Capsicum chinense* Jacq.): Proline accumulation, ions dynamics and sodium root-shoot partition and compartmentation. *Front. Plant Sci.* **2014**, *5*, 605. [CrossRef]

57. Kaya, C.; Tuna, A.L.; Okant, A.M. Effect of foliar applied kinetin and indole acetic acid on maize plants grown under saline conditions. *Turk. J. Agric. For.* **2010**, *34*, 529–538.

58. Habib, S.H.; Kausar, H.; Saud, H.M. Plant Growth-Promoting Rhizobacteria Enhance Salinity Stress Tolerance in Okra through ROS-Scavenging Enzymes. *BioMed Res. Int.* **2016**, *2016*, 6284547. [CrossRef]

59. Bremer, E.; Kramer, R. Responses of microorganisms to osmotic stress. *Annu. Rev. Microbiol.* **2019**, *73*, 313–334. [CrossRef]

60. Nautiyal, C.S.; Srivastava, S.; Chauhan, P.S.; Seem, K.; Mishra, A.; Sopory, S.K. Plant growth-promoting bacteria *Bacillus amyloliquefaciens* NBRISN13 modulates gene expression profile of leaf and rhizosphere community in rice during salt stress. *Plant Physiol. Biochem.* **2013**, *66*, 1–9. [CrossRef] [PubMed]

61. Krasensky, J.; Jonak, C. Drought, salt, and temperature stress-induced metabolic rearrangements and regulatory networks. *J. Exp. Bot.* **2012**, *63*, 1593–1608. [CrossRef] [PubMed]

62. Wang, Q.; Dodd, I.C.; Belimov, A.A.; Jiang, F. Rhizosphere bacteria containing 1-aminocyclopropane-1-carboxylate deaminase increase growth and photosynthesis of pea plants under salt stress by limiting Na+ accumulation. *Funct. Plant Biol.* **2016**, *43*, 161–172. [CrossRef] [PubMed]

63. Prado, F.E.; Boero, C.; Gallardo, M.; González, J.A. Effect of NaCl on germination, growth, and soluble sugar content in *Chenopodium quinoa* wild seeds. *Bot. Bull. Acad. Sin.* **2000**, *41*, 27–343.

64. Upadhyay, S.K.; Singh, D.P. Effect of salt-tolerant plant growth-promoting rhizobacteria on wheat plants and soil health in a saline environment. *Plant Biol. (Stuttg.)* **2015**, *17*, 288–293. [CrossRef] [PubMed]

65. Moghaddam, M.; Farhadi, N.; Panjtandoust, M.; Ghanati, F. Seed germination, antioxidant enzymes activity and proline content in medicinal plant *Tagetes minuta* under salinity stress. *Plant Biosyst.* **2019**. [CrossRef]

66. Wang, C.J.; Yang, W.; Wang, C.; Gu, C.; Niu, D.D.; Liu, H.X.; Wang, Y.P.; Guo, J. Induction of drought tolerance in cucumber plants by a consortium of three plant growth-promoting rhizobacterium strains. *PLoS ONE* **2012**, *7*, e52565. [CrossRef]

10

PGPR Modulation of Secondary Metabolites in Tomato Infested with *Spodoptera litura*

Bani Kousar [1], Asghari Bano [2,*] and Naeem Khan [3,*]

[1] Department of Plant Sciences Faculty of Biological Sciences, Quaid-i-Azam University, Islamabad 45320, Pakistan; bani.kousar@gmail.com
[2] Department of Biosciences, University of Wah, Wah Cantt 47040, Pakistan
[3] Department of Agronomy, Institute of Food and Agricultural Sciences, University of Florida, Gainesville, FL 32608, USA
* Correspondence: banoasghari@gmail.com (A.B.); naeemkhan@ufl.edu (N.K.)

Abstract: The preceding climate change demonstrates overwintering of pathogens that lead to increased incidence of insects and pest attack. Integration of ecological and physiological/molecular approaches are imperative to encounter pathogen attack in order to enhance crop yield. The present study aimed to evaluate the effects of two plant growth promoting rhizobacteria (*Bacillus endophyticus* and *Pseudomonas aeruginosa*) on the plant physiology and production of the secondary metabolites in tomato plants infested with *Spodoptera litura* (Fabricius) (Lepidoptera: Noctuidae). The surface sterilized seeds of tomato were inoculated with plant growth promoting rhizobacteria (PGPR) for 3–4 h prior to sowing. Tomato leaves at 6 to 7 branching stage were infested with *S. litura* at the larval stage of 2nd instar. Identification of secondary metabolites and phytohormones were made from tomato leaves using thin-layer chromatography (TLC) and high performance liquid chromatography (HPLC) and fourier-transform infrared spectroscopy (FTIR). Infestation with *S. litura* significantly decreased plant growth and yield. The PGPR inoculations alleviated the adverse effects of insect infestation on plant growth and fruit yield. An increased level of protein, proline and sugar contents and enhanced activity of superoxide dismutase (SOD) was noticed in infected tomato plants associated with PGPR. Moreover, p-kaempferol, rutin, caffeic acid, p-coumaric acid and flavonoid glycoside were also detected in PGPR inoculated infested plants. The FTIR spectra of the infected leaf samples pre-treated with PGPR revealed the presence of aldehyde. Additionally, significant amounts of indole-3-acetic acid (IAA), salicylic acid (SA) and abscisic acid (ABA) were detected in the leaf samples. From the present results, we conclude that PGPR can promote growth and yield of tomatoes under attack and help the host plant to combat infestation via modulation in IAA, SA, ABA and other secondary metabolites.

Keywords: *Spodoptera litura* (Fabricius) (Lepidoptera: Noctuidae); *Solanum lycopersicum* L.; secondary metabolites; plant insect interactions

1. Introduction

Lycopersicon esculentum Mill. (Tomato) is one of the widely used vegetables cultivated all over the world. It is the important source of vitamin C and vitamin A [1], lycopene (carotenoids), pro-vitamin A, β-carotene and flavonoids [2]. In the recent years, its yield is significantly reduced by the infestation of leaf caterpillars.

Leaf caterpillar *S. litura* (Fabricius) (Lepidoptera: Noctuidae), also known as tropical armyworm, is among the main pests of cultivated crops that can cause significant damage to tomato crop. To this date, *S. litura* has infected about 290 plant species, belonging to 99 families [3,4]. It grows throughout the year, and mounts nearly 7 to 8 generations per year. The larvae of *S. litura* feed initially on plant leaves and latterly feed on almost every part of the plant. The larvae can cause 12 to 23% damage to tomatoes in the monsoon and 9.4 to 27.4% in winter [5]. This insect had shown strong resistance to all conventional and some new chemically synthesized insecticides [6,7]. To combat this notorious insect attack, one can develop new insect resistant cultivars. The main drawbacks of the new cultivars' development are time and expenses. Alternatively, the use of plant growth promoting rhizobacteria having biocontrol properties is a sustainable and eco-friendly approach.

Rhizosphere bacteria form a close association with the roots of plants, they nourish on the soil nutrients and root-exudates of plants; in return they protect the host against the biotic and abiotic stresses and help in host growth [8,9]. Plant growth promoting rhizobacteria (PGPR) boost plant growth directly through the production of phytohormones and indirectly as biocontrol agents [10]. PGPR employs different mechanisms to promote plant growth and control phyto-pathogens. One of the widely recognized mechanisms is the production of inhibitory allelo-chemicals, the production of antibiotics, siderophore, lytic enzymes and the induction of systemic resistance (ISR) in host plants against a broad spectrum of pathogens [11]. Induced systemic resistance (ISR) protects the plant against a broad range of diseases [12,13], triggered by a wide variety of beneficial microbes [14].

PGPR consortium of *S. marcescens*, *B. amyloliquefaciens*, *P. putida*, *P. fluorescens* and *B. cereus* significantly increased the number of fruit/plant [15]. The three bacterial species viz. *B. amyloliquefaciens*, *B. subtilis* and *B. brevis* have significantly improved the activity of defense related enzymes in tomato plants infected with bacterial canker [16]. Several bacterial species (*Pseudomonas*, *Azotobacter*, *Azospirillum*, *Pseudomonas* + *Azotobacter*, *Pseudomonas* + *Azospirillum*, *Azotobacter* + *Azospirillum* and *Pseudomonas* + *Azotobacter* + *Azospirillum*) played a key role in nutrient uptake by tomato plants. Also, the rhizospheric bacteria significantly improved shoot and root dry weights, enhanced and modulated production of secondary metabolites [17] and induced resistance to various diseases [18]. *Pseudomonas aeruginosa* is an aerobic, gram-negative rod-shaped bacterium of *Pseudomonadaceae* [19] that was reported to have antifungal activity against *Fusarium moniliforme* [20]. Both *Pseudomonas aeruginosa* and *Bacillus endophyticus* were catalase and oxidase positive, solubilize phosphorus and produce bacteriocin. These bacterial strains showed significant ($p < 0.05$) increase in dry matter production, plant height and root length of maize [21]. They were found positive for the production of antibiotics [22] and had a protruding impact on plant metabolism and plant defense against environmental stresses [23,24].

The present investigation was based on the hypothesis that rhizobacteria isolated from stressed habitats can induce tolerance to plants against environmental stresses in a much better way than those from normal conditions [25]. The rhizobacteria *Bacillus endophyticus* strainY5 (Accession no. JQ792035) and *Pseudomonas aeruginosa* JYR (Accession no JQ792038) were isolated from the semiarid areas of Yousaf wala Sahiwal (15% soil moisture) and arid areas of Jhang (9% soil moisture), where maize is grown as a main crop. Soil sampling was done at the tasseling stage of maize. The role of those two PGPRs used as bioinoculants was studied on growth and yield of tomato (*Solanum lycopersicum* L.) infested with *S. litura*.

2. Materials and Methods

2.1. Plant Material

The experiment was conducted in the green house of Quaid-i-Azam University, Islamabad. Seeds of *Solanum lycopersicum* L. cv. Rio Grande was obtained from the National Agricultural Research Centre (NARC) Islamabad. Prior to sowing the seeds were surface sterilized with 70% ethanol for 2–3 min, followed by shaking in 10% clorox for 2–3 min. The seeds were finally washed with autoclaved distilled water to remove the traces of treated chemicals [13].

2.2. Preparation of Inocula and Method of Inoculation

Fresh cultures (24 h old) of *Bacillus endophyticus* and *Pseudomonas aeruginosa* were used to inoculate Luria-Bertani (LB) broth, incubated on a rotary shaker for 48 h at 28 °C. The cultures were centrifuged at 3000 rpm for 10 min. Supernatant was discarded, and the pellet containing the bacterial cells was suspended in the autoclaved distilled water to adjust the optical density ($\lambda = 1$) at 660 nm. The inoculum prepared was found to have 10^6 cells/mL. Sterilized seeds were soaked in the bacterial inoculum for 3 to 4 h. The seeds soaked in autoclaved distilled water for the same period were treated as a control [5].

2.3. Growing Conditions and the Treatments

Seeds were sown in pots containing autoclaved sand and soil mixed in 1:3 ratio [26]. Pots were kept in the greenhouse of Quaid-i-Azam University using randomized complete block design with four replicates per treatment. The growing conditions were: photoperiod 16 h, temp 22–28 °C and humidity 60–80%.

The treatments included: Tomato seeds uninoculated uninfested control (C); Tomato seeds inoculated with *Bacillus endophyticus* (T1); Tomato seeds inoculated with *Pseudomonas aeruginosa* (T2); plants infested with *S. litura* (T3); Tomato seeds inoculated with *Bacillus endophyticus* and latterly infested the leaves at 6 to 7 branching stage with *S. litura* (T4); Tomato seeds inoculated with *Pseudomonas aeruginosa* and infested the leaves at 6 to 7 branching stage with *S. litura* (T5).

The tropical armyworm was obtained from the Insectary department, National Agricultural Research Centre (NARC), Islamabad. The leaves of tomato seedlings at 6 to 7 branching stage were infested with larvae of *S. litura* at larval stage of 2nd instar. The larvae were starved for 2 h prior to infestation.

2.4. Height and Weight of Plants and Weight of Tomato Fruit

At the time of harvesting, four plants were marked from each treatment to measure the average height (cm) of the plant and their fresh and dry weights were recorded. After 180 days of sowing, the red ripened fruits were harvested and their fresh weight was measured [27].

2.5. Physiological and Biochemical Attributes of Plants

The physiological and biochemical parameters of leaves were measured after insect infestation.

2.5.1. Leaf Protein Content

Protein content of fresh leaves of tomato plant was estimated following the method of Lowry et al. [28], using Bovine Serum Albumin (BSA) as a standard. Fresh leaves (0.1 g) were grinded in 1 mL of phosphate buffer (pH 6.8) and centrifuged for 10 min at 3000 rpm. The supernatant (0.1 mL) was poured into the test tube and a total volume of 1 mL was made with distilled water. A mixture of 50 mL of Na_2CO_3, NaOH and Na-K tartrate and 1mL of $CuSO_4.5H_2O$ was added. After shaking for 10 min, 0.1 mL of Folin phenol reagent was added. The absorbance of each sample was recorded at 650 nm after 30 min incubation. The concentration of protein was determined using the following formula:

$$Protein \left(\frac{mg}{g} \right) = \frac{K - value \times dilution\ factor \times absorbance}{weight\ of\ sample}$$

K value = 19.6

Dilution factor = 2

Weight of leaf sample = 100 mg

2.5.2. Chlorophyll and Carotenoids Content

Estimation of chlorophyll contents was made according to the method of Arnon [29]. The tomato leaves (0.05 g) were grinded in 10 mL dimethyl sulfoxide (DMSO). The tubes were incubated at 65 °C for 4 h and then the optical density of the sample was recorded at 665 nm and 645 nm. The carotenoids content was determined following the method of Lichtenthaler and Welburn [30].

$$Chloropyll\ a\left(\frac{mg}{g}\right) = 1.07(OD_{663}) - 0.09(OD_{645})$$

$$Chloropyll\ b\left(\frac{mg}{g}\right) = 1.77(OD_{645}) - 0.28(OD_{663})$$

$$Carotenoids\left(\frac{mg}{g}\right) = Absorbance\ (OD_{663}) \times 4$$

2.5.3. Proline Content of Leaves (µg/g)

Free proline content in tomato plant leaves was estimated following the method of Bates et al. [31]. Fresh plant leaf (0.5 g) was grounded in 3% sulfosalicylic acids and kept overnight at 4 °C. The extract was centrifuged at 3,000 rpm for 10 min. The supernatant was mixed with acidic ninhydrin and boiled for 1 h. The solution was then cooled and toluene was added. The absorbance of the toluene layer was recorded at 520 nm against toluene blank. The content of free proline was estimated on fresh weight basis following the formula:

$$Proline\left(\frac{\mu g}{g}\right) = \frac{K - value \times dilution\ factor \times absorbance}{leaf\ weight}$$

Value of K= 17.52
Dilution factor= 2
Weight of leaf sample= 100 mg

2.5.4. Sugar Estimation

The colorimetric determination of total sugar (simple sugar, oligosaccharides and reducing sugar) was done following the method of Dubois et al. [32]. Fresh tomato leaves (500 mg) were grinded with 10 mL of distilled water in autoclaved mortar and pestle, centrifuged at 3000 rpm for 5 min. To the supernatant (100 µL), 1 mL of 80% (w/v) phenol and 5 mL concentrated sulfuric acid was added. The mixture was heated in a water bath till boiling and then incubated for 4 h at room temperature. The absorbance of each sample was finally measured at 420 nm.

$$Sugar\left(\frac{mg}{g}\right) = \frac{K - value \times dilution\ factor \times absorbance}{leaf\ weight}$$

Value of K = 20
Dilution factor = 10
Weight of leaf sample = 500mg

2.5.5. Superoxide Dismutase (SOD) Assay

The SOD activity was estimated following the method of Beauchamp and Fridovich [33]. The activity of Superoxide dismutase was expressed as units/100 g F.W. Superoxide dismutase was calculated by the following formula:

$$R4 = R_3 - R_2$$

SOD activity = R_4/A

R_1 = O.D of Reference, R_2 = O.D of Blank, R_3 = O.D of Sample

$A = R_1$ (50/100)

2.5.6. Determination of Indole acetic acid (IAA), Gibberellic acid (GA) and Abscisic acid (ABA) Contents

The extraction and purification for above mentioned phytohormones were made following the method of Kettner and Doerffling [34]. Plant leaves (1g) were grinded in 80% methanol at 4 °C with butylated hydroxytoluene (BHT) used as antioxidant. The extract was centrifuged and the supernatant was reduced by using a rotary thin film evaporator (RFE). The aqueous phase was partitioned 4 times at pH 2.5–3 with $\frac{1}{2}$ volume of ethyl acetate. The ethyl acetate was evaporated by a rotary thin film evaporator. The residue was re-dissolved in 1 mL of methanol (100%) and examined on HPLC (LC-8A Shimadzu, C-R4A Chromatopac; SCL-6B system controller) using UV detector and C-18 column (39 × 300 mm). The wavelength used for the detection of IAA was 280 nm and for GA was 254 nm. For ABA, the samples were injected onto a C_{18} column and eluted at 254 nm with a linear gradient of methanol (30–70%), containing 0.01% acetic acid, at a flow rate of 0.8 mL min^{-1} [35].

2.5.7. Determination of Salicylic Acid (SA) Content of Leaves

Enyedi et al. [36] and Seskar et al. [37] method was employed for salicylic acid detection. After crushing the fresh leaves (1 g) of tomato in 10 mL of 80% methanol at 4 °C. The sample was kept for 3 days with subsequent change in methanol after 24 h. The methanol was then evaporated using RFE and the residue was dissolved again in methanol, filtered and subjected to high-performance liquid chromatography (HPLC) (Agilent Technologies USA) equipped with S-1121 dual piston solvent delivery system and S-3210 UV/VIS diode array detector. Detection of SA was done at 280 nm by co-chromatography with 2-hydroxybenzoic acid as standard. The peak areas were recorded and calculated with SRI peak simple chromatography data acquisition and integration software (SRI instruments, Torrance, CA, USA).

2.5.8. Measurement of Shoot and Root Fresh and Dry Weights and Root Area

Shoots of 4 plants per treatment were cut at the base and weighed immediately by using the electronic balance, to measure the fresh weight of shoot. The chopped shoot was then dried at 70 °C for 72 h and dry weight was measured. The roots of the same plants were washed thoroughly with running tap water to remove soil debris. The water was absorbed on filter paper and weighed to measure the fresh weight of the root. The same root samples were used for determination of root dry weight after drying in the oven till constant weight was obtained [13]. The root area was calculated by using root law Software, Washington State University [38].

2.6. Thin Layer Chromatography of Methanolic Extract of Tomato Leaves

Leaves of tomato plant were harvested 24 h after infestation (80 DAS); shade dried at room temperature and grinded to fine powder. Powdered leaves (20 g) was extracted in 400 mL methanol for 72 h.The methanolic extract was dried using rotary evaporator (RFE), the residue (3 mg) was dissolved in 500 μL methanol and collected in eppendorf tube and stored at −4 °C.

Extract was spotted on a TLC plate (20 × 20 cm) coated with silica gel HF (250-350 nm). The mobile phase used was chloroform: methanol (95:5 v/v). The bands, representing various compounds were visualized under UV (254 nm and 380 nm) [39]. The Rf value of each band was calculated and identification of the compound from each band at specific Rf was made from the literature documented.

2.7. FTIR Spectroscopy

All spectra were obtained with the help of an OMNI-sampler attenuated total reflectance (ATR) accessory on a Nicolet FTIR spectrophotometer followed by the method of Lu et al. [40] and Liu et al. [41] with some modifications. Small amount of TLC eluent corresponding to the Rf value of major bands were placed directly on the germanium piece of the infrared spectrometer with constant pressure applied and data of infrared absorbance, collected over the wave number ranged from 4000 cm^{-1} to 675 cm^{-1} and computerized for analyses by using the Omnic software [42].

2.8. Statistical Analysis of Data

The data was subjected to analysis of variance using Statistix 8.1 software. The differences among various treatment means were compared using the least significant differences test (LSD) at $p \leq 0.05$ probability level (Table S1).

3. Results

3.1. Plant Growth Attributes

The plant spread, which is a measurement of plant width, was significantly (31%) higher in PGPR treated plants under unstressed condition over control (Figure 1). Insect infestation decreased the plant spread by 41%, the decrease was ameliorated by the PGPRs and the value was even greater than the control. The plant height was significantly increased in PGPR inoculated plants (Figure 1). The insect infestation significantly reduced ($p \leq 0.05$) the height of the plant by 40%, and root area by 50% of the control (Figure 1). The PGPR inoculated plants alleviated the inhibitory effects of insect infestation on plant height and root area such that the root area was significantly higher than the control. Both the shoot and root fresh weights were significantly (44% and 34%) increased in PGPR inoculated plants (Figures 2 and 3). Infestation with the insect decreased the fresh weights of both the root and shoot, the shoot fresh weight was more adversely affected. The PGPR inoculation had ameliorated the insect-induced decrease in the root and shoot fresh weight.

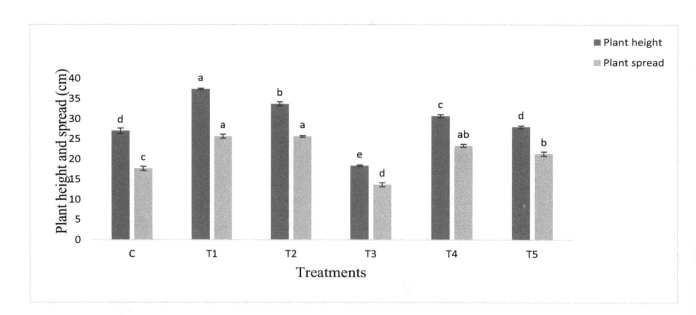

Figure 1. Mean plant height and plant spread (cm) of tomato under control and infested conditions. Data are means of four replicates along with standard error bars. Different letters on the bar represent significant differences ($p < 0.05$) among treatments.

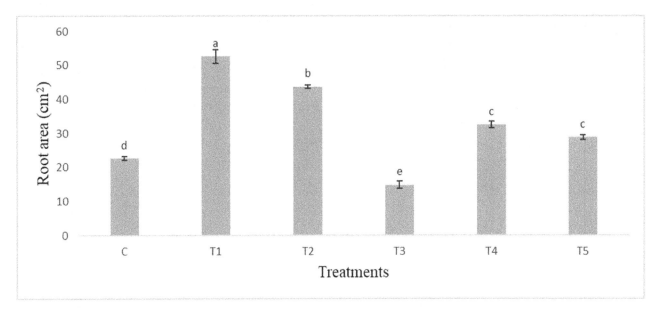

Figure 2. Root area (cm^2) of tomato plant infested with *S. litura* and under control condition. Data are means of four replicates along with standard error bars. Different letters are indicating significant differences ($p < 0.05$) among treatments.

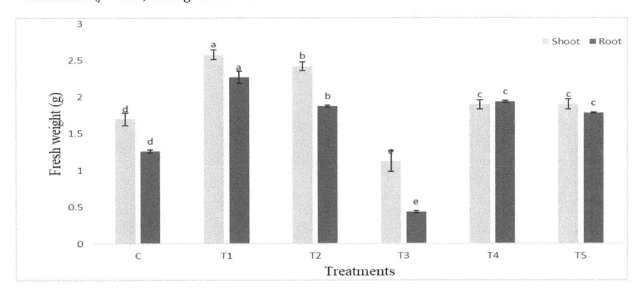

Figure 3. Fresh weight of shoot and root (g) of tomato plant infested with *S. litura* and under control condition. Data are means of four replicates along with standard error bars. Different letters are indicating significant differences ($p < 0.05$) among treatments.

C-uninoculated uninfested control, T1-Seeds inoculated with *Bacillus endophyticus*, T2-Seeds inoculated with *Pseudomonas aeruginosa*, T3-Plants infested with *S. litura*, T4-Seeds inoculated with *Bacillus endophyticus* and plants infested with *S. litura*, T5-Seeds inoculated with *Pseudomonas aeruginosa* and plants infested with *S. litura*.

C-uninoculated uninfested control, T1-Seeds inoculated with *Bacillus endophyticus*, T2-Seeds inoculated with *Pseudomonas aeruginosa*, T3-Plants infested with *S. litura*, T4-Seeds inoculated with *Bacillus endophyticus* and plants infested with *S. litura*, T5-Seeds inoculated with *Pseudomonas aeruginosa* and plants infested with *S. litura*.

C-uninoculated uninfested control, T1-Seeds inoculated with *Bacillus endophyticus*, T2-Seeds inoculated with *Pseudomonas aeruginosa*, T3-Plants infested with *S. litura*, T4-Seeds inoculated with

Bacillus endophyticus and plants infested with *S. litura*, T5-Seeds inoculated with *Pseudomonas aeruginosa* and plants infested with *S. litura*.

The dry weight of root and shoot was also higher ($p \leq 0.05$) in PGPR inoculated plants (Figure 4). The root was more responsive and the % increase in root weight was greater. The leaves were almost eaten by the insect and the shoot weight was significantly decreased to 81% whereas root weight was decreased by 38% over the control.

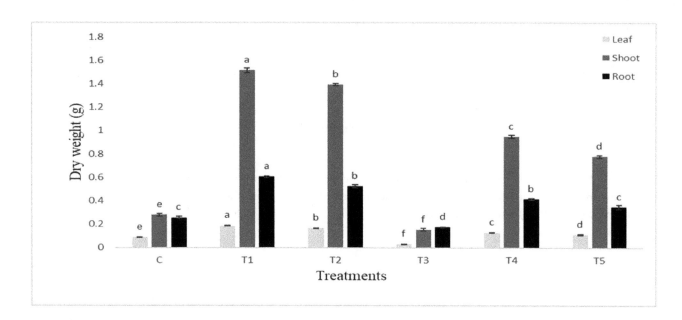

Figure 4. Dry weight of leaf, shoot and root (g) of tomato plant infested with *S. litura* and under control condition. Data are means of four replicates along with standard error bars. Different letters are indicating significant differences ($p < 0.05$) among treatments.

C-uninoculated uninfested control, T1-Seeds inoculated with *Bacillus endophyticus*, T2-Seeds inoculated with *Pseudomonas aeruginosa*, T3-Plants infested with *S. litura*, T4-Seeds inoculated with *Bacillus endophyticus* and plants infested with *S. litura*, T5-Seeds inoculated with *Pseudomonas aeruginosa* and plants infested with *S. litura*.

3.2. Physiological Parameters

The proline production was lower ($p \leq 0.05$) in the untreated control plants (Figure 5). Under unstressed conditions the PGPR treatments stimulated proline content of leaves by 18% over control. Similar percent of increase was recorded in plants infested with *S. litura*. Both the PGPR inoculated plants infested with *S. litura* exhibited marked increase in proline content of leaves over infested plants. The maximum (59%) increase was recorded in the *Bacillus endophyticus* inoculated plants infested with *S. litura*. Chlorophyll a, b and carotenoids followed the similar pattern of response to PGPR and *S. litura* infestation (Figure 6). The response of PGPR was higher ($p \leq 0.05$) particularly for carotenoids content. Both the protein and the sugar contents were higher ($p \leq 0.05$) in PGPR inoculated plants (Figure 7) under unstressed conditions. *Pseudomonas aeruginosa* showed maximum (1.4 fold) increase in sugar content over infested plants. The infestation with *S. litura* had increased sugar and protein contents significantly higher than the control. The inoculated plants receiving insect infestation exhibited up to 2.25 fold increase in sugar content as compared to that of infested plants.

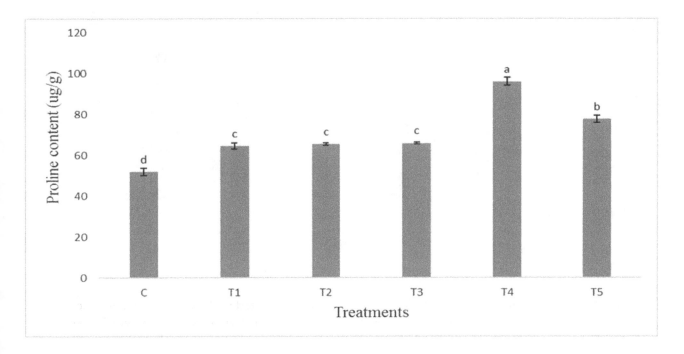

Figure 5. Proline content (μg/g) of tomato leaves infested with *S. litura* and under control condition. Data are means of four replicates along with standard error bars. Different letters are indicating significant differences ($p < 0.05$) among treatments.

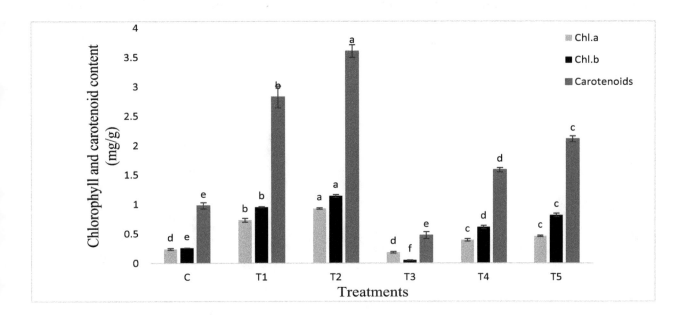

Figure 6. Chlorophylls and carotenoids content (mg/g) of tomato leaves infested with *S. litura* and under control condition. Data are means of four replicates along with standard error bars. Different letters are indicating significant differences ($p < 0.05$) among treatments.

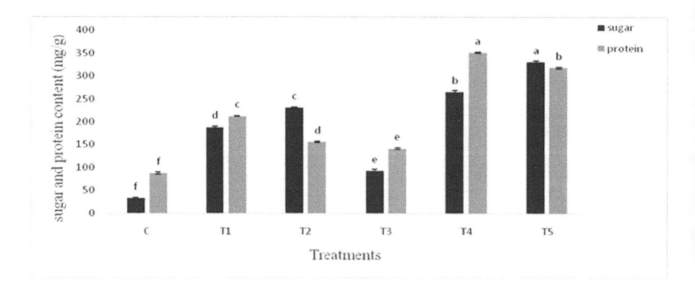

Figure 7. Sugar and protein content (mg/g) of tomato leaves infested with *S. litura* and under control condition. Data are means of four replicates along with standard error bars. Different letters are indicating significant differences (*p* < 0.05) among treatments.

The weight of tomato fruit was about 35% greater in plants inoculated with *Bacillus endophyticus* while *Pseudomonas aeruginosa* inoculated plants exhibited 44% increase over control. There was 26% decrease in the weight of tomato fruit in infested plants (Figure 8). The PGPR inoculated plants ameliorated the inhibitory effect of the insect and showed up to 78% increase in the fruit weight over infested plants.

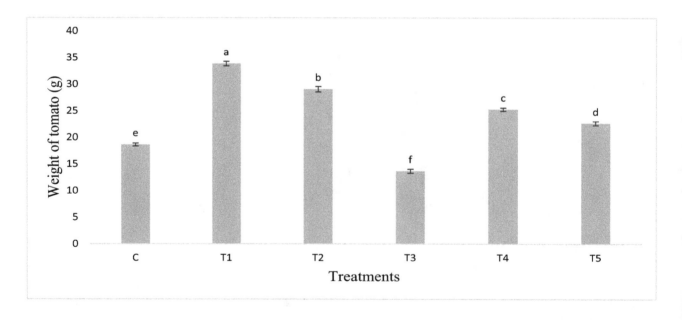

Figure 8. Weight of tomato fruits/plant (g) infested with *S. litura* and under control condition. Data are means of four replicates along with standard error bars. Different letters are indicating significant differences (*p* < 0.05) among treatments.

The infestation with insects enhanced the SOD activity. The SOD activity was three fold higher in leaves of plants inoculated with *Bacillus endophyticus* (T1). Plants inoculated with *Pseudomonas aeruginosa* (T2) on infestation further augmented SOD (Figure 9).

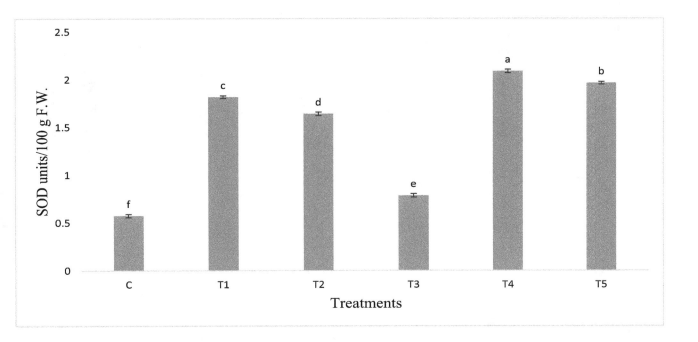

Figure 9. Superoxide dismutase (SOD) activity in tomato leaves infested with *S. litura* and under control condition. Data are means of four replicates along with standard error bars. Different letters are indicating significant differences ($p < 0.05$) among treatments.

3.3. Phytohormones Contents of Leaves

The data in Figure 10 revealed that uninoculated uninfested control leaves of tomato had traceable amounts of Salicylic acid. Insect infestation produced very little amounts of SA. Both the PGPR produced significantly higher amounts of SA in plants, *Pseudomonas* sp. being more efficient. The SA was 1.8 folds greater than infested plant leaves. In *Pseudomonas* inoculated plants, this was further augmented and significantly higher (3.6 fold) SA was recorded in infested plant leaves pretreated with *Pseudomonas aeruginosa*. IAA was not detected in the control and insect infested plants but both the PGPR produced significant amount of IAA in the leaves of inoculated plant which was further augmented and up to 449 µg IAA/g leaves was detected in the leaves of plants infested with *S. litura* and pretreated with *Pseudomonas aeruginosa* (Figure 10). Insect infestation increased the GA content of leaves significantly over control. Several fold increases in GA production were recorded in both the PGPR inoculated plants: *Pseudomonas aeruginosa* being most efficient. Both the PGPR inoculated plants overcame the insect infestation induced decrease in GA content (Figure 10).

The ABA content was significantly lower in the infested plant leaves as compared to control. *Bacillus endophyticus* inoculation showed significantly higher ABA production under controlled conditions and the value was several times greater than control in the inoculated plant infested with *S. litura*.

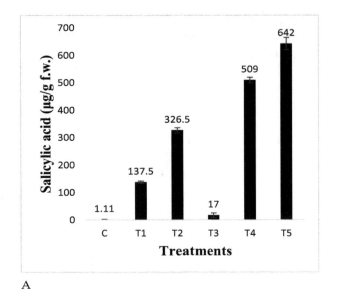

A

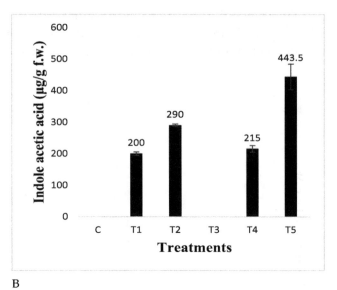

B

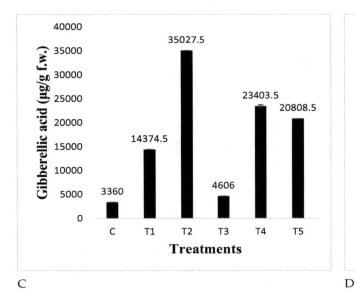

C

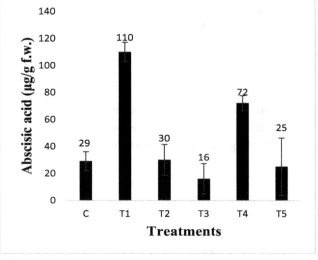

D

Figure 10. Phytohormone content in the leaves of tomato plants infested with *S. litura* and under control condition. (**A**): Salicylic acid; (**B**): Indole acetic acid; (**C**): Gibberellic acid; (**D**): Abscisic acid.

3.4. Detection of Secondary Metabolites from Extract of Tomato Leaves

Thin layer chromatography of tomato leaf extract showed 29 bands of different colors under UV light (Table 1). Calculated Rf values of leaf extract were compared with Rf values of standard compounds ferulic acid (0.72), salicylic acid (0.60), o-coumeric acid (0.74), trans-cinnamic acid (0.74), caffeic acid (0.85), p-coumaric acid (0.77).

The un-inoculated non infested control plant leaves extract contained caffeic acid (Rf 0.85) and quercetin (Rf 0.88). This was in contrast to *Bacillus endophyticus* inoculated plant leaves which exhibited some unidentified compounds at Rf 0.50 in addition to myricitin (Rf 0.73) o-coumaric (Rf 0.74) whereas, *Pseudomonas aeruginosa* inoculated plants showed the presence of flavonoids, ferulic acid, o-coumaric, kaempferol-7-neoheps-eridiside-glycosides in addition to some unidentified compounds of low polarity. Infestation with *S. litura* resulted in the production of caffeic acid and o-coumaric acid in addition to low and high polarity unidentified compounds. PGPR inoculated plants on infestation produced salicylic acid, rutin and kaempferol in addition to p-Coumaric acid and some unidentified compounds.

Table 1. Putative secondary metabolites identified on the basis of the Rf values in the extract of tomato leaves of different treatments.

Treatments	Rf Values	Color	Compounds
Control	0.85	Red	Caffeic acid
	0.85	Red	Quercitin
T1	0.50	Red	Unidentified
	0.73	Red	Myricitin
	0.79	Red	o-coumaric acid
T2	0.21	Red	Flavonoid-glycoside
	0.39	Red	Unidentified
	0.50	Red	Unidentified
	0.71	Red	Ferulic acid
	0.79	Red	o-coumaric acid
T3	0.55	Yellow	Kampferol-7-neoheps-eridiside
	0.14	Red	Unidentified
	0.41	Red	Unidentified
	0.84	Red	Caffeic acid
	0.76	Yellow	p-Cumaric acid
T4	0.16	Red	Unidentified
	0.23	Red	Unidentified
	0.43	Red	Rutin
	0.60	Red	Salicylic acid
	0.82	Red	Kaempferol
	0.76	Yellow	p-Cumaric acid
T5	0.16	Red	Unidentified
	0.23	Red	Unidentified
	0.43	Red	Rutin
	0.60	Red	Salicylic acid
	0.82	Red	Kaempferol
	0.77	Yellow	p-Cumaric acid

3.5. Fourier Transform Infrared Spectrometry (FTIR) of Tomato Leaves

The data presented in Table 2 revealed that control plant (uninfested and uninoculated) leaves extract had shown the presence of amines and amides with N-H stretch and bend, aliphatic amines stretching with C-N, alkenes with C-H bend and alkyl halides with C-Cl stretch. Plants inoculated with *Pseudomonas aeruginosa* exhibited an additional bonding indicating the presence of alkynes (at frequency of 638.51 with –C≡C–H: C–H bend) which were absent in uninoculated un-infested plants. While extract of plant leaves infested with *S. litura* exhibited alkanes with C–H stretch. In addition to amines and amides with N–H stretch, aliphatic amines stretching with C-N, alkenes with C-H bend and alkyl halides with C-Cl stretching. This was in contrast to plant leaves extract previously inoculated with *Bacillus endophyticus* (T4) or *Pseudomonas aeruginosa* and infested with *S. litura* (T5) which exhibited higher frequency of N-H stretch and =C-H bend and additional bonding indicating the presence of aldehyde and amine with H–C=O: C–H stretch, N–H bend (at frequencies of 2827.91 and 1630.78) as compared to plant extract infested with *S. litura* (T5).

Table 2. Fourier-transform infrared spectroscopy (FTIR) of thin-layer chromatography (TLC) eluent of tomato leaves under different treatments.

Treat.	Frequency	Bond	Functional Group	Characteristics of Peak
C	3408.9	N–H- stretch	1°, 2° amines/amides	Medium
	1632.5	N–H- bend	1° amines	Medium
	1068.3	C–N- stretch	aliphatic amines	Medium
	967.6	=C–H- bend	Alkenes	Strong
	799.5	C–Cl stretch	alkyl halides	Medium
T1	3412.90	N–H stretch	1°, 2° amines, amides	Medium
	1633.25	N–H- bend	1° amines	Medium
	1066.14	C–N- stretch	aliphatic amines	Strong
	967.48	=C–H- bend	Alkenes	Medium
	799.21	C–Cl stretch	alkyl halides	Broad, strong
T2	3410.75	N–H- stretch	1°, 2° amines, amides	Medium
	1633.24	C–N- stretch	aliphatic- amines	Strong
	1068.66	C–N- stretch	aliphatic amines	Strong
	967.13	=C–H- bend	Alkenes	Medium
	799.17	C–Cl- stretch	alkyl halides	Broad, strong
	638.51	–C≡C–H:C–H- bend	Alkynes	Broad, strong
T3	3376.84	N–H stretch	1°, 2° amines, amides	Medium
	2922.79	C–H- stretch	Alkanes	Medium
	1071.40	C–N- stretch	aliphatic amines	Strong
	966.62	=C–H- bend	Alkenes	Medium
	799.60	C–Cl- stretch	alkyl halides	Broad, strong
T4	3412.33	N–H- stretch	1°, 2° amines, amides	Medium
	2827.91	H–C=O: C–H- stretch	Aldehydes	Medium
	1630.78	N–H- bend	1° amines	Medium
	1066.63	C–N- stretch	aliphatic amines	Strong
	967.01	=C–H- bend	Alkenes	Medium
	799.50	C–Cl- stretch	alkyl halides	Broad, strong
T5	3405.01	N–H- stretch	1°, 2° amines, amides	Medium
	1632.25	N–H- bend	1° amines	Medium
	1066.10	C–N- stretch	aliphatic amines	Strong
	967.23	=C–H- bend	Alkenes	Medium
	799.56	C–Cl- stretch	alkyl halides	Broad, strong

Values are mean of 4 replications per treatment. Small amount of TLC eluent corresponding to the Rf-value of key bands were placed directly on the germanium piece of the infrared spectro-meter with persistent pressure and the infrared absorbance was collected over the wave number ranged from 4000 cm^{-1} –675 cm^{-1} and computerized for analyses by using the Omnic software.

4. Discussion

This paper evaluates the effect of PGPR as a growth promoter as well as a biocontrol agent. Although the study deals with tomatoes only in one season and also limited by the lack of behavioral study of the insect.

The growth parameters of tomato were considerably amplified after PGPR inoculation, under uninfested condition; the *Pseudomonas aeruginosa* being more effective. Of note, the effectivity of PGPR were higher under infested conditions and produce higher proline as osmoregulant, more defense hormone e.g., SA and higher level of growth promoting hormone, e.g., IAA contents and for inducing antioxidant enzyme, SOD. The PGPR inoculation not only overcame the infestation induced decrease in root and shoot weight but also increased the root and shoot weight. The PGPR effect was more pronounced on shoot dry weight. Similar results were reported by Avis et al. [43] and Babalola [44]. Shannag and Abadneh [45] reported that fresh and dry weight of shoot and root were decreased by *Aphis fabae Scopoli* in Faba Bean as compared to its respective control. Yadav et al. [46]

reported a marked increase in shoot and root dry weight in chickpea treated with PGPR. The increased weight of tomato fruit was correlated with the number of flowers, branches, plant biomass concomitant with the osmotic balance and alleviating oxidative stress. Both the PGPR were effective and significantly enhanced (≥35%) the fruit fresh weight. This increase in fruit fresh weight by the PGPR may be attributed to the fact that PGPR significantly improves the root growth and plant vigor which lead to enhanced fruit production. Fabro et al. [47] reported that tomato plants treated with PGPR showed the highest number of branches when compared to infested control. PGPR has positive effects on tomato fruit quality attributes, particularly on size and texture [48]. Widnyana [49] reported that inoculation of tomato plants with *Pseudomonas* and *Bacillus* sp. speeded up the plant growth and yield and protection against plant pathogens. Similar results that PGPR inoculation enhances the plant growth, yield and fruit weight were also reported by Almaghrabi et al. [50] and Murphy et al. [51].

Results revealed that PGPR alleviated the osmotic imbalance by increasing, proline content in the insect infested plants. The different PGPR behaved differently, both for proline production and antioxidant enzymes. It is demonstrated that *Pseudomonas aeruginosa* combats osmotic stress in infested plants through increase in sugar content as osmoregulant whereas, *Bacillus endophyticus* enhances proline content to combat osmotic imbalance. The osmotic stress is one of the secondary stresses caused by insect infestation. Proline is revealed to be an osmoregulant that accumulates in plants under a wide range of stress conditions [52,53]. It is well known that free proline accumulation in vascular plants demonstrated stresses including pathogen attack [54,55]. The accumulation of cellular osmolytes such as proline, sugar alcohols, glucosinolates etc. and soluble sugars and the expression of antioxidant systems help plants in sustaining cellular function, crucial for physiological stability of plants under stress. Ullah et al. [56] indicated that application of PGPR to plants displayed substantial increase in proline content as compared to untreated plants. Phenolics are produced by many plant species for protection against biotic or abiotic stress growth conditions and their accumulation correlates with antioxidant capacity of plants in a number of species [57,58].

Compatible solutes are used for osmotic adjustment under adverse environmental conditions [59,60]. The soluble carbohydrates in plants attacked by a fungal pathogen, as well as proportions of individual sugars, may be variously modified, both by plant regulatory mechanisms and by pathogen interference. There are several causes for quantitative and qualitative changes of sugars at the infection site. The level of sugars is reduced by their consumption for both energy and structural purposes, their uptake by the pathogen, while in autotrophic tissues it happens due to the inhibition of photosynthesis [61]. The results further demonstrate the PGPR induced changes in chlorophyll and carotenoids in normal and insect infested plants and the *P. aeruginosa* being most effective. Similar results were reported by Wang et al. [62] that PGPR isolates increased chlorophyll content significantly in tomatoes. Inoculation with *Pseudomonas* B-25 resulted in greater synthesis of chlorophyll than the diseased control. Botha et al. [63] showed that *Diuraphis noxia* feeding caused decrease in chlorophyll content in Tugela and decreased levels of chlorophyll a upon infestation [64,65].

It was demonstrated that the *P. aeruginosa* adjust osmotic stress following infestation by stimulation in antioxidant SOD activity. The PGPR effectively enhanced the SOD activity to scavenge the ROS and prevent oxidative stress in plant cells. The observed enhancement in PGPR induced SOD activity in infested plants is a mechanism to combat insect induced oxidative stress. Recently it has been reported by Sharma and Mathur [66] that PGPR alone and/or in association with fungi significantly enhanced the antioxidant enzyme activities in *Brassica juncea* infested with *Spodoptera litura* that lead to enhanced immune system against herbivory. Similarly, Zhao et al. [67] reported Aphid resistance in plants infested with *B. tabaci* nymphs, associated with enhanced antioxidant activities. They concluded that this resistance probably acted via interactions with SA-mediated defense responses.

PGPR promoted growth by nutrient acquisition and by producing bioactive compounds [68,69]. They also improve the nutrient uptake in plants by modulating plant hormones level, thereby increasing root proliferation [70]. However, response of the 2 PGPR differed substantially; *Bacillus endophyticus* exhibited lower IAA but higher GA than that of *Pseudomonas aeruginosa*. PGPR can control plant disease directly, through the production of antagonistic compounds, and indirectly, through the elicitation of a plant defense response [71]. Fernandez-Aunion et al. [72] also reported that PGPR enhances plant growth by synthesis of bioactive compounds and activating plant defense system.

While the PGPR-elicited ISR has been studied extensively in the model plant *Arabidopsis*, it is not well characterized in crop plants. The induction of ISR was investigated by *Bacillus cereus* strain BS107 against *Xanthomonas axonopodis* pv. *vesicatoria* in pepper leaves. Choudhary and Jobri [73] demonstrated the induction of ISR elicited by *Bacillus* spp. against several fungal bacterial and viral pathogens including root knot nematodes. Yang et al. [74] reported genetic evidence of the priming effect of a *rhizobacterium* on the expression of defense genes involved in ISR in pepper. A stronger negative effect of the PGPR on the performance of leaf folder larvae was noted in rice and found that combined treatment of PGPR is more effective than individually. Several plant secondary compounds such as glucosinolates and cyanogenic glycosides yield toxic products after hydrolysis by enzymes stored and liberated during attack by chewing insects [75,76].

It is demonstrated from the present findings that SA and ABA are both involved in inducing tolerance to plants, but the mechanism of inducing tolerance against the insect varied among the PGPR used. For example, *Bacillus endophyticus* inoculation ameliorated the adverse effects of insect infestation by significantly increasing SA and ABA many folds higher than infested plants whereas, *Pseudomonas aeruginosa* ameliorated the infestation by increasing SA higher than the former strain. Salicylic acid is the integral part of signal transduction pathways initiating resistance to disease and infection [77–79]. Plant defense in response to microbial attack is controlled by signaling molecules including SA, JA and ethylene [80]. SA is an important director of pathogen stimulated systemic acquired resistance (SAR), whereas JA and ET are compulsory for rhizobacteria-mediated induced systemic resistance (ISR) [81]. Branch et al. [82] found that SA is a vital constituent of motioning the induced resistance to root-knot nematodes [83].

Plant phenolics comprises a wide array of secondary metabolites including flavonoids, Cinnamic acid, Kaempferol, Coumaric acid as well as salicylic acid synthesized to provide resistance to plants. Their number, type and concentration increase under insect attack [84] and appear to be stimulated following PGPR application. *Pseudomonas aeruginosa* produced both flavonoid glycoside and kaempferol in addition to coumaric acid whereas, *Bacillus endophyticus* had only myricitin and lack kaempferol and coumaric acid but on infestation both produced similar bioactive metabolites in plant. The chromatographic separation of leaf extract also revealed the presence of bands corresponding to Rf value of SA as well as phenolic compounds e.g., Kaempferol and coumaric acid thereby demonstrating the induction of ISR by PGPR in the inoculated plants infested with *S. litura*. Generally, the role of phenolic compounds in defence is related to their antibiotic, antinutritional or unpalatable properties. Besides their involvement in plant- animal or plant-microbe interactions; plant phenolics also play a key role as antioxidants and stress signaling [85–87]. Hammerschmidt [88] reported that phenolic metabolites are related to the resistance phenomenon of plants against their enemies.

The FTIR spectrum was used to identify the functional group of the active components based on the peak value in the region of infrared radiation. Production of alkynes and aldehyde in plants inoculated with *Bacillus endophyticus* and *Pseudomonas aeruginosa* on infestation with *S. litura* demonstrate the PGPR induced defense strategy against insects. Similar results reported by Panda and Khush [89] that chemical derived substances e.g., alkanes, aldehydes, ketones, waxes are involved in host-plant resistance to insects' function as a protective layer to save the plant. Whereas, Shavit [90] reported that inoculation of tomato plants with *P. fluorescens* WCS417r enhanced the performance of the phloem feeding insect *Bemicia tabaci*. Previously, the FTIR has been applied to classify the actual structure of certain plant secondary metabolites [91]. FTIR is one of the extensively used approaches to categorize the chemical ingredients and clarify the compounds structures [92]. The FTIR of the leaves extract revealed the presence of additional peaks of aldehyde in the FTIR of leaves of infested plants pretreated with PGPR. Chehab et al. [93] reported that aldehydes play a positive role in plant defense. Plants defend themselves from pathogens attack by producing secondary metabolites and proteins [94,95].

5. Conclusions

The *Bacillus endophyticus* and *Pseudomonas aeruginosa* can be used to combat oxidative and osmotic stresses induced by *S. litura* infestation. Both the PGPRs combat insects induced adverse effects on plant growth and productivity through the production of phenolics, SA and ABA. *Bacillus endophyticus* was more effective in the improved defense strategy induction through the modulation of phytohormones and secondary metabolites. These PGPR are more effective under uninfested conditions and can be implicated as bioinoculant to endure the plants to cope better with insect infestation. Since there is alteration in the functional group and presence of aldehyde predominantly detected in plants treated with PGPR and infested with the insect armyworm. Further investigations using nuclear magnetic resonance (NMR) and liquid chromatography-mass spectrometry (LC-MS) are needed to unveil the secondary metabolites produced in PGPR inoculated plants versus uninoculated insect infested plants. Finally, an integrated approach of molecular mechanism of PGPR induced defense in plants against pests and parasites needs thorough investigation.

Author Contributions: Conceptualization, A.B.; Methodology, B.K. and N.K.; Software, B.K. and N.K.; Validation, B.K., A.B. and N.K.; Formal Analysis, N.K.; Investigation, B.K. and A.B.; Resources, A.B.; Data Curation, N.K.; Writing—Original Draft Preparation, B.K.; Writing—Review & Editing, A.B. and N.K; Visualization, B.K.; Supervision, A.B.; Project Administration, A.B. All authors have read and agreed to the published version of the manuscript.

References

1. Tyssandier, V.; Feillet-Coudray, C.; Caris-Veyrat, C.; Guilland, J.C.; Coudray, C.; Bureau, S.; Reich, M.; Amiot-Carlin, M.J.; Bouteloup-Demange, C.; Boirie, Y.; et al. Effect of tomato product consumption on the plasma status of antioxidant microconstituents and on the plasma total antioxidant capacity in healthy subjects. *J. Am. Coll. Nutr.* **2004**, *23*, 148–156. [CrossRef]

2. Friedman, M. Anticarcinogenic, cardioprotective, and other health benefits of tomato compounds lycopene, α-tomatine, and tomatidine in pure form and in fresh and processed tomatoes. *J Agr Food Chem.* **2013**, *61*, 9534–9550. [CrossRef]

3. Wu, C.J.; Fan, S.Y.; Jiang, Y.H.; Yao, H.H.; Zhang, A.B. Inducing gathering effect of taro on *Spodoptera lituraFabricius*. *Chin. J. Eco.* **2004**, *23*, 172–174. [CrossRef]

4. Ahmad, M.; Ghaffar, A.; Rafiq, M. Host plants of leaf worm, Spodoptera litura (Fabricius) (Lepidoptera: Noctuidae) in Pakistan. *Asian J. Agric. Biol.* **2013**, *1*, 23–28.

5. Khan, N.; Bano, A.; Zandi, P. Effects of exogenously applied plant growth regulators in combination with PGPR on the physiology and root growth of chickpea (*Cicer arietinum*) and their role in drought tolerance. *J. Plant Interact.* **2018**, *13*, 239–247. [CrossRef]

6. Sayyed, A.H.; Ahmad, M.; Saleem, M.A. Cross-resistance and genetics of resistance to indoxacarb in *Spodoptera litura* (*Lepidoptera: Noctuidae*). *J. Econ. Entomol.* **2008**, *101*, 472–479. [CrossRef]

7. Saleem, M.A.; Ahmad, M.; Aslam, M.; Sayyed, A.H. Resistance to selected organochlorine, organophosphate, carbamates and pyrethroid insecticides in *Spodoptera litura* (Lepidoptera: Noctuidae) from Pakistan. *J. Econ. Entomol.* **2008**, *101*, 1667–1675. [CrossRef]

8. Bais, H.P.; Weir, T.L.; Perry, L.G.; Gilroy, S.; Vivanco, J.M. The role of root exudates in rhizosphere interactions with plants and other organisms. *Annu. Rev. Plant Biol.* **2006**, *57*, 233–266. [CrossRef] [PubMed]

9. Khan, N.; Bano, A. Growth and Yield of Field Crops Grown Under Drought Stress Condition Is Influenced by the Application of PGPR. In *Field Crops: Sustainable Management by PGPR*; Springer Nature: Berlin/Heidelberg, Germany, 2019; pp. 337–349.

10. Nihorimbere, V.; Ongena, M.; Smargiassi, M.; Thonart, P. Beneficial effect of the rhizosphere microbial community for plant growth and health. *Biotechnol. Agron. Soc. Environ.* **2011**, *15*, 327.

11. Compant, S.; Duffy, B.; Nowak, J.; Clément, C.; Barka, E.A. Use of plant growth-promoting bacteria for biocontrol of plant diseases: Principles, mechanisms of action, and future prospects. *Appl. Environ. Microbiol.* **2013**, *71*, 4951–4959. [CrossRef]

12. War, A.R.; Paulraj, M.G.; War, M.Y.; Ignacimuthu, S. Differential defensive response of groundnut to *Helicoverpa armigera* (Hubner) (Lepidoptera: Noctuidae). *J. Plant Interact.* **2011**, *7*, 45–55. [CrossRef]

13. Khan, N.; Bano, A.; Rahman, M.A.; Rathinasabapathi, B.; Babar, M.A. UPLC-HRMS-based untargeted metabolic profiling reveals changes in chickpea (*Cicer arietinum*) metabolome following long-term drought stress. *Plant Cell Environ.* **2019**, *42*, 115–132. [CrossRef] [PubMed]

14. Naseem, H.; Ahsan, M.; Shahid, M.A.; Khan, N. Exopolysaccharides producing rhizobacteria and their role in plant growth and drought tolerance. *J. Basic Microbiol.* **2018**, *58*, 1009–1022. [CrossRef] [PubMed]

15. Vaikuntapu, P.R.; Dutta, S.; Samudrala, R.B.; Rao, V.R.; Kalam, S.; Podile, A.R. Preferential promotion of Lycopersicon esculentum (Tomato) growth by plant growth promoting bacteria associated with tomato. *Indian J. Microbiol.* **2014**, *54*, 403–412. [CrossRef]

16. Passari, A.K.; Upadhyaya, K.; Singh, G.; Abdel-Azeem, A.M.; Thankappan, S.; Uthandi, S.; Hashem, A.; Abd_Allah, E.F.; Malik, J.A.; Alqarawi, A.S.; et al. Enhancement of disease resistance, growth potential, and photosynthesis in tomato (*Solanum lycopersicum*) by inoculation with an endophytic actinobacterium, Streptomyces thermocarboxydus strain BPSAC147. *PLoS ONE* **2019**, *14*, e0219014. [CrossRef]

17. Moradali, M.F.; Ghods, S.; Rehm, B.H. *Pseudomonas aeruginosa* lifestyle: A paradigm for adaptation, survival, and persistence. *Front. Microbiol.* **2017**, *7*, 39. [CrossRef]

18. Khan, N.; Zandi, P.; Ali, S.; Mehmood, A.; Adnan Shahid, M.; Yang, J. Impact of salicylic acid and PGPR on the drought tolerance and phytoremediation potential of Helianthus annus. *Front. Microbiol.* **2018**, *9*, 2507. [CrossRef]

19. Quais, M.K.; Munawar, A.; Ansari, N.A.; Zhou, W.W.; Zhu, Z.R. Interactions between brown planthopper (Nilaparvata lugens) and salinity stressed rice (Oryza sativa) plant are cultivar-specific. *Sci. Rep.* **2020**, *10*, 1–14. [CrossRef]

20. Farooq, U.; Bano, A. Screening of indigenous bacteria from rhizosphere of maize (Zea mays L.) for their plant growth promotion ability and antagonism against fungal and bacterial pathogens. *J. Anim. Plant Sci.* **2013**, *23*, 1642–1652.

21. Naz, R.; Bano, A.; Yasmin, H.; Samiullah, H.; Farooq, U. Antimicrobial potential of the selected plant species against some infectious microbes used. *J. Med. Plants Res.* **2011**, *5*, 5247–5253.

22. Egamberdieva, D.; Wirth, S.J.; Alqarawi, A.A.; Abd_Allah, E.F.; Hashem, A. Phytohormones and beneficial microbes: Essential components for plants to balance stress and fitness. *Front. Microbiol.* **2017**, *8*, 2104. [CrossRef] [PubMed]

23. Khan, N.; Bano, A.; Ali, S.; Babar, M.A. Crosstalk amongst phytohormones from planta and PGPR under biotic and abiotic stresses. *Plant. Growth Regul.* **2020**, *90*, 189–203. [CrossRef]

24. Kudoyarova, G.; Arkhipova, T.N.; Korshunova, T.; Bakaeva, M.; Loginov, O.; Dodd, I.C. Phytohormone mediation of interactions between plants and non-symbiotic growth promoting bacteria under edaphic stresses. *Front. Plant Sci.* **2019**, *10*, 1368. [CrossRef] [PubMed]

25. Kumar, A.; Patel, J.S.; Meena, V.S.; Ramteke, P.W. Plant growth-promoting rhizobacteria: Strategies to improve abiotic stresses under sustainable agriculture. *J. Plant Nutr.* **2019**, *42*, 1402–1415. [CrossRef]

26. Mahmoudi, T.R.; Yu, J.M.; Liu, S.; Pierson III, L.S.; Pierson, E.A. Drought-Stress Tolerance in Wheat Seedlings Conferred by Phenazine-Producing Rhizobacteria. *Front. Microbiol.* **2019**, *10*, 1590. [CrossRef]

27. Baliyan, S.P.; Rao, M.S. Evaluation of tomato varieties for pest and disease adaptation and productivity in Botswana. *Int. J. Agric. Food Res.* **2013**, *2*, 2. [CrossRef]

28. Lowry, O.H.; Rosebrough, N.J.; Farr, A.L.; Randall, R.J. Protein measurement with the Folin phenol reagent. *J. Biol. Chem.* **1951**, *193*, 265–275.

29. Arnon, D.I. Copper enzymes in isolated chloroplasts. Polyphenoloxidase in Beta vulgaris. *Plant Physiol.* **1949**, *24*, 1–15. [CrossRef]

30. Lichtenthaler, H.K.; Wellburn, A.R. Determinations of total carotenoids and chlorophylls a and b of leaf extracts in different solvents. *Biochem. Soc. Trans.* **1983**, *11*, 591–592. [CrossRef]

31. Bates, L.S.; Waldren, R.P.; Teare, I.D. Rapid determination of free proline for water-stress studies. *Plant Soil* **1973**, *39*, 205–207. [CrossRef]

32. Dubois, S.M.; Giles, K.A.; Hamilton, J.K.; Rebers, P.A.; Smith, F. Calorimetric method for determination of sugar and related substance. *Anal. Chem.* **1956**, *28*, 350. [CrossRef]

33. Beauchamp, C.; Fridovich, I. Superoxide dismutase: Improved assays and an assay applicable to acrylamide gels. *Anal. Biochem.* **1971**, *44*, 276–287. [CrossRef]

34. Kettner, J.; Dörffling, K. Biosynthesis and metabolism of abscisic acid in tomato leaves infected with Botrytis cinerea. *Planta* **1995**, *196*, 627–634. [CrossRef]

35. Hansen, H.; Dörffling, K. Root-derived trans-zeatin riboside and abscisic acid in drought-stressed and rewatered sunflower plants: Interaction in the control of leaf diffusive resistance? *Funct. Plant Biol.* **2003**, *30*, 365–375. [CrossRef]

36. Enyedi, A.J.; Raskin, I. Induction of UDP-glucose: Salicylic acid glucosyltransferase activity in tobacco mosaic virus-inoculated tobacco (*Nicotiana tabacum*) leaves. *Plant Physiol.* **1993**, *101*, 1375–1380. [CrossRef]

37. Sesker, M.; Shulaev, V.; Raskin, I. Endogenous methyl salicylate in pathogen-inoculated tobacco plants. *Plant Physiol.* **1998**, *116*, 387–392. [CrossRef]

38. Khan, N.; Bano, A. Effects of exogenously applied salicylic acid and putrescine alone and in combination with rhizobacteria on the phytoremediation of heavy metals and chickpea growth in sandy soil. *Int. J. Phytoremediat.* **2018**, *20*, 405–414. [CrossRef]

39. Subramanian, S.; Ramakrishnan, N. Chromatographic finger print analysis of Naringi crenulata by HPTLC technique. *Asian Pac. J. Trop. Biomed.* **2011**, *1*, S195–S198. [CrossRef]

40. Lu, X.; Wang, J.; Al-Qadiri, H.M.; Ross, C.F.; Powers, J.R.; Tang, J.; Rasco, B.A. Determination of total phenolic content and antioxidant capacity of onion (*Allium cepa*) and shallot (*Allium oschaninii*) using infrared spectroscopy. *Food Chem.* **2011**, *129*, 637–644. [CrossRef]

41. Liu, Y.; Hu, T.; Wu, Z.; Zeng, G.; Huang, D.; Shen, Y.; He, X.; Lai, M.; He, Y. Study on biodegradation process of lignin by FTIR and DSC. *Environ. Sci. Pollut. Res.* **2014**, *21*, 14004–14013. [CrossRef]

42. Lammers, K.; Arbuckle-Keil, G.; Dighton, J. FT-IR study of the changes in carbohydrate chemistry of three New Jersey pine barrens leaf litters during simulated control burning. *Soil Bio Biochem.* **2009**, *41*, 340–347. [CrossRef]

43. Avis, T.J.; Gravel, V.; Antoun, H.; Tweddell, R.J. Multifaceted beneficial effects of rhizosphere microorganisms on plant health and productivity. *Soil Biol. Biochem.* **2008**, *40*, 1733–1740. [CrossRef]

44. Babalola, O.O. Beneficial bacteria of agricultural importance. *Biotechnol. Lett.* **2010**, *32*, 1559–1570. [CrossRef] [PubMed]

45. Shannag, H.K.; Abadneh, J.A. Influence of Black Bean Aphids, *Aphis fabae Scopoli*. On growth rates of Faba Bean. *World J. Agric. Sci.* **2007**, *3*, 344–349.

46. Yadav, J.; Jay, P.V.; Kavindra, N.T. Plant growth promoting activities of fungi and their effect on chickpea plant growth. *Asian J. Biol. Sci.* **2011**, *4*, 291–299. [CrossRef]

47. Ali, N.; Hadi, F. Phytoremediation of cadmium improved with the high production of endogenous phenolics and free proline contents in Parthenium hysterophorus plant treated exogenously with plant growth regulator and chelating agent. *Environ Sci Pollut.* **2015**, *22*, 13305–13318. [CrossRef]

48. Mena-Violante, H.G.; Olalde-Portugal, V. Alteration of tomato fruit quality by root inoculation with plant growth-promoting rhizobacteria (PGPR): Bacillus subtilis BEB-13bs. *Sci. Hortic.* **2007**, *113*, 103–106. [CrossRef]

49. Widnyana, I.K. PGPR (Plant Growth Promoting Rizobacteria) Benefits in Spurring Germination, Growth and Increase the Yield of Tomato Plants. In *Recent Advances in Tomato Breeding and Production*; IntechOpen: London, UK, 2018.

50. Almaghrabi, O.A.; Massoud, S.I.; Abdelmoneim, T.S. Influence of inoculation with plant growth promoting rhizobacteria (PGPR) on tomato plant growth and nematode reproduction under greenhouse conditions. *Saudi J. Biol. Sci.* **2013**, *20*, 57–61. [CrossRef]

51. Murphy, J.F.; Reddy, M.S.; Ryu, C.M.; Kloepper, J.W.; Li, R. Rhizobacteria-mediated growth promotion of tomato leads to protection against Cucumber mosaic virus. *Phytopathol* **2003**, *93*, 1301–1307. [CrossRef]

52. Lu, H.F.; Cheng, C.G.; Tang, X.; Hu, Z.H. Spectrum of *Hypericum* and *Triadenum* with reference to their identification. *J. Integr. Plant Biol.* **2004**, *46*, 401–406. [CrossRef]

53. Fabro, G.; Kovács, I.; Pavet, V.; Szabados, L.; Alvarez, M.E. Proline accumulation and AtP5CS2 gene activation are induced by plant-pathogen incompatible interactions in Arabidopsis. *Mol. Plant-Microbe Interact.* **2004**, *17*, 343–350. [CrossRef] [PubMed]

54. Janardan, Y.; Verma, J.P.; Tiwari, K.N. Effect of plant growth promoting rhizobacteria on seed germination and plant growth chickpea (*Cicer arietinum* L.) under in vitro conditions. *Biol. Forum* **2010**, *2*, 15–18.

55. Isah, T. Stress and defense responses in plant secondary metabolites production. *Biol. Res.* **2009**, *52*, 39. [CrossRef] [PubMed]

56. Ullah, S.; Ashraf, M.; Asghar, H.N.; Iqbal, Z.; Ali, R. Review Plant growth promoting rhizobacteria-mediated amelioration of drought in crop plants. *Soil Environ.* **2019**, *38*, 1–20. [CrossRef]

57. Abideen, Z.; Qasim, M.; Rasheed, A.; Adnan, M.Y.; Gul, B.; Khan, M.A. Antioxidant activity and polyphenolic content of *Phragmites karka* under saline conditions. *Pak. J. Bot.* **2015**, *47*, 813–818.

58. Welbaum, G.E.; Sturz, A.V.; Dong, Z.; Nowak, J. Managing soil microorganisms to improve productivity of agro-ecosystems. *Crit. Rev. Plant Sci.* **2004**, *23*, 175–193. [CrossRef]

59. Ullah, S.; Asema, M.; Asghari, B. Effect of PGPR on growth and performance of *Zea mays*. *Res. J. Agric. Environ. Manag.* **2013**, *2*, 434–447.

60. Willcox, J.K.; Catignani, G.L.; Lazarus, S. Tomatoes and cardiovascular health. *Crit. Rev. Food Sci. Nutr.* **2003**, *43*, 1–18. [CrossRef]

61. Chen, Y.; Cao, C.; Guo, Z.; Zhang, Q.; Li, S.; Zhang, X.; Gong, J.; Shen, Y. Herbivore exposure alters ion fluxes and improves salt tolerance in a desert shrub. *Plant Cell Environ.* **2020**, *43*, 400–419. [CrossRef]

62. Berger, S.; Sinha, A.K.; Roitsch, T. Plant physiology meets phytopathology: Plant primary metabolism and plant–pathogen interactions. *J. Exp. Bot.* **2007**, *58*, 4019–4026. [CrossRef]

63. Wang, C.J.; Yang, W.; Wang, C.; Gu, C.; Niu, D.D.; Liu, H.X.; Wang, Y.P.; Guo, J.H. Induction of drought tolerance in cucumber plants by a consortium of three plant growth-promoting rhizobacterium strains. *PLoS ONE* **2012**, *7*, e52565. [CrossRef] [PubMed]

64. Botha, A.M.; Lacock, L.; van Niekerk, C.; Matsioloko, M.T.; du Preez, F.B.; Loots, S.; Venter, E.; Kunert, K.J.; Cullis, C.A. Is photosynthetic transcriptional regulation in Triticum aestivum L. cv. 'TugelaDN' a contributing factor for tolerance to Diuraphis noxia (Homoptera: Aphididae)? *Plant Cell Rep.* **2006**, *25*, 41–54. [CrossRef] [PubMed]

65. Burd, J.D.; Elliott, N.C. Changes in chlorophyll a fluorescence induction kinetics in cereals infested with Russian wheat aphid (Homopetra: Aphididea). *J. Econ. Entomol.* **1996**, *89*, 1332–1337. [CrossRef]

66. Ni, X.; Quisenberry, S.S.; Heng-Moss, T.; Markwell, J.; Higley, L.; Baxendale, F.; Sarath, G.; Klucas, R. Dynamic change in photosynthetic pigments and chlorophyll degradation elicited by cereal aphid feeding. *Entomol. Exp. Appl.* **2002**, *105*, 43–53. [CrossRef]

67. Sharma, G.; Mathur, V. Modulation of insect-induced oxidative stress responses by microbial fertilizers in Brassica juncea. *FEMS Microbiol. Ecol.* **2020**, *96*, 40. [CrossRef]

68. Zhao, Z.H.; Hui, C.; Hardev, S.; Ouyang, F.; Dong, Z.; Ge, F. Responses of cereal aphids and their parasitic wasps to landscape complexity. *J. Econ. Entomol.* **2014**, *107*, 630–637. [CrossRef]

69. Mehboob, I.; Zahir, Z.A.; Arshad, M.; Tanveer, A.; Khalid, M. Comparative effectiveness of different Rhizobium sp. for improving growth and yield of maize (*Zea mays* L.). *Soil Environ.* **2012**, *31*, 37–46.

70. Zhang, S.; Moyne, A.L.; Reddy, M.S.; Kloepper, J.W. The role of salicylic acid in induced systemic resistance elicited by plant growth-promoting rhizobacteria against blue mold of tobacco. *Biol. Control.* **2002**, *25*, 288–296. [CrossRef]

71. Bobby, M.N.; Wesely, E.G.; Johnson, M. FT-IR studies on the leaves of *Albizia lebbeck* benth. *Int. J. Pharm. Pharm. Sci.* **2012**, *4*, 293–296.

72. Mansour, M.F. Nitrogen containing compounds and adaptation of plants to salinity stress. *Biol. Plant.* **2000**, *43*, 491–500. [CrossRef]

73. Fernandez-Aunión, C.; Hamouda, T.B.; Iglesias-Guerra, F.; Argandoña, M.; Reina-Bueno, M.; Nieto, J.J.; Aouani, M.E.; Valrgas, C. Biosynthesis of compatible solutes in rhizobial strains isolated from Phaseolus vulgaris nodules in Tunisian fields. *BMC Microbiol.* **2010**, *10*, 192. [CrossRef] [PubMed]

74. Choudhary, D.K.; Johri, B.N. Interactions of Bacillus spp. and plants–with special reference to induced systemic resistance (ISR). *Microbiol. Res.* **2009**, *164*, 493–513. [CrossRef] [PubMed]

75. Yang, J.; Yen, H.E. Early salt stress effects on the changes in chemical composition in leaves of ice plant and Arabidopsis. A Fourier transform infrared spectroscopy study. *Plant Physiol.* **2000**, *130*, 1032–1042. [CrossRef] [PubMed]

76. Liu, H.; Sun, S.; Lv, G.; Chan, K.C. Study on Angelica and its different extracts by Fourier transform infrared spectroscopy and two-dimensional correlation IR spectroscopy. *Spectrochim. Acta Part A* **2006**, *64*, 321–326. [CrossRef]

77. Pineda, A.; Zheng, S.J.; Van Loon, J.J.; Pieterse, C.M.; Dicke, M. Helping plants to deal with insects: The role of beneficial soil-borne microbes. *Trends Plant Sci.* **2010**, *15*, 507–514. [CrossRef]

78. Schoonhoven, L.M.; Van Loon, B.; van Loon, J.J.; Dicke, M. *Insect-Plant Biology*; Oxford University Press on Demand: Oxford, UK, 2005.

79. Natikar, P.K.; Balika, R.A. Tobacco caterpillar Spodoptera litura (Fabricus) toxicity, ovicidal action, oviposition deterrent activity, ovipositional preference and its management. *Biochem. Cell. Arch* **2015**, *15*, 383–389.

80. Brimecombe, M.J.; De Leij, F.A.; Lynch, J.M. Rhizodeposition and microbial populations. In *the Rhizosphere Biochemistry and Organic Substances at the Soil-Plant Interface*; Pinton, R., Veranini, Z., Nannipieri, P., Eds.; Taylor and Francis: New York, NY, USA, 2007; pp. 73–110.

81. Tsukanova, K.A.; Ch botar, V..; Meyer, J.J.M.; Bibikova, T.N. Effect of plant growth-promoting Rhizobacteria on plant hormone homeostasis. *South Afri. J. Bot.* **2017**, *113*, 91–102. [CrossRef]

82. Mendes, R..; Garbeva, P..; Raaijmakers, J.M. The Rhizosphere Microbiome: Significance of Plant Beneficial, Plant Pathogenic, and Human Pathogenic Microorganisms. *FEMS Microbiol. Rev.* **2013**, *37*, 634–663. [CrossRef]

83. Branch, C.; Hwang, C.F.; Navarre, D.A.; Williamson, V.M. Salicylic acid is part of the Mi-1-mediated defense response to root-knot nematode in tomato. *Mol. Plant-Microbe Interact.* **2004**, *17*, 351–356. [CrossRef]

84. Ton, J.; Van-Pelt, J.A.; Van Loon, L.C.; Pieterse, C.M. Differential effectiveness of salicylate-dependent and jasmonate/ethylene-dependent induced resistance in *Arabidopsis*. *Mol. Plant-Microbe Interact.* **2002**, *15*, 27–34. [CrossRef]

85. Dixit, G.; Srivastava, A.; Rai, K.M.; Dubey, R.S.; Srivastava, R.; Verma, P.C. Distinct defensive activity of phenolicsandphenylpropanoid pathway genesin different cotton varieties toward chewing pests. *Plant. Signal. Behav.* **2020**, *15*, 1747689. [CrossRef] [PubMed]

86. Lattanzio, V.; Lattanzio, V.M.; Cardinali, A. Role of phenolics in the resistance mechanisms of plants against fungal pathogens and insects. *Phytochem. Adv. Res.* **2006**, *661*, 23–67.

87. Jacobo-Velázquez, D.A.; González-Agüero, M.; Cisneros-Zevallos, L. Cross-talk between signaling pathways: The link between plant secondary metabolite production and wounding stress response. *Sci. Rep.* **2015**, *5*, 8608. [CrossRef] [PubMed]

88. Weber, D.; Egan, P.A.; Muola, A.; Ericson, L.E.; Stenberg, J.A. plant resistance does not compromise parasitoid-based biocontrol of a strawberry pest. *Sci. Rep.* **2020**, *10*, 1–10. [CrossRef]

89. Hammerschmidt, R. Phenols and plant–pathogen interactions: The saga continues. *Physiol. Mol. Plant Pathol.* **2005**, *66*, 77–78. [CrossRef]

90. Panda, N.; Khush, G.A. *Host Plant Resistance to Insects*; CAB International: Wallingford, UK, 1995.

91. Shavit, R.; Lalzar, M.O.; Saul, B.; Morin, S. Inoculation of tomato plants with rhizobacteria enhances the performance of the phloem-feeding insect *Bemisia tabaci*. *Front. Plant Sci.* **2013**, *4*, 1941–1947. [CrossRef]

92. Subramanion, L.J.; Zakaria, Z.; Sreenivasan, S. Phytochemicals screening, DPPH free radical scavenging and xanthine oxidase inhibitiory activities of *Cassia fistula* seeds extract. *J. Med. Plants Res.* **2011**, *5*, 1941–1947.

93. Khan, N.; Bano, A. Modulation of phytoremediation and plant growth by the treatment with PGPR, Ag nanoparticle and untreated municipal wastewater. *Int. J. phytoremediat.* **2016**, *18*, 1258–1269. [CrossRef]

94. Mintenig, S.M.; Int-Veen, I.; Löder, M.G.; Primpke, S.; Gerdts, G. Identification of microplastic in effluents of waste water treatment plants using focal plane array-based micro-Fourier-transform infrared imaging. *Water Res.* **2017**, *108*, 365–372. [CrossRef]

95. Chehab, E.W.; Eich, E.; Braam, J. Thigmomorphogenesis: A complex plant response to mechano-stimulation. *J. Exp. Bot.* **2009**, *60*, 43–56. [CrossRef]

11

Nematicidal Evaluation and Active Compounds Isolation of *Aspergillus japonicus* ZW1 against Root-Knot Nematodes *Meloidogyne incognita*

Qiong He [1], Dongya Wang [1], Bingxue Li [1], Ambreen Maqsood [1,2]🆔 and Haiyan Wu [1,*]🆔

[1]   Guangxi Key Laboratory of Agric-Environment and Agric-Products Safety, Agricultural College of Guangxi University, Nanning 530004, China; heqiong3344@163.com (Q.H.); wdy15677131171@163.com (D.W.); 18437958381@163.com (B.L.); ambreenagrarian@gmail.com (A.M.)

[2]   Department of Plant Pathology, Faculty of Agriculture and Environmental Sciences, The Islamia University of Bahawalpur, Bahawalpur 63100, Pakistan

*   Correspondence: whyzxb@gmail.com

Abstract: The root-knot nematode is one of the most damaging plant-parasitic nematodes worldwide, and the ecofriendly alternative approach of biological control has been used to suppress nematode populations. Here the nematicidal activity of *Aspergillus japonicus* ZW1 fermentation filtrate against *Meloidogyne incognita* was evaluated in vitro and in greenhouse, and the effects of *A. japonicus* ZW1 fermentation filtrate on seed germination and the active compound of *A. japonicus* ZW1 fermentation filtrate were determined. The 2-week fermentation filtrate (2-WF) of *A. japonicus* ZW1 exhibited markedly inhibitory effects on egg hatching, and 5% 2-WF showed potential nematicidal activities on second-stage juveniles (J2s); the mortality of J2s was 100% after 24 h exposure. The internal contents of nematodes were degraded and remarkable protruded wrinkles were present on the body surface of J2s. The nematicidal activity of the fermentation was stable after boiling and was not affected by storage time. A germination assay revealed that 2-WF did not have a negative effect on the viability and germination of corn, wheat, rice, cowpeas, cucumbers, soybeans, or tomato seeds. The pot-grown study confirmed that a 20% fermentation broth solution significantly reduced root galls and egg numbers on tomatoes, and decreased galls and eggs by 47.3% and 51.8% respectively, over Czapek medium and water controls. The active compound from the *A. japonicus* ZW1 fermentation filtrate was isolated and identified as 1,5-Dimethyl Citrate hydrochloride ester on the basis of nuclear magnetic resonance (NMR) and LC-MS (liquid chromatograph-mass spectrometer) techniques. Thus, fermentation of *A. japonicus* ZW1 could be considered a potential new biological nematicide for the control of *M. incognita*.

Keywords: biocontrol *Aspergillus japonicus*; root-knot nematode; fermentation filtrate; biological control; seed germination

1. Introduction

Root-knot nematodes (*Meloidogyne* spp.) are economically important worldwide pathogens causing considerable damage to many crops, including cucumbers, tomatoes, rice [1–4], and even cotton [5,6]. *Meloidogyne incognita* is an important species of root-knot nematodes worldwide due to its direct impact on crop yields [7–9]. Specifically, it is capable of causing an estimated yield loss of 5–43% within vegetable crops cultivated in tropical and subtropical areas [10] and estimated $100 billion loss per year worldwide [11].

Due to their short life cycle and high reproduction rates, these root-knot nematodes have been particularly challenging to control. Previously, chemical nematicides are efficiently used to suppress nematode populations, such as fenamiphos, sebufos, dazomet, and carbofuran [12]; however, these have been found to be harmful to both the eco-environment and human health due to their toxic effects. Thus, as a result of these negative impacts and the significant economic losses which can result from nematodes, new and alternative biological control options are urgently needed [13]. Therefore, the use of biological agents to suppress the population of plant-parasitic nematodes could provide an alternative strategy to sustainably manage plant-parasitic nematodes. Using biofumigation instead of harmful fumigants (like synthetic nematicide methyl bromide) to control nematodes is an increasingly feasible method of parasitic nematode management [14]. Plants such as *Melia azedarach* have been found to be potential sources of biofumigation plant material to control *Meloidogyne* spp. on tomato [15]. Moreover, microbial agents for the control of plant-parasitic nematodes is also a potential method; such as bacteria [16,17], fungi [18,19] and actinomycetes [20], which are nematophagous or antagonistic for root-knot nematodes. Specifically, *Arthrobotrys irregularis*, *Pochonia chlamydosporium*, *Paecilomyces lilacinus*, *Myrothecium verrucaria*, bacteria *Pasteuria usgae*, *Bacillus firmus*, *Burkholderia cepacia*, *Pseudomonas fluorescens*, and *Streptomyces avermitilis* [21,22] have been commercially used in many countries for the control of plant-parasitic nematodes. Some potential microbial sources were constantly obtained, volatiles from beneficial bacteria (*Bacillus* sp., *Paenibacillus* sp. and *Xanthomonas* sp.) can control *M. graminicola* second-stage juveniles (J2s) on rice and significantly reduced infection of susceptible rice [23]. Co-inoculation of *Streptomyces* spp. strains KPS-E004 and KPS-A032 showed success in suppressing root-knot nematode [24].

In our previous study, *A. japonicus* ZW1 culture filtrate was shown to have marked nematicidal activity against *M. incognita*. As a result, the main objective of this work was to evaluate the potential biological control of *A. japonicus* ZW1 against root-knot nematodes including: (1) the nematicidal activity of *A. japonicus* ZW1 fermentation filtrate on eggs and J2s within pot and in vitro experiments; (2) electron microscopic evaluation of J2 bodies after treatment with 2-week fermentation filtrate (2-WF); (3) effect of boiling and storage time on nematicidal activity stability of the fermentation filtrate; and (4) evaluation for the effect of *A. japonicus* ZW1 fermentation filtrate on the germination of various crop seeds.

2. Materials and Methods

2.1. Nematode Preparation

Tomato seeds (cv. Xin Bite 2 F1) were sourced from Yashu Garden Seeds Co., Ltd., (Guangzhou, China) and were used to generate seedlings for culturing the *M. incognita*. For the nematodes culture, one-month-old tomato seedlings were transplanted into pots ($7 \times 7 \times 8$ cm) with second stage juveniles of root-knot nematode-infected peat moss (Gui Yu Xin Nong Technology Co., Ltd., Nanning, China) and maintained at 25 °C with a 14 h light (22000 Lux) and 10 h dark photoperiod treatment within a GXZ-280C incubator (Jiangnan Instrument Factory, Ningbo, China). Tomato roots were collected 35 days after inoculation and were gently rinsed with tap water. Eggs were then extracted with 1% NaOCl [25] and hatched at 25 °C using the modified Baermann funnel method [26]. Eggs were put in 30 ìm pore sieves, nested in petri dishes (6 cm-diameter) containing 3 mL distilled water, and the fresh J2s in water were then collected on the day of experiment and used for subsequent experimentation.

2.2. Fermentation Filtrate Preparation

A. japonicus ZW1 from soil was deposited in the China Center for Type Culture Collection (accession number CCTCC No. M 2014641) and GenBank (accession number KR708636.1). One cm^2 potato dextrose agar (PDA) with a fresh culture of *A. japonicus* ZW1 (cultured 3-5 days at 25 °C) was inoculated in triangular flasks with 100 mL Czapek medium ($NaNO_3$ 0.2 g, KCl 0.05 g, $FeSO_4$ 0.001 g, K_2HPO_4 0.1 g, $MgSO_4$ 0.05 g, Sucrose 3.0 g, H_2O 100 mL) and incubated in a MQD-S2R shaker (Minquan Instrument Co., Ltd., Shanghai, China) at 150 rpm and 25 °C [27] for 3 consecutive weeks, with 10 triangular flasks replicates per week. Czapek medium without inoculation was used as a negative control. At the end of the 3-week period, fermentation broth from a total of 30 conical flasks was then filtered using 0.45 μm Millipore filters (Whatman, Clifton, NJ, USA) and 1-week fermentation filtrate (1-WF), 2-WF, and 3-week fermentation filtrate (3-WF) were prepared. The concentration of 2.5% (i.e., fermentation filtrate volume: sterilized water volume = 1:39), 5% (1:19), 10% (1:9), 20% (1:4) and 50% (1:1) of 1-week fermentation filtrate(1-WF), 2-week fermentation filtrate (2-WF), and 3-week fermentation filtrate (3-WF) were used and 20% Czapek medium and sterilized water were used as control.

2.3. Effect of Fermentation Filtrate on Meloidogyne Incognita Egg Hatching

Fresh eggs were treated with 2.5%, 5%, 10%, 20%, and 50% 1-WF, 2-WF, and 3-WF; and also 20% Czapek medium and sterilized water as controls. The specific experimental conditions were as follows: approximately 100 eggs and 200 μL of different concentrations of fermentation filtrate were dispensed into each well of 96-well plate, with 4 replicates for each treatment. Additionally, all experiments were performed in triplicate. The initial number of eggs was counted, and the hatched J2s were recorded using an inverted microscope (Ti-S, Nikon Instruments Inc., Tokyo, Japan) at 0, 3, 6, 9, 12, 15 d after exposure in the dark at 25 °C. The cumulative hatching rate was calculated using the following formula: cumulative hatching rate = (the number of hatched J2s)/(the initial number of eggs) × 100%.

2.4. Nematicidal Activity of Fermentation Filtrate on Meloidogyne Incognita J2s

Approximately 60 fresh J2s were contained in each well of a 96-well plate and treated with 200 μL of 2.5%, 5%, 10%, 20%, and 50% 1-WF, 2-WF, and 3-WF, 20% Czapek medium and sterilized water. The number of dead nematodes were counted using a Ti-S Nikon microscope (Nikon Instruments Inc., Tokyo, Japan) at 6, 12, 24, 48 h after treatment with the solutions and pictures were taken at each time point except for 48 h. It wax determined whether he bodies of dead J2s were straight and lacking movement even after mechanical prodding [28,29]. The test was conducted at 25 °C in the dark and the experiment was replicated 4 times. J2 mortality was calculated for each well as follows: mortality = (the number of dead J2s/total J2s) × 100%. This experiment was performed a total of three times.

2.5. Scanning Electron Microscopy Observations

J2s were treated with 10% 2-WF for 10 h and subsequently analyzed with scanning electron microscopy (SEM) using the approach as described below [30,31]. In preparation for the microscopic evaluations, J2 specimens were fixed in 2.5% glutaraldehyde with 0.1 M phosphate buffer (pH 7.2) at 4 °C overnight and subsequently washed 3 times in 0.1 M phosphate buffer. Afterwards, they were then fixed in 1% osmium tetroxide for 2 h, washed 3 times in 0.1 M phosphate buffer again, dehydrated in a graded series of ethanol, critical point dried with Quorum K850 critical dryers (Emitech, East Sussex, England, UK) and finally sputter coated with MSP-2S gold-palladium (IXRF, Austin, TX, USA). Prepared J2 specimens were observed using a SU8100 scanning electron microscope (Hitachi, Tokyo, Japan) operating at 3.0 kV accelerating voltage.

2.6. Transmission Electron Microscopy Observations

The technical approach was very similar to the aforementioned method described for 'scanning electron microscopy observations'; however, after J2s were dehydrated with ethanol, they were subsequently embedded in Araldite (Sigma-Aldrich, Sigma-Aldrich LLC., Darmstadt, Germany). To enable evaluation of the specimens, ultrathin sections (70 nm) were obtained using an EM UC7 ultramicrotome (Leica, Wetzlar, Germany) with a Diatome Ultra 45° diamond knife (Diatome Ltd., Helmstrasse Nidau, Switzerland). Sectioned samples were then stained with uranyl acetate and lead citrate using carbon film copper 500 mesh [30,32]. Sections of the J2 bodies were then observed using an HT7700 transmission electron microscope (Hitachi, Tokyo, Japan) operating at an 80.0 kV accelerating voltage.

2.7. Greenhouse Experiment

Thirty day old (3–4 leaf stage) healthy tomato seedlings (cv. Xin Bite 2 F1) were transplanted in a pot (785 cm^3) containing 250 g autoclaved and dried peat moss. A total of 2000 fresh J2s were inoculated in each pot at 3 days after transplanting. Subsequently, 130 mL of 20% and 50% 2-WF were used in this experiment and applied in pots. 20% of Czapek medium and tap water were utilized as controls. A randomized design with 6 replicates for each treatment group was used for the pot experiment and all materials were maintained after inoculation at 25 °C in a greenhouse with a 14 h light and 10 h dark photoperiod. Thirty-five days after transplantation, tomato roots were collected and gently washed with tap water to remove residual materials. Plant height, root fresh weight, and the total number of galls and eggs per plant root system were determined. The eggs were extracted separately from plants with a 1% NaOCl method as previously described [25] and were subsequently collected in beakers with water. Afterwards, 50 μL of a well-mixed egg suspension solution were transferred to a counting dish to enable egg count determination. Eggs were counted three times and the total number of eggs in the entire suspension was calculated. This experiment was repeated twice.

2.8. Effect of Boiling and Storage Time on Nematicidal Activity Stability of Fermentation Filtrate

Two-hundred mL of fresh 2-WF was dispensed into two 100 mL beakers respectively. One of the beakers was boiled in a microwave oven at 100 °C, whereas the second beaker was maintained at room temperature. The fermentation filtrate from two beakers were diluted to 10% and sterilized water was used as a control. Nematicidal activity was then conducted as described above and the experiment was triplicated.

For the analysis of storage time, the experiment was set up for 1-, 2-, and 3-week old 2-WF at 4 °C and 25 °C in dark, respectively; with 4 replicates for each treatment. After storage, the 2-WF solution was filtered through a sterile 0.45 μm polyethersulfone filter (Whatman, Clifton, NJ, USA) and subsequently diluted to a 10% solution in sterilized water. Sterilized water alone was used as a negative control. The nematicidal activity was measured as described above and this experiment was repeated 3 times.

2.9. Evaluation of the Strain Fermentation Filtrate on The Germination of Crop Seeds

In this study, the effect of 2-WF of *A. japonicus* ZW1 was evaluated on seed germination of various crops, e.g., from corn (Qingnong 13), wheat (Mianmai 41), cowpeas (Shanlv), cabbage (Green column), cucumbers (Liaoning 8), rice (Teyou 09103), tomatoes (Hongyingguo 808), and soybeans (Ludou 4). First, healthy seeds were surface sterilized with 2% NaOCl for 3 min and subsequently rinsed 5 times with sterilized water [33]. Seeds were treated with 10% and 20% 2-WF in triplicates across 3 independent experiments, with sterilized water used as a negative control. The sterilized crop seeds were then exposed to the fermentation filtrate in a moist chamber and incubated for several days in the dark at room temperature (25 °C). Sprouted seeds were counted every day until the seed germination rate no longer changed. The seed germination rate was calculated as: (number of germinated seed/total tested seeds) × 100%.

2.10. Isolation and Structural Determination of Aspergillus Japonicus ZW-1 Nematicidal Metabolites

Eight litre of *A. japonicus* ZW-1 2-week fermentation broth was filtered through 8 layers of muslin gauze, then concentrated to 500 mL using rotary evaporation (Hei-VAP Core ML G3, Instruments GmbH & Co. Heidolph, KG, Schwabach, Germany) at 55 °C. The crude extract (15.6 g) from *A. japonicus* ZW-1 fermentation broth was extracted with 1-butanol and evaporated at 40 °C until dry, dissolved in methanol (MeOH) and chromatographed on methylated sephadex LH20 (Beijing Solarbio Science & Technology Co., Ltd., Beijing, China) using MeOH as eluent to give two fractions, the two fractions were dissolved in distilled water to make 2.0 mg mL^{-1} aqueous solution for activity assay. One fraction showed activity against J2. This active fraction was dissolved in the chloroform, at which point white crystals formed. The solution was filtered through cotton which was then washed 20 times using chloroform and dried at room temperature to get the purified active compound.

The chemical structures of the active compound were determined using nuclear magnetic resonance (NMR) analysis and high-resolution electrospray ionization mass spectrometry (HR-ESI-MS) analysis. ^{1}H nuclear magnetic resonance (NMR) and ^{13}C NMR spectra were acquired in MeOH with a Bruker AVANCE III HD600 spectrometer (Bruker Corporation, Faellanden, Switzerland) at 600 MHz for ^{1}H NMR spectra and 125 MHz for ^{13}C NMR spectra using tetramethylsilane as the internal standard. HR-ESI-MS analysis was performed using a Waters E2695 model ion trap mass spectrometer (Waters, Milford, MA, USA) [34]. The nematicidal activity of active compounds at different concentrations (1.25, 1.00, 0.75, 0.50, 0.25 mg mL^{-1}) was measured as described above and this experimental approach was repeated 3 times. Sterilized water was used as a control.

2.11. Statistical Analysis

Data were analyzed using SPSS 19.0. software (SPSS Inc. Chicago, IL, USA) and statistical significance was calculated using a one-way analysis of variance (ANOVA). The means of different parameters for each treatment group were compared among each other using a Fisher's protected least significant difference (LSD) test at $p < 0.05$. All figures for statistical analyses were made using Sigma Plot 10.0 (SPSS Inc., Chicago, IL, USA).

3. Results

3.1. Effect of Fermentation Filtrates on Hatching of Meloidogyne Incognita Eggs

The fermentation filtrate of *A. japonicus* ZW1 at various concentrations and different time points showed significant nematicidal activity against cumulative hatching rate of eggs. The cumulative hatching rate of eggs increased over time in the 1-WF, 2-WF, and 3-WF treatments (Figure 1). In relative comparison to 1-WF, *M. incognita* eggs exhibited higher sensitivity to 2-WF and 3-WF. Fifteen days after incubation, the cumulative hatching rates in 20% and 50% 1-WF were 71.1% and 30.1%, respectively, and were significantly lower in comparison to 2.5%, 5%, and 10% 1-WF and controls ($p < 0.05$). For the 2-WF treated samples, cumulative hatching rates in 5%, 10%, 20%, and 50% 2-WF were 42.5%, 36.0%, 24.3%, and 6.4%, respectively, 15 d after incubation. These values were significantly lower than that of the 2.5% 2-WF and control treatments ($p < 0.05$). Cumulative hatching rates in 5%, 10%, 20%, and 50% 3-WF treatments were 53.0%, 42.2%, 34.6%, and 21.2%, respectively, 15 d after incubation. These results were significantly lower than that of the 2.5% 2-WF and control treatments ($p < 0.05$).

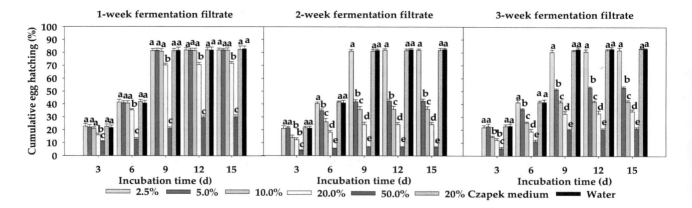

Figure 1. Cumulative *Meloidogyne incognita* eggs hatching rates in *Aspergillus japonicus* ZW1 fermentation filtrate. The bars represent the standard error. The same letter is not significantly different ($p < 0.05$) according to a Fisher's protected least significant difference (LSD) test.

3.2. Nematicidal Activity of Fermentation Filtrates on Meloidogyne Incognita J2s

The time of culturing influenced the nematicidal activity of the fermentation filtrate on J2s (Figure 2). In comparison to the 1-WF and control treatments, the mortality of J2s was higher in 2-WF and 3-WF treatments at different time points post incubation. In the 1-WF treatment, the mortality of J2s was less than 3.3% and no significant difference was observed after treatment for a 6 to 48 h period. Conversely, application of 2-WF and 3-WF resulted in a significantly higher mortality of J2s at different concentrations of the fermentation filtrates as compared to the controls ($p < 0.05$). When investigating 50% 2-WF and 3-WF, the mortality of J2s reached 100% after a 6 h incubation period. After the 48 h incubation period, the mortality of 2.5% 2-WF and 3-WF treatments reached 56.1% and 56.8%, respectively, and were all significantly higher than the controls ($p < 0.05$). From a morphological perspective, treatment with 2-WF resulted in differences in the J2 when compared to the controls (Figure 3). Specifically, microscopic observations revealed that the bodies of J2s in the 2-WF treatment were either straight or arched without movements at 6 h post-incubation (Figure 3, A2). However, bubbles (Figure 3, Bu) appeared in the body of J2s over time and protruded wrinkles (Figure 4, Wr) on the body surface and areas of intensive cytoplasmic vacuolization were observed (such as damaged areas; Figure 5, Da) at 10 h post-exposure to treatment with 2-WF.

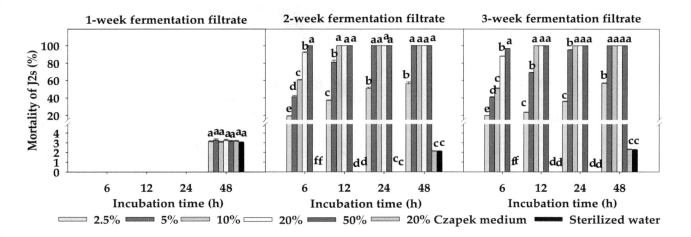

Figure 2. The mortality of *Meloidogyne incognita* J2s in *Aspergillus japonicus* ZW1 fermentation filtrate. Means with the same letter in each group designate no significant differences ($p < 0.05$) based on analysis with a Fisher's protected LSD test.

3.3. Greenhouse Experiment

Treatment with fermentation broth of *A. japonicus* ZW1 resulted in a significant reduction in the number of root galls and eggs per plant as compared to controls (Table 1). The number of root galls and eggs were 8.2 and 3488.9 per plant in the 50% fermentation broth treatment, respectively; whereas 16.8 and 6020 were observed per plant in the 20% fermentation broth treatment, respectively. In both treatments, the number of root galls and eggs was significantly lower than what was observed in controls ($p < 0.05$). The 50% fermentation broth decreased root galls by 78.6% and eggs by 69.4% per plant in comparison to treatment with the Czapek medium control (38.4 root galls and 11413.3 eggs) and 79.9% root galls and 72.0% eggs per plant compared with the tap water control (40.8 root galls and 12480.0 eggs, respectively), and root galls and eggs from the 20% fermentation broth treatment decreased by 56.3% and 47.3% per plant compared with the Czapek medium control (38.4 root galls and 11413.3 eggs, respectively), and 58.8% and 51.8% compared with the tap water control (40.8 root galls and 12480.0 eggs, respectively).

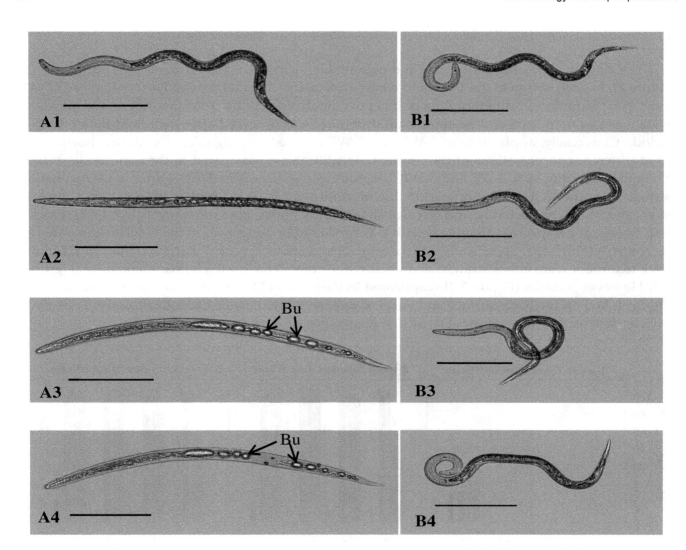

Figure 3. Morphology of second-stage juveniles of *Meloidogyne incognita* treated with 10% 2-week fermentation filtrate (2-WF) of *Aspergillus japonicus* ZW1. **A1–A4** were treated with 10% 2-WF; **B1–B4** were treated with sterilized water; **A1** and **B1** were treated at 0 h; **A2** and **B2** were treated at 6 h; **A3** and **B3** were treated at 12 h; and **A4** and **B4** were treated at 24 h. **Bu**: bubbles. Scale bars of **A1–A4** and **B1–B4** were 100 µm.

Table 1. Effect of *Aspergillus japonicus* ZW1 fermentation broth on the formation of galls and eggs on roots and the growth of tomato plants infected with *Meloidogyne incognita*.

Treatments	Plant Height (cm)	Fresh Root Weight (g)	Root Galls per Plant	Egg Number per Plant
50% Fermentation Broth	26.6 ± 0.6 a	0.6 ± 0.3 a	8.2 ± 1.7 c	3488.9 ± 155.6 d
20% Fermentation Broth	26.5 ± 0.6 a	0.9 ± 0.2 a	16.8 ± 1.4 b	6020.0 ± 214.9 c
Czapek Medium Control	26.9 ± 0.5 a	0.7 ± 0.1 a	38.4 ± 4.3 a	11413.3 ± 338.9 b
Tap Water Control	26.4 ± 0.6 a	0.8 ± 0.2 a	40.8 ± 3.8 a	12480.0 ± 200.4 a

Values represent means ± standard error of six replicate plants per treatment using the combination of two different experiments. Means with the same letter were not significantly different ($p < 0.05$) according to a Fisher's protected LSD test.

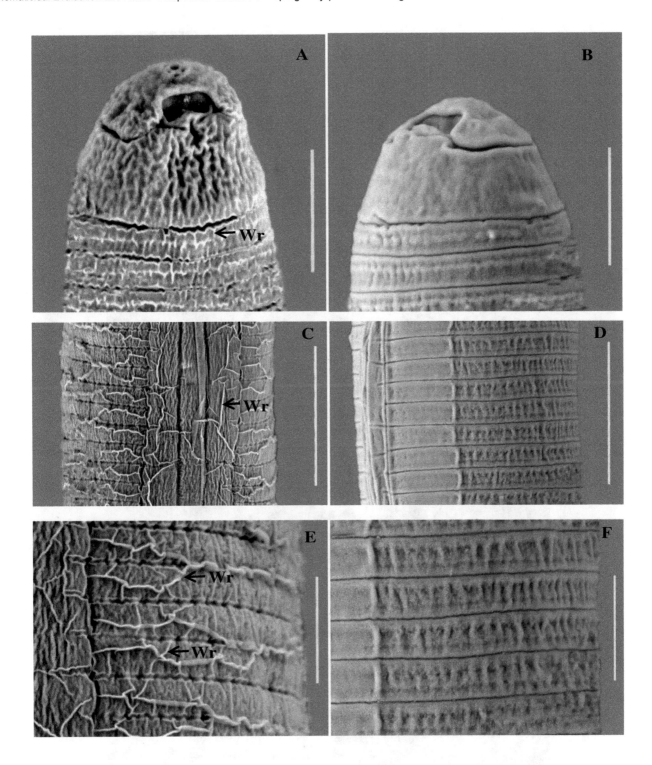

Figure 4. Visualization of the effect of 10% 2-WF of *Aspergillus japonicus* ZW1 on the morphology of *Meloidogyne incognita* J2s with scanning electron microscopy. (**A,C,E**) J2s treated with 10% 2-WF. (**B,D,F**) J2s treated with sterilized water. (**A,B**) Head region of J2. (**C–F**) The lateral field of J2. Scale bars of (**A,B,E,F**) and (**C,D**) were 2 and 5 μm, respectively. **Wr**: protruded wrinkles (black arrow).

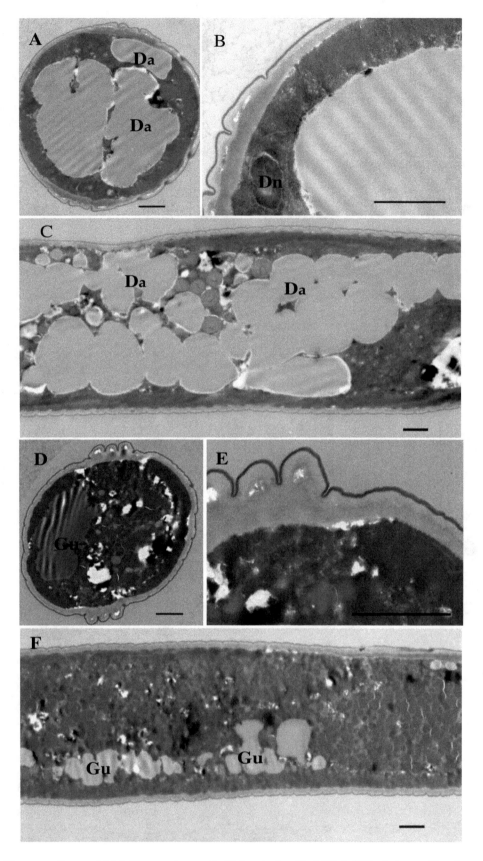

Figure 5. Cross-sections of *Meloidogyne incognita* J2 treated with 10% 2-WF of *Aspergillus japonicus* ZW1. (**A–C**) J2s treated with *A. japonicus* ZW1 fermentation filtrate. (**D–F**) J2s treated with sterilized water. Scale bars of **A, B, C, D, E**, and **F** were 2 μm. **Da**: damaged and area. **Gu**: gut. **Dn**: destructed nuclei.

3.4. Effect of Boiling and Storage Time on the Nematicidal Activity of Fermentation Filtrate

The mortality of J2s in fresh and boiled 10% 2-WF did not display any significant differences (Table 2). After a 48 h incubation period, the mortality of J2 reached 100.0% in both fermentation filtrates and was significantly higher than what was observed in the sterilized water treatment ($p < 0.05$).

Table 2. Nematicidal activity of the boiled fermentation filtrate of *Aspergillus japonicus* ZW1 on *Meloidogyne incognita* J2s.

Treatment with 10% 2-WF	Incubation Time (h)			
	6	12	24	48
Untreated	44.9 ± 5.6 a	91.2 ± 3.3 a	91.9 ± 3.4 a	99.2 ± 0.8 a
Boiled	40.5 ± 4.7 a	93.6 ± 2.4 a	96.6 ± 2.0 a	99.0 ± 1.0 a
Sterilized Water	0.0 ± 0.0 b	0.0 ± 0.0 b	0.0 ± 0.0 b	0.1 ± 0.1 b

Values represent means ± standard deviation of three replicates. Means with the same letter are not significantly different ($p < 0.05$) according to a Fisher's protected LSD test.

No significant difference was observed in the mortality of J2s exposed to different storage conditions of 10% 2-WF (Table 3). Specifically, they all reached 100% mortality after a 48 h incubation period, which was higher than the sterilized water treatment ($p < 0.05$).

Table 3. Mortality of *Meloidogyne incognita* J2s in *Aspergillus japonicus* ZW1 fermentation filtrate under different storage conditions.

Treatments with 10% 2-WF	Storage Time	Incubation Time (h)			
		6	12	24	48
	1-week	58.9 ± 5.3 a	99.4 ± 0.6 a	100.0 ± 0.0 a	100.0 ± 0.0 a
4 °C	2-week	58.3 ± 2.2 a	98.8 ± 0.7 a	100.0 ± 0.0 a	100.0 ± 0.0 a
	3-week	60.8 ± 2.2 a	100.0 ± 0.0 a	100.0 ± 0.0 a	100.0 ± 0.0 a
	1-week	62.1 ± 1.8 a	100.0 ± 0.0 a	100.0 ± 0.0 a	100.0 ± 0.0 a
25 °C	2-week	55.6 ± 3.3 a	99.0 ± 0.6 a	100.0 ± 0.0 a	100.0 ± 0.0 a
	3-week	58.4 ± 4.4 a	98.7 ± 0.8 a	99.6 ± 0.4 a	100.0 ± 0.0 a
Sterilized Water	–	0.0 ± 0.0 b	0.0 ± 0.0 b	0.0 ± 0.0 b	1.4 ± 0.8 b

Values represent the means ± standard error of four replicates; means with the same letter are not significantly different ($p < 0.05$) according to a Fisher's protected LSD test.

3.5. Effect of Fermentation Filtrate on Seed Germination

The 20% and 10% 2-WF did not influenced the germination of corn, rice, tomato, cowpea, and cucumber seeds (Table 4). Two days after incubation with 10% 2-WF, the wheat seed germination rate was 85.4% and was significantly higher than what was observed in the control ($p < 0.05$). After an extended period of time beyond the 48-h time period, this value did not increase any further. For soybean seeds treated with 10% 2-WF, germination was significantly lower than what was observed in sterilized water ($p < 0.05$) at day 1; however, there were no statistically significant differences 2–5 days post-incubation across 20% and 10% 2-WF and sterilized water treatments. For cabbage seeds, germination in 20% 2-WF was significantly lower than what was observed in 10% 2-WF and control treatments ($p < 0.05$).

Table 4. Seed germination (%) in different concentrations of 2-week *Aspergillus japonicus* ZW1 fermentation filtrate.

Seeds	Treatments	Incubation Time (d)					
		1	2	3	4	5	6
Wheat	20%	64.6 ± 4.5 a	78.1 ± 1.8 ab	78.1 ± 1.8 ab	78.1 ± 1.8 ab	78.1 ± 1.8 ab	–
	10%	63.5 ± 5.5 a	85.4 ± 3.8 a	85.4 ± 3.8 a	85.4 ± 3.8 a	85.4 ± 3.8 a	–
	Sterilized Water	60.4 ± 4.6 a	72.5 ± 1.3 b	73.5 ± 0.9 b	73.5 ± 0.9 b	73.5 ± 0.9 b	–
Corn	20%	10.4 ± 1.0 a	83.3 ± 1.0 a	88.5 ± 2.8 a	88.5 ± 2.8 a	88.5 ± 2.8 a	–
	10%	11.5 ± 4.5 a	81.3 ± 6.5 a	90.6 ± 4.8 a	90.6 ± 4.8 a	90.6 ± 4.8 a	–
	Sterilized Water	9.4 ± 4.8 a	78.1 ± 4.8 a	89.6 ± 2.1 a	89.6 ± 2.1 a	89.6 ± 2.1 a	–
Rice	20%	3.0 ± 1.8 a	94.0 ± 1.8 a	97.0 ± 1.8 a	97.0 ± 1.8 a	97.0 ± 1.8 a	–
	10%	4.0 ± 2.7 a	96.0 ± 4.0 a	97.0 ± 3.0 a	97.0 ± 3.0 a	97.0 ± 3.0 a	–
	Sterilized Water	1.0 ± 1.0 a	90.9 ± 3.0 a	93.9 ± 1.8 a	93.9 ± 1.7 a	93.9 ± 1.8 a	–
Tomato	20%	0.0	48.1 ± 2.9 a	73.1 ± 3.2 a	81.8 ± 3.3 a	87.5 ± 1.8 a	87.5 ± 1.8 a
	10%	0.0	51.6 ± 2.7 a	69.5 ± 3.6 a	79.5 ± 4.6 a	89.3 ± 3.3 a	91.3 ± 3.2 a
	Sterilized Water	0.0	46.6 ± 8.2 a	72.8 ± 4.2 a	85.0 ± 4.3 a	90.0 ± 2.5 a	91.0 ± 1.6 a
Soybean	20%	33.3 ± 2.8 ab	90.6 ± 4.8 a	100.0 ± 0.0 a	100.0 ± 0.0 a	100.0 ± 0.0 a	–
	10%	26.0 ± 5.2 b	86.5 ± 1.0 a	96.8 ± 1.8 a	99.0 ± 1.0 a	99.0 ± 1.0 a	–
	Sterilized Water	41.8 ± 3.8 a	90.6 ± 3.6 a	100.0 ± 0.0 a	100.0 ± 0.0 a	100.0 ± 0.0 a	–
Cowpea	20%	56.3 ± 3.6 a	99.0 ± 1.0 a	100.0 ± 0.0 a	100.0 ± 0.0 a	100.0 ± 0.0 a	–
	10%	64.3 ± 7.3 a	100.0 ± 0.0 a	100.0 ± 0.0 a	100.0 ± 0.0 a	100.0 ± 0.0 a	–
	Sterilized Water	65.6 ± 3.6 a	99.0 ± 1.0 a	99.0 ± 1.0 a	99.0 ± 1.0 a	99.0 ± 1.0 a	–
Cucumber	20%	100.0	100.0	100.0	100.0	100.0	–
	10%	100.0	100.0	100.0	100.0	100.0	–
	Sterilized Water	100.0	100.0	100.0	100.0	100.0	–
Cabbage	20%	65.1 ± 4.6 b	72.1 ± 6.9 b	72.1 ± 6.9 b	72.1 ± 6.9 b	72.1 ± 6.9 b	–
	10%	89.0 ± 1.2 a	95.1 ± 1.8 a	96.1 ± 2.5 a	96.1 ± 2.5 a	96.1 ± 2.5 a	–
	Sterilized Water	77.9 ± 4.0 a	95.0 ± 3.7 a	95.0 ± 3.7 a	95.0 ± 3.7 a	95.0 ± 3.7 a	–

Values represent the means ± standard error of four replicates; means with the same letter were not significantly different ($p < 0.05$) according to a Fisher's protected LSD test.

3.6. Structural Confirmation of Nematicidal Substance from 2-WF

The active compound was a pale-yellow crystal, which can dissolve easily in water. The ^{1}H NMR spectrum in MeOH exhibited signals due to two methyl groups at δ 3.68 (each 3H, s, 7, 8-CH$_3$), 2.95, 2.85 (each 2H, AB system, d, J = 12.0 Hz, 2, 4-CH$_2$). The ^{13}C NMR and heteronuclear multiple-quantum correlation spectra revealed two carbonyl carbons at δ$_C$ 175.00 (s, C-6), 170.46 (s, C-1, C-5), two methoxy groups 72.84 (s, C-3), 50.78 (q, C-7, C-8), 42.63 (t, C-2, C-4). The electrospray ionization mass spectrometry (ESI-MS) data of active compound was identified the molecular formula of C$_8$H$_{12}$O$_7$ by the [M]$^-$ ion signal at *m/z* 219 [M]$^-$. The structure of the active compound was determined to be 1,5-Dimethyl Citrate hydrochloride ester (C$_8$H$_{12}$O$_7$ HCl, Figure 6) by the analysis of its spectroscopic data and comparison with the values in the literature [35].

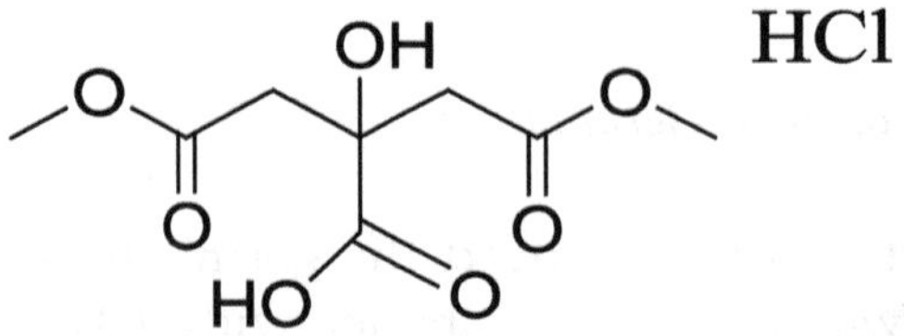

Figure 6. Chemical structures of active compound from *Aspergillus japonicus* ZW1 fermentation filtrate.

3.7. Effect of 1,5-Dimethyl Citrate Hydrochloride Ester on Meloidogyne Incognita J2s

1,5-Dimethyl Citrate hydrochloride ester had a strong toxic activity against J2s at low concentrations, and J2s mortality increased with the duration of exposure in different concentration of 1,5-Dimethyl Citrate hydrochloride ester (Table 5). There were significant differences in mortality between concentrations and control after exposure ($p < 0.05$). The mortality of J2s in concentrations of 1.25, 1.00, 0.75, 0.50, and 0.25 mg mL^{-1} of 1,5-Dimethyl Citrate hydrochloride ester were 91.7%, 57.7%,

36.9%, 20.8%, and 3.3% respectively at 48 h after exposure, which were significantly higher than that of sterilized water ($p < 0.05$).

Table 5. Mortality (%) of *Meloidogyne incognita* J2s in different concentrations of active compound from *Aspergillus japonicus* ZW-1 fermentation filtrate.

Concentration mg/mL	Incubation Time (h)			
	6	12	24	48
1.25	63.4 ± 0.9 a	72.9 ± 0.5 a	78.8 ± 0.6 a	91.7 ± 0.5 a
1.00	39.9 ± 0.7 b	44.4 ± 0.6 b	47.1 ± 0.4 b	57.7 ± 0.5 b
0.75	23.3 ± 0.8 c	31.4 ± 0.3 c	34.1 ± 0.7 c	36.9 ± 0.7 c
0.50	2.0 ± 0.3 d	4.8 ± 0.1 d	7.9 ± 0.2 d	20.8 ± 0.7 d
0.25	0.0 ± 0.0 e	0.0 ± 0.0 e	1.6 ± 0.1 e	3.3 ± 0.1 e
Sterilized Water	0.0 ± 0.0 e	0.0 ± 0.0 e	0.0 ± 0.0 f	0.0 ± 0.0 f

Values represent the means ± standard error of four replicates; means with the same letter each column were not significantly different ($p < 0.05$) according to a Fisher's protected LSD test.

Nematicidal activity of 1,5-Dimethyl Citrate hydrochloride ester was evaluated by comparing the median lethal concentrations (LC50) for different concentrations on *M. incognita* J2s under different exposure times. The concentrations at which 50% of the dead *M. incognita* J2s (LC50) were 1.0373, 0.9646, 0.9397, and 0.7614 mg mL^{-1} 1,5-Dimethyl Citrate hydrochloride ester for 6, 12, 24, and 48 h respectively. The LC50 values were decreasing with the enhanced of exposure time (Table 6).

Table 6. Toxicity of active compound to *Meloidogyne incognita* J2s at different treatment durations.

Exposure Time (h)	Slope (±SE)	Correlation Coefficient	LC50 (95%CI)	LC90 (95%CI)
6	4.8790(±0.2118)	0.9881	1.0373 (0.9112–1.1808)	1.5283 (1.2756–1.8312)
12	5.1225(±0.2843)	0.9800	0.9646 (0.8229–1.1308)	1.4059 (1.1282–1.7520)
24	5.1099(±0.1618)	0.9760	0.9397 (0.7922–1.1145)	1.9421 (1.4234–2.6498)
48	5.4928(±0.2180)	0.9596	0.7614 (0.6261–0.9260)	1.5469 (1.0971–2.1811)

LC-lethal concentration expressed in mg/mL active compound with 95% confidence intervals (CI). SE, standard error.

4. Discussion

In general, the management of parasitic nematodes is a challenging process and current control strategies are mostly dependent upon the application of nematicides [36]. However, many effective nematicides have been restricted for usage and have been banned from the market in recent years due to environmental concerns [37]. Biological options are gaining attention as promising new tools due to their environmentally-friendly and non-toxic characteristics. The potential for using microbes in controlling plant-parasitic nematodes has been documented [38] and effective microbes have been obtained from soil, plants, and the surface of nematodes [39–41]. *Aspergillus* spp. are very common in soil and are lethal to the nematode population; *A. niger* and *A. candidus* were the potential fungal agents to be used against plant-parasitic nematodes [35,42,43]. The results of this study indicated that fermentation of the *A. japonicus* ZW1 from soil was found to not only inhibit egg hatching but was also toxic to nematodes in vitro. The 2-WF was shown to be more toxic to J2s than 1-WF and 3-WF; this effect showed the presence of more active compounds in 2-WF, worth previous characterization. The similar behavior of several fungi and bacteria were also studied against plant parasitic nematodes. Among them a culture filtrate of the rhizosphere bacterium *Pseudoxanthomonas japonensis* isolated from soil exhibited strong nematicidal activity against the *M. incognita* [30]; a metabolite of *Xylaria grammica* KCTC 13121BP isolated from lichen showed strong J2 killing and egg-hatching inhibitory effects [44];

and a culture medium of *Stenotrophomonas maltophilia* and *Rhizobium nepotum* isolated from the surface of nematodes reduced the pathogenicity of wild pine wood nematodes [39].

Natural products have many limitations, such as natural laccases, which have poor stability of enzymatic activity [45]. As a result, it was important to determine and assess if the novel environmentally-friendly nematicides could be stable for practical and durable application opportunities. Consequently, in our present study, we were interested to determine the durability of the novel biological filtrates. Importantly, the toxic activity of the *A. japonicus* ZW1 fermentation filtrate was not effected by boiling, storage time (1-, 2-week, and 3-week) and warm/cold conditions (25 °C and 4 °C). Usually, the surface coating of nematodes was considered to play an important role in the external protection of nematode bodies, sensing, and communication [46,47].

The microbes and plant produced several acidic metabolites or proteinases that specifically degraded the outer membrane of host cells during primary infection [42,48,49]. In our study, wrinkles on the surface of the body of J2s in 2-WF were observed with scanning electron microscopy, and internal bubbles appeared in their body over time. Additionally, other prominent changes such as intensive cytoplasmic vacuolization areas were observed using transmission electron microscopy; suggesting that the activity of compounds produced by *A. japonicus* ZW1 targeted the skin of nematodes and changed its permeability [50]. Previous research showed acidoid (acetic acid) damage the nuclei of cells and led to intensive cytoplasmic vacuolization areas in the body of J2 *M. incognita* [28]. Nematicidal metabolites from the endophytic fungus *Chaetomium globosum* YSC5 significantly reduced the reproduction of *M. javanica* as well [51]. In our present study, nematicidal compound 1,5-Dimethyl Citrate hydrochloride ester from *A. japonicus* ZW1, first isolated and identified on the basis of NMR, LC-MS techniques, was different with the nematicidal compounds produced by *A. niger* (oxalic acid) and *A. candidus* (Citric acid and 3-hydroxy-5-methoxy-3-(methoxycarbonyl)-5-oxopentanoic acid). *M. incognita* J2 mortality reached 100% at 1 day, and egg hatching was suppressed by 95.6% at 7 days after treated with 2 mmol L^{-1} (180 µg m L $^{-1}$) oxalic acid [42]. 3-hydroxy-5-methoxy-3-(methoxycarbonyl)- 5-oxopentanoic acid was an isomer of 1,5-Dimethyl Citrate, which increased the mean percentage of immobile *Ditylenchus destructor* by 50% at a concentration of 50 mg mL^{-1} after exposure for 72 h [35]. In our study, *M. incognita* J2 treated with 1,5-Dimethyl Citrate hydrochloride ester, mortality reached 91.7% at 48 h after exposure to 1.25 mg mL^{-1} concentration, the LC50 was 0.7614 mg mL^{-1}, which exhibited the most potent toxic activity against the J2 of *M. incognita*. However, the interesting thing was that in in vitro bioassay, fermentation of the strain exhibited better nematicidal effects, and the mortality of J2s reached 100% after exposed to 5% concentration (approximately 100 µg m L $^{-1}$ 1,5-Dimethyl Citrate hydrochloride ester) *A. japonicus* ZW1 fermentation filtrate at 24 h. Our speculation is that the nematicidal effect originated 1,5-Dimethyl Citrate hydrochloride ester combined with some other compounds produced by *A. japonicus* ZW1. Thus, we still need further study to find and proved other nematicidal activity compounds by metabonomics analysis.

No effect on the seed germination of corn, wheat, rice, cowpeas, cucumbers, soybeans, and tomatoes was observed for the 10% and 20% 2-WF treatments. In whole pot experiments, treatment with the fermentation broth of the strain suppressed root galls and egg populations for tomatoes. As a result, these results suggested that *A. japonicus* ZW1 produced and excreted metabolites that were toxic to root-knot nematodes but did not exert negative effects on seed germination. Thus, *A. japonicus* showed desirable, effective, and safe biocontrol properties against *M. incognita* for both in vitro and greenhouse conditions. Taken together, these observations suggest that the fermentation filtrate of *A. japonicus* ZW1 is safe for use as a biological control fungus against root-knot nematodes. However, further studies are warranted and necessary to evaluate the in vivo efficacy of the strain against root-knot nematodes or other plant-parasitic nematodes.

5. Conclusions

A. japonicus ZW1 fermentation filtrate exhibited a potential biocidal activity on *M. incognita* in vitro and in vivo. The *A. japonicus* ZW1 2-week fermentation filtrate exhibited markedly inhibitory effects on egg hatching and nematicidal activities on J2s followed by 3-week fermentation filtrate. The *A. japonicus* ZW1 filtrate penetrated the body wall of *M. incognita* and caused intensive cytoplasmic vacuolization with remarkable protruded wrinkles appearing on the body surface of the J2s. Moreover, the nematicidal activity of the fermentation was stable after a boiling treatment and was not affected by storage time. *A. japonicus* ZW1 fermentation filtrate had no negative effect on the viability and germination of corn, wheat, rice, cowpeas, cucumbers, soybeans, and tomato seeds. The main active compound of 1,5-Dimethyl Citrate hydrochloride ester was first isolated and identified from the *A. japonicus* ZW1 fermentation filtrate. Finally, this work highlights the relevance of *A. japonicus* ZW1 fermentation filtrate as a potential new biological nematicide resource for the control of *M. incognita*.

Author Contributions: Conceptualization, Q.H., A.M. and H.W.; methodology, Q.H., A.M.; software, D.W.; B.L. and A.M.; validation, Q.H.; formal analysis, Q.H., A.M., D.W. and B.L.; investigation, A.M. and H.W.; resources, H.W.; data curation, Q.H., A.M. and B.L.; writing—original draft preparation, Q.H.; writing—review and editing, Q.H., A.M., D.W., and H.W.; and supervision, H.W. All authors have read and agreed to the published version of the manuscript.

References

1.　Fanelli, E.; Cotroneo, A.; Carisio, L.; Troccoli, A.; Grosso, S.; Boero, C.; Boero, C.; Capriglia, F.; Luca, F.D. Detection and molecular characterization of the rice root-knot nematode *Meloidogyne graminicola* in Italy. *Eur. J. Plant Pathol.* **2017**, *149*, 467–476. [CrossRef]

2.　Kayani, M.Z.; Mukhtar, T.; Hussain, M.A. Effects of southern root knot nematode population densities and plant age on growth and yield parameters of cucumber. *Crop Prot.* **2017**, *92*, 207–212. [CrossRef]

3.　Besnard, G.; Thi-Phan, N.; Ho-Bich, H.; Dereeper, A.; Nguyen, H.T.; Quénéhervé, P.; Aribi, J.; Bellafiore, S. On the close relatedness of two rice-parasitic root-knot nematode species and the recent expansion of *Meloidogyne graminicola* in Southeast Asia. *Genes* **2019**, *10*, 175. [CrossRef]

4.　Bozbuga, R.; Dasgan, H.Y.; Akhoundnejad, Y.; Imren, M.; Günay, O.C.; Toktay, H. Effect of *Mi* gene and nematode resistance on tomato genotypes using molecular and screening assay. *Cytol. Genet.* **2020**, *54*, 154–164. [CrossRef]

5.　Alves, G.C.S.; Ferri, P.H.; Seraphin, J.C.; Fortes, G.A.C.; Rocha, M.R.; Santos, S.C. Principal Response Curves analysis of polyphenol variation in resistant and susceptible cotton after infection by a root-knot nematode (RKN). *Physiol. Mol. Plant Pathol.* **2016**, *96*, 19–28. [CrossRef]

6.　Lopes, C.M.L.; Cares, J.E.; Perina, F.J.; Nascimento, G.F.; Mendona, J.S.F.; Moita, A.W.; Castagnone-Sereno, P.; Carneiro, R.M.D.G. Diversity of *Meloidogyne incognita* populations from cotton and aggressiveness to *Gossypium* spp. accessions. *Plant Pathol.* **2019**, *68*, 816–824. [CrossRef]

7.　Trudgill, D.L.; Blok, V.C. Apomictic, polyphagous root-knot nematodes: Exceptionally successful and damaging biotrophic root pathogens. *Annu. Rev. Phytopathol.* **2001**, *39*, 53–77. [CrossRef]

8.　Onkendi, E.M.; Kariuki, G.M.; Marais, M.; Moleleki, L.N. The threat of root-knot nematodes (*Meloidogyne* spp.) in Africa: A review. *Plant Pathol.* **2014**, *63*, 727–737. [CrossRef]

9.　Janati, S.; Houari, A.; Wifaya, A.; Essarioui, A.; Mimouni, A.; Hormatallah, A.; Sbaghi, M.; Dababat, A.A.; Mokrini, F. Occurrence of the root-knot nematode species in vegetable crops in Souss region of Morocco. *Plant Pathol. J.* **2018**, *34*, 308–315.

10.　Seid, A.; Fininsa, C.; Mekete, T.; Decraemer, W.; Wesemael, W.M.L. Tomato (*Solanum lycopersicum*) and root-knot nematodes (*Meloidogyne* spp.)–a century-old battle. *Nematology* **2015**, *17*, 995–1009. [CrossRef]

11.　Mukhtar, T.; Hussain, M.A.; Kayani, M.Z.; Aslam, M.N. Evaluation of resistance to root-knot nematode (*Meloidogyne incognita*) in okra cultivars. *Crop Prot.* **2014**, *56*, 25–30. [CrossRef]

12.　Patel, B.K.; Patel, H.R. Effect of physical, cultural and chemical methods of management on population dynamics of phytonematodes in bidi tobacco nursery. *Tob. Res.* **1999**, *25*, 51–60.

13. Sikora, R.A. Management of the antagonistic potential in agricultural ecosystems for the biological control of plant parasitic nematodes. *Annu. Rev. Phytopathol.* **1992**, *30*, 245–270. [CrossRef]

14. Brennan, R.J.B.; Glaze-Corcoran, S.; Wick, R.; Hashemi, M. Biofumigation: An alternative strategy for the control of plant parasitic nematodes. *J. Integr. Agric.* **2020**, *19*, 1680–1690. [CrossRef]

15. Ntalli, N.; Monokrousos, N.; Rumbos, C.; Kontea, D.; Zioga, D.; Argyropoulou, M.D.; Menkissoglu-Spiroudi, U.; Tsiropolos, N.G. Greenhouse biofumigation with *Melia azedarch* controls *Meloidogyne* spp. and enhances soil biological activity. *J. Pest Sci.* **2017**, *91*, 29–40. [CrossRef]

16. Stirling, G.R.; Wong, E.; Bhuiyan, S. *Pasteuria*, a bacterial parasite of plant-parasitic nematodes: Its occurrence in Australian sugarcane soils and its role as a biological control agent in naturally-infested soil. *Australas. Plant Pathol.* **2017**, *46*, 563–569. [CrossRef]

17. Viljoen, J.J.F.; Labuschagne, N.; Fourie, H.; Sikora, R.A. Biological control of the root-knot nematode *Meloidogyne incognita* on tomatoes and carrots by plant growth-promoting rhizobacteria. *Trop. Plant Pathol.* **2019**, *44*, 284–291. [CrossRef]

18. Hussain, M.; Zouhar, M.; Rysanek, P. Suppression of *Meloidogyne incognita* by the entomopathogenic fungus *Lecanicillium muscarium*. *Plant Dis.* **2018**, *102*, 977–982. [CrossRef]

19. Hussain, M.; Maòasová, M.; Zouhar, M.; Rysanek, P. Comparative virulence assessment of different nematophagous fungi and chemicals against northern root-knot nematodes, *Meloidogyne hapla*, on carrots. *Pak. J. Zool.* **2020**, *52*, 199–206. [CrossRef]

20. Nimnoi, P.; Pongsilp, N.; Ruanpanun, P. Monitoring the efficiency of *Streptomyces galilaeus* strain KPS-C004 against root knot disease and the promotion of plant growth in the plant-parasitic nematode infested soils. *Biol. Control* **2017**, *114*, 158–166. [CrossRef]

21. Dong, L.Q.; Zhang, K.Q. Microbial control of plant-parasitic nematodes: A five-party interaction. *Plant Soil* **2006**, *288*, 31–45. [CrossRef]

22. Li, J.; Zou, C.G.; Xu, J.P.; Ji, X.L.; Niu, X.M.; Yang, J.K.; Huang, X.W.; Zhang, K.Q. Molecular mechanisms of nematode-nematophagous microbe interactions: Basis for biological control of plant-parasitic nematodes. *Annu. Rev. Phytopathol.* **2015**, *53*, 67–95. [CrossRef]

23. Bui, H.X.; Hadi, B.A.R.; Oliva, R.; Schroeder, N.E. Beneficial bacterial volatile compounds for the control of root-knot nematode and bacterial leaf blight on rice. *Crop Prot.* **2020**, *135*, 104792. [CrossRef]

24. Nimnoi, P.; Ruanpanun, P. Suppression of root-knot nematode and plant growth promotion of chili (*Capsicum flutescens* L.) using co-inoculation of *Streptomyces* spp. *Biol. Control* **2020**, *145*, 104244. [CrossRef]

25. Hussey, R.S.; Barker, K.R. A comparison of methods of collecting inocula of *Meloidogyne* ssp. including a new technique. *Plant Dis. Rep.* **1973**, *57*, 1025–1028.

26. Wu, H.Y.; Silva, J.O.; Becker, J.S.; Becker, J.O. Fluazaindolizine mitigates plant-parasitic nematode activity at sublethal dosages. *J. Pest Sci.* **2020**. [CrossRef]

27. Hahn, M.H.; Mio, L.L.M.D.; Kuhn, O.J.; Duarte, H.D.S.S. Nematophagous mushrooms can be an alternative to control *Meloidogyne javanica*. *Biol. Control* **2019**, *138*, 104024. [CrossRef]

28. Choi, I.H.; Kim, J.; Shin, S.C.; Park, I.K. Nematicidal activity of monoterpenoids against the pine wood nematode (*Bursaphelencus xylophilus*). *Russ. J. Nematol.* **2007**, *15*, 35–40.

29. Hajji-Hedfi, L.; Larayedh, A.; Hammas, N.C.; Regaieg, H.; Horrigue-Raouani, N. Biological activities and chemical composition of *Pistacia lentiscus* in controlling *Fusarium* wilt and root-knot nematode disease complex on tomato. *Eur. J. Plant Pathol.* **2019**, *155*, 281–291. [CrossRef]

30. Holland, R.J.; Williams, K.L.; Khan, A. Infection of *Meloidogyne javanica* by *Paecilomyces lilacinus*. *Nematology* **1999**, *1*, 131–139. [CrossRef]

31. Janssen, T.; Karssen, G.; Topalović, O.; Coyne, D.; Bert, W. Integrative taxonomy of root-knot nematodes reveals multiple independent origins of mitotic parthenogenesis. *PLoS ONE* **2017**, *12*. [CrossRef] [PubMed]

32. Ntalli, N.; Ratajczak, M.; Oplos, C.; Menkissoglu-Spiroudi, U.; Adamski, Z. Acetic acid, 2-undecanone, and (e)-2-decenal ultrastructural malformations on *Meloidogyne incognita*. *J. Nematol.* **2016**, *48*, 248–260. [CrossRef] [PubMed]

33. Zhao, J.; Liu, W.; Liu, D.; Lu, C.; Zhang, D.; Wu, H.; Dong, D.; Meng, L. Identification and evaluation of *Aspergillus tubingensis* as a potential biocontrol agent against grey mould on tomato. *J. Gen. Plant Pathol.* **2018**, *84*, 148–159. [CrossRef]

34. Hu, Y.; Li, J.; Li, J.; Zhang, F.; Wang, J.; Mo, M.; Liu, Y. Biocontrol efficacy of *Pseudoxanthomonas japonensis*

against *Meloidogyne incognita* and its nematostatic metabolites. *FEMS Microbiol. Lett.* **2019**, *366*, fny287. [CrossRef]

35. Shemshura, O.N.; Bekmakhanova, N.E.; Mazunina, M.N.; Meyer, S.L.F.; Rice, C.P.; Masler, E.P. Isolation and identification of nematode-antagonistic compounds from the fungus *Aspergillus candidus*. *FEMS Microbiol. Lett.* **2016**, *363*, fnw26. [CrossRef]

36. Desaeger, J.A.; Watson, T.T. Evaluation of new chemical and biological nematicides for managing *Meloidogyne javanica* in tomato production and associated double-crops in Florida. *Pest Manag. Sci.* **2019**, *75*, 3363–3370. [CrossRef]

37. Sissell, K. EPA bans carbofuran residues; sued over endosulfan. *Chem. Week* **2008**, *170*, 29.

38. Liang, L.M.; Zou, C.G.; Xu, J.Q.; Zhang, K.Q. Signal pathways involved in microbe-nematode interactions provide new insights into the biocontrol of plant-parasitic nematodes. *Philos. Trans. R. Soc. B* **2019**, *374*, 20180317. [CrossRef]

39. Liu, K.C.; Zeng, F.L.; Ben, A.L.; Han, Z.M. Pathogenicity and repulsion for toxin-producing bacteria of dominant bacteria on the surface of American pine wood nematodes. *J. Phytopathol.* **2017**, *165*, 580–588. [CrossRef]

40. Liu, M.J.; Hwang, B.S.; Zhi, J.C.; Li, W.J.; Park, D.J.; Seo, S.T.; Seo, S.T.; Kim, C.J. Screening, isolation and evaluation of a nematicidal compound from actinomycetes against the pine wood nematode, *Bursaphelenchus xylophilus*. *Pest Manag. Sci.* **2019**, *75*, 1585–1593. [CrossRef]

41. Ponpandian, L.N.; Rim, S.O.; Shanmugam, G.; Jeon, J.; Park, Y.H.; Lee, S.K.; Bae, H. Phylogenetic characterization of bacterial endophytes from four *Pinus* species and their nematicidal activity against the pine wood nematode. *Sci. Rep.* **2019**, *9*, 12457. [CrossRef] [PubMed]

42. Jang, J.Y.; Choi, Y.H.; Shin, T.S.; Kim, T.H.; Shin, K.S.; Park, H.W.; Kim, Y.; Kim, H.; Choi, G.J.; Jang, K.S.; et al. Biological control of *Meloidogyne incognita* by *Aspergillus niger* F22 producing oxalic acid. *PLoS ONE* **2016**, *11*. [CrossRef] [PubMed]

43. Jin, N.; Liu, S.M.; Peng, H.; Huang, W.K.; Kong, L.A.; Wu, Y.H.; Chen, Y.P.; Ge, F.Y.; Jian, H.; Peng, D.L. Isolation and characterization of *Aspergillus niger* NBC001 underlying suppression against *Heterodera glycines*. *Sci. Rep.* **2019**, *9*, 591. [CrossRef] [PubMed]

44. Kim, T.Y.; Jang, J.Y.; Yu, N.H.; Chi, W.J.; Bae, C.H.; Yeo, J.H.; Park, A.R.; Hur, J.S.; Park, H.W.; Park, J.Y.; et al. Nematicidal activity of grammicin produced by *Xylaria grammica* KCTC 13121BP against *Meloidogyne incognita*. *Pest Manag. Sci.* **2018**, *74*, 384–391. [CrossRef] [PubMed]

45. Kang, C.; Ren, D.; Zhang, S.; Zhang, X.; He, X.; Deng, Z.; Huang, C.; Guo, H. Effect of polyhydroxyl compounds on the thermal stability and structure of laccase. *Pol. J. Environ. Stud.* **2019**, *28*, 3253–3259. [CrossRef]

46. Spiegel, Y.; McClure, M.A. The surface coat of plant-parasitic nematodes: Chemical composition, origin, and biological role—A review. *J. Nematol.* **1995**, *27*, 127–134.

47. Curtis, R.H.C. Plant-nematode interactions: Environmental signals detected by the nematode's chemosensory organs control changes in the surface cuticle and behaviour. *Parasite* **2008**, *15*, 310–316. [CrossRef]

48. Djian, C.; Pijarowski, L.; Ponchet, M.; Arpin, N.; Favre-Bonvin, J. Acetic acid: A selective nematicidal metabolite from culture filtrates of *Paecilomyces Lilacinus* (Thom) Samson and *Trichoderma Longibrachiatum* Rifai. *Nematologica* **1991**, *37*, 101–112.

49. Phiri, A.M.; Pomerai, D.D.; Buttle, D.J.; Behnke, J.M.B. Developing a rapid throughput screen for detection of nematicidal activity of plant cysteine proteinases: The role of *Caenorhabditis elegans* cystatins. *Parasitology* **2014**, *141*, 164–180. [CrossRef]

50. Jatala, P. Biological control of plant-parasitic nematodes. *Annu. Rev. Phytopathol.* **1986**, *24*, 453–489. [CrossRef]

51. Khan, B.; Yan, W.; Wei, S.; Wang, Z.Y.; Zhao, S.S.; Cao, L.L.; Rajput, N.A.; Ye, Y.H. Nematicidal metabolites from endophytic fungus *Chaetomium globosum* YSC5. *FEMS Microbiol. Lett.* **2019**, *366*, fnz169. [CrossRef] [PubMed]

Relevance of Plant Growth Promoting Microorganisms and their Derived Compounds, in the Face of Climate Change

Judith Naamala[ID] and Donald L. Smith *

Department of Plant Science, McGill University, Lakeshore Road, Ste. Anne de Bellevue, 21111, Montreal, QC H9X3V9, Canada; naamala.judith@mail.mcgill.ca
* Correspondence: Donald.Smith@McGill.Ca

Abstract: Climate change has already affected food security in many parts of the world, and this situation will worsen if nothing is done to combat it. Unfortunately, agriculture is a meaningful driver of climate change, through greenhouse gas emissions from nitrogen-based fertilizer, methane from animals and animal manure, as well as deforestation to obtain more land for agriculture. Therefore, the global agricultural sector should minimize greenhouse gas emissions in order to slow climate change. The objective of this review is to point out the various ways plant growth promoting microorganisms (PGPM) can be used to enhance crop production amidst climate change challenges, and effects of climate change on more conventional challenges, such as: weeds, pests, pathogens, salinity, drought, etc. Current knowledge regarding microbial inoculant technology is discussed. Pros and cons of single inoculants, microbial consortia and microbial compounds are discussed. A range of microbes and microbe derived compounds that have been reported to enhance plant growth amidst a range of biotic and abiotic stresses, and microbe-based products that are already on the market as agroinputs, are a focus. This review will provide the reader with a clearer understanding of current trends in microbial inoculants and how they can be used to enhance crop production amidst climate change challenges.

Keywords: plant growth promoting microorganisms; climate change; abiotic stress; biotic stress

1. Introduction

The world is at a point where we can no longer prevent all of the effects of climate change (because some of it is already here), but can only slow its further progress. The purpose of this paper is therefore to give the reader an understanding of why plant growth promoting organisms, or their products, are relevant, amidst climate change challenges, by showing how they can be used to mitigate the effects of climate change on crop production. The paper also highlights the various ways in which this approach can be used, and the role that inoculant formulation plays in maintaining the efficacy, durability and handling of microbial inoculants. The major drivers of climate change are human driven [1–3]. Burning of fossil fuels for energy, agriculture and industrialisation all contribute to emission of greenhouse gases (GHGs), such as: methane, carbon dioxide and nitrous oxide (N_2O). Agriculture is a major contributor to greenhouse gas emissions [4,5], especially through the use of N based fertilizers, methane emissions from animals and animal manure, deforestation to acquire more land for crop production, etc. According to the intergovernmental panel on climate change (IPCC) report on GHG emissions, energy consumption contributes about 35%, agriculture, forestry and related land use 24%, industry 21% and transport 14% [6]. The greenhouse gases then trap heat radiating from the earth's surface, causing global warming. Unfortunately, climate change also adversely affects agriculture [6,7], especially because, along with increases in global temperature, comes the increased prevalence of biotic and abiotic stresses that are detrimental to agriculture production, such as: pests, pathogens, nutrient deficiencies, salinity and weather extremes [1,8–10], some of which may encourage the further use of chemicals to correct, while there is little that can be done about others such as high temperatures and floods. Unmanaged, such factors affect plant growth and render arable land unproductive. This puts us in a challenging situation, especially because world population is growing so that there is a need to increase food production [5], both through increasing yield per unit area and reclaiming more land for crop production [11]. Therefore, while we strive hard to hold greenhouse gas emissions to 'bearable' levels, there is also a need for sustainable approaches that will ensure increased food production in the face of climate change. The use of agrochemicals has boosted crop productivity and contributed to food security, especially in developed countries. However, shortcomings related to their improper and continuous use, such as: increased greenhouse gas emissions (which is a major contributor to global warming), surface and ground water contamination, residual contamination of crop harvest, which poses health concerns to both humans and animals, as well as high costs related to their use. These circumstances have created a need for a more ecofriendly and sustainable approach for enhancing crop productivity in the face of climate change [11–13].

Several approaches have been suggested; the use of plant growth promoting microorganisms and compounds that they produce is perhaps the most promising [14]. The holobiont refers to plants and their associated microbes, which probably coexisted since the colonization of land by the first terrestrial plants [15–17]. This association is referred to as the holobiont [18], and it is dynamic, with the plant asserting a great influence on the nature of phytomicrobiome, especially in its rhizosphere [19], which is mainly attributed to the composition of their root exudates. The rhizosphere, endosphere and phyllosphere may be comprised of pathogenic, neutral and beneficial microbes, in relation to the plant [18,20]. Microbes that exert beneficial effects on the plant are termed plant growth promoting microorganisms (PGPM). These microbes may inhabit the rhizosphere, rhizoplane, phyllosphere, endosphere, etc. [19] For decades, PGPM such as rhizobia, mycorrhizae and plant growth promoting bacteria (PGPR, first defined by Kloepper and Schroth, in 1978) have been reported to enhance plant growth under stressed and non-stressed conditions. The use of microbial inoculants is an old practice [21] that has recently gained more prominence during the last three decades. Much research has been done on rhizobia, and currently a lot is being done on plant growth promoting rhizobacteria and PGPR derived compounds. The ability of microbes to suppress plant pathogens, as well as mitigate the effect of abiotic stress on plants, has been investigated by many researchers, and the findings are promising.

Although they occur naturally in the rhizosphere, and plant tissue, PGPM populations are often insufficient to induce desired effects, hence, it is recommendable to isolate them from their natural environments and multiply their populations before reintroduction into the soil or onto the plant as microbial inoculants [14]. Products in the form of microbe-produced compounds are currently gaining popularity among researchers, although they are less well known among farmers, in comparison to microbial cell inoculants, packaged as either single microbial strains or consortia, which have been commercialised for quite some time [21,22]. Microbe based inoculants are generally from the bacteria (such as Bacillus and Rhizobia) and fungi (especially Trichordema) subgroups [19,22,23], although some groups of archea have also been reported to enhance plant growth. Microbially produced compounds, such as lipochitooligosaccharides (LCO), as plant growth enhancers, on the other hand, are only gaining attention recently, which may explain their lesser availability on the agro-input market. Figure 1 below summarizes some of the mechanisms PGPM employ to mitigate the effects of biotic and abiotic stress on plants, which are later discussed in detail.

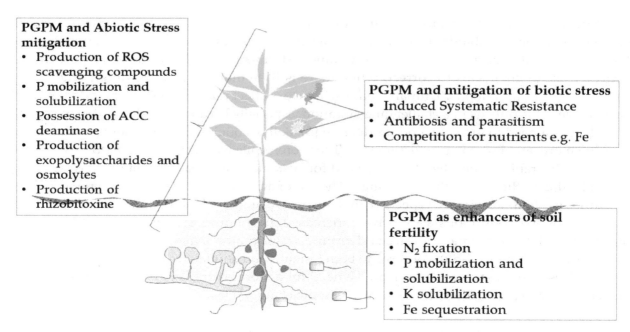

PGPM and Abiotic Stress mitigation
- Production of ROS scavenging compounds
- P mobilization and solubilization
- Possession of ACC deaminase
- Production of exopolysaccharides and osmolytes
- Production of rhizobitoxine

PGPM and mitigation of biotic stress
- Induced Systematic Resistance
- Antibiosis and parasitism
- Competition for nutrients e.g. Fe

PGPM as enhancers of soil fertility
- N$_2$ fixation
- P mobilization and solubilization
- K solubilization
- Fe sequestration

Figure 1. Mechanisms employed by plant growth promoting microorganisms (PGPM) to mitigate effects of biotic and abiotic stress on plants.

2. PGPM as Enhancers of Soil Fertility

For proper growth and development, plants need enough supply of essential macro (Nitrogen, Phosphorus, Potassium, Magnesium, Calcium, etc.) and micro (iron, manganese, boron, zinc, molybdenum, copper) nutrients. N, P and K are the most limiting as far as plant growth is concerned. Unfortunately, with climate change comes abiotic stresses like high temperature, drought and salinity, which influence the biogeochemical transformation of nutrients like P, K, and N, making them either available or less available for plant uptake [24–26]. While the lack of bioavailable macro and microelements is natural in the soil, it could be worsened by climate change. Nitrogen, phosphorus and potassium as the most plant growth limiting elements and their biogeochemical cycle, are affected by temperature and rainfall amongst other abiotic factors, which happen to be affected by climate change. Processes like decomposition, mineralisation, immobilisation, etc. are largely influenced by temperature and rainfall. Processes like soil erosion should also be noted, which is majorly due to run off and wind affect soil fertility as the nutrient rich topsoil is washed away.

Alkalinity affects the availability of Fe, Cu, Zn and Mn, while very low pH is associated with Al toxicity. Processes such as mineralization and nitrogen fixation are affected by moisture, temperature and pH, because they are driven by soil microorganisms like rhizobia, nitrifying bacteria, etc., and enzymes [24,26], which are also affected by abiotic stress. A study by DaMatta et al. [27] showed a decrease in leaf N content of *Coffea canephora* due to water stress. For PGPM technology to be relevant, amidst climate change, it is paramount that stress tolerant strains are identified and used. At the same time, the availability of these nutrients is essential, because they play a key role in minimizing the effects of other abiotic stresses like drought, salinity and high temperature on crops. The roles N, P, K, Ca, Mg and Fe play in the mitigation of abiotic stress have been reported by many researchers [27–33]. For instance, N and P have been reported to minimize the effects of drought stress [24–26,34,35]. K plays a major role in drought stress as well, since it is involved in the opening and closing of the stomata. Agricultural soils have been degraded due to continuous and intense cropping. Agricultural practices like continuous cropping, especially monocropping of non-leguminous crops, without application of fertiliser, is one way of depleting soils of nutrients [36]. This is a common practice of many smallholder farmers, especially in sub-Saharan Africa, due to the inaccessibility and cost of fertiliser [37]. Climate change is only further degrading the situation, because factors such as high temperatures, drought, flooding, salinity, extreme pH, etc. may cause changes in the physiochemical properties of essential soil nutrients such as N, Fe, P and K, thereby limiting their mobility and/or affect their availability for plant uptake, while enhancing the accumulation of toxic elements such as aluminum (Al^{3+}). The role of stress tolerant beneficial microbes in maintaining/increasing crop production amidst climate change challenges cannot be ignored. In order to reclaim land that has been abandoned due to inadequate nutrients for crop growth, considering the financial and environmental costs related to synthetic fertilisers, stress tolerant plant growth promoting organisms can be a cheaper and sustainable approach. With the need to reclaim more land for crop production, emphasis on enhancing soil fertility is inevitable, because nutrients can enhance plant tolerance to abiotic stress. Therefore, there is a need to address the issue using more sustainable approaches. With limited alternatives, and research output so far, microbial inoculants are a promising approach to enhance soil fertility, particularly in conjunction with the various challenges associated with climate change. Microbial inoculants may be defined as formulations comprised of microorganisms, such as bacteria and fungi, as the active ingredients, which once applied on plants, can enhance their growth [19,22,38]. They may also enhance plant quality through the increased concentration of essential nutrients such as proteins [14], and valuable metabolites such as flavonoids, phenolics, alkaloids and carotenoids [23]. Microbial inoculants may also enhance soil biodiversity and properties such as soil structure [22]. As biofertilizers, microbial inoculants enhance the availability and uptake of essential plant nutrients, such as: nitrogen, phosphorus, iron, zinc, and potassium [11–13], which, if lacking or available in inadequate quantities, could limit plant growth.

2.1. Nitrogen Fixation

Some free-living and symbiotic bacteria fix atmospheric dinitrogen into plant usable forms, initially ammonium, through biological nitrogen fixation. Symbionts such as *Rhizobia*, *Bradyrhizobium*, *Sinorhizobium*, *Frankia*, *Actinobacteria* and *Bukholderia* form specialized structures called nodules on their host plants, where they obtain nourishment and shelter, and in turn, fix nitrogen [38,39]. The process is referred to as symbiotic nitrogen fixation and it occurs in both legumes and non-leguminous plants, although that of legumes is the most studied. Communication in the form of molecular signals from both the microbe and host plant, as well as a complex of enzymes (e.g., nitrogenase) and genes (nif and/or symbiotic genes), are involved in the process of nitrogen fixation. On the other hand, free-living nitrogen fixing bacteria such as *Azotobacter* do not need to occupy plant tissue to fix nitrogen. Because of its high energy requirement, plants tend to prefer applied N fertilizer to biological nitrogen fixation (BNF), hence, for effectiveness, synthetic N should not be used along with biological nitrogen fixing organisms, because the plant may suppress the nitrogen fixing symbiosis. Where a starter dose of synthetic N is necessary, it should be applied cautiously, because high N supply can have an inhibitory effect on nodulation (nodule dry weight and number of nodules) and nitrogenase activity [24,40,41]. Arbuscular mycorrhizal fungi, through their hyphae, can enhance the acquisition of soil N by the plant [42], although there are wide variabilities as to the degree of this, whose causes are not yet known [22]. The efficiency and effectiveness of nitrogen fixing bacteria varies among and within plant species, and, in the agricultural context, are largely limited to members of the fabaceae family. Other crops can benefit from the symbiosis by including legumes in crop rotation regimes. There is also a need for more research on how to extend such modifications to non-leguminous plants. Approaches such as genetic engineering to enable non-legume nitrogen fixation and enhance effective communication with N fixing microorganisms can be further researched. Although genetic engineering is questionable, especially its ecological impact, some of the questions are likely from a lack of adequate information on the technology. Extensive research to address most of the questions can be very helpful.

2.2. Phosphate Mobilisation and Solubilisation

Although phosphorus is an abundant element in most soils, it frequently occurs in forms unavailable for plant use. The application of external sources of P fertiliser, such as single super phosphate, diammonium phosphate, etc., can help meet plants' P requirements, but this too may be immobilised shortly after application, making it largely unavailable for plant uptake [43]. mobilization (chemical solubilization and mineralization), which results in plant available forms of the respective nutrients and solubilization, which is a more general term and does not necessarily result in readily plant available forms. For instance, the solubilization of organic P does not necessarily mean that the P is already plant available, as it may still be bound in unavailable organic forms (e.g., phytates). PGPM may enhance soil phosphorus availability for plant uptake through solubilisation and/or mobilisation of inorganic phosphorus. A PGPM may possess both or either mechanisms. The terms, phosphorus solubilisation and phosphorus mobilisation are often used synonymously by many researchers, although they are not necessarily the same thing. P solubilisation is the broader term, which may entail P mobilisation. Goldstein and Krishnaraj [44] described phosphate solubilising microorganisms as those that convert sparing soluble organic or mineral P, into soluble orthophosphate, in a way that significantly increases P availability to a specific plant or plant population within the microorganism's native soil ecosystem.

The same author defined phosphate mobilising microorganisms as those that convert sparingly soluble organic or mineral P, into soluble orthophosphate P, in a way that significantly contributes to pool of available orthophosphate (Pi) in the native soil ecosystem. Phosphorus solubilising bacteria, such as: *Pseudomonas*, *Bacillus*, *Burkholderia* and *Rhizobium*, and some fungal species solubilise inorganic phosphates from sparingly soluble forms such as: tricalcium phosphate, dicalcium phosphate and aluminum phosphate, to forms such as hydrogen phosphate (HPO_4^{-2}), or dihydrogen phosphate ($H_2PO_4^{-1}$), which plants can utilise [37,42–46] through the production of low molecular weight organic acid anions, such as gluconate, lactate, glycolate and oxalate. Phosphorus mobilisers, on the other hand, produce enzymes (such as phosphatase, phytase and phosphonoacetate hydrolase) that chelate cations, bind phosphates and dephosphorylate organic phosphates [22,24]. Dephosphorylation is catalyzed by hydrolase enzymes such as phosphonoacetate, which some PGPM can produce. For instance, ectomycorrhiza and ericoid mycorrhizal fungi produce extracellular acid phosphatases and phytases, which catalyse the mineralisation of P from organic complexes in the soil [42,47]. Other fungal species, such as *Aspergillus niger*, also produce organic acids which aid the process of P solubilisation [48,49]. Through the possession of hyphae, some mycorrhizae such as arbuscular mycorrhizae can deliver up to 80% of the phosphorus taken up by the host plant [27,40].

2.3. Sequestering of Iron

Some PGPM, like *Pseudomonas fluorescens* and *Rhizobia meliloti*, sequester iron through the production of siderophores, which can be grouped into four, namely: hydroxamates, catecholates, carboxylates and pyoverdines [50]. Currently, about 500 siderophores have been reported by researchers. Although plants cannot absorb Fe^{3+}, siderophores have a high affinity for Fe^{3+}, which results in an iron-siderophore complex that is then absorbed by plants [51], into their tissues, hence, aiding plants in meeting their iron requirements [14,17,43,52]. In 2013, study findings of Radzki et al. [53] showed an increase in iron content at 12 weeks for iron deficient tomato plants, following inoculation of siderophore producing bacteria, evidence that microbial siderophores can be a source of iron for plants. A study by Sharma and Johri [54] also showed an increase in maize plant growth following inoculation with siderophore producing PGPR. The uptake of Fe-microbial siderophore complexes by strategy II plants, via ligand exchange, between ferrated microbes and a phyto siderophore, was also reported by Yehuda et al. [55] It should also be noted that some plant species can also produce siderophores which bind Fe^{3+}, to form a complex that can be taken up by the plant with the aid of ligands. Production of siderophores is also a benefit in the context of biocontrol in a sense that potential plant pathogens, especially fungal pathogens, are outcompeted for iron sources, which may lead to their death, or ineffectiveness.

2.4. Potassium Solubilisation

Microbes such as: *Arthrobacter* sp. *Bacillus edaphicus*, *Bacillus circulans* and *Bacillus mucilaginosus* convert sparingly soluble and mineral potassium to soluble forms available for plant use [56]. Through the release of H^+ and organic anions, such as citrate, malate and oxalate, arbuscular mycorrhiza can also increase the solubility of mineral K [57], thereby increasing the availability of potassium anions for plant uptake, although the increase in K^+ availability is sometimes related to the increase in phosphorus availability [22,58].

Some PGPR can also directly influence plant growth through the production of phytohormones such as auxins and gibberellins, which enhance plant growth when plant phytohormones are at suboptimal concentrations [37]. They may also produce enzymes which regulate hormone concentration in plant tissue. For instance, some plant growth promoting microorganisms can produce an enzyme, ACC deaminase, which breaks down ACC, a precursor of ethylene, into an alpha keto butyrate and ammonium, hence lowering ethylene concentration in plant tissues [56,59–62]. With more research and proper manipulation, PGPM, with the ability to enhance plant growth, may not necessarily fully replace chemical fertilizer, but lower their use, directly and indirectly, through increasing the plants' nutrient uptake efficiency from applied chemical fertilizers [63]. Manipulations such as developing an effective consortium of microbes that are able to make available key elements in the soil would greatly reduce the need to use chemical fertilizers. For instance, rhizobial species require iron for good growth, and in their nitrogenase complex, hence co-inoculation of rhizobia and siderophore producing PGPM could enhance nodulation and nitrogen fixation [64]. Use of biofertilizers can lower the need to burn fossil fuels for fertilizer production, and the associated contribution to greenhouse gas emissions.

3. PGPM and Control of Plant Pests and Diseases

With global warming comes new species of pests, weeds and pathogens currently prevalent in warmer environments. The use of chemicals to suppress such plant growth inhibitors is effective but with negative outcomes related to improper use, cost, and increasing evolution of tolerance to the chemical. The antagonist properties of biocontrols against such plant growth suppressors have been reported by many researchers, and the results are promising. A diversity of PGPM with biocontrol properties has been identified by researchers, conferring benefits to a variety of crop species [65–73]. Berendsen [74] showed that plants, when exposed to pathogen attack, can recruit specific plant growth promoting microbes with biocontrol activity, against the pathogen in question. It is believed that manipulating plant recruited PGPM for inoculant production could be more effective in controlling targeted pathogens, than PGPM isolated from places with no pathogen attack. Biocontrols have the potential to minimise the use of industrially manufactured chemicals in agricultural production. This would mean a decline in burning of fossil fuels, and hence a reduction in greenhouse gas emissions. This is because some pesticides are synthesized in laboratories using hydrocarbons like petroleum, which is a fossil fuel. Reduction in their use can mean a reduction in burning fossil fuels, hence less CO_2 emission to the atmosphere. It would also reduce effects on non-targeted members in the ecosystems, which are sometimes affected by chemical use.

Biocontrols may act directly to inhibit growth of biotic agents through hyper parasitism and production of bioactive substances, such as: antibiotics, hydrogen cyanide and phenazines [73], or indirectly through competition for nutrients and active sites on plants, as well as inducing the plant's systemic resistance against the harmful biotic factor [22,38]. Siderophore producing PGPM tend to outcompete other microorganisms for iron sources, which causes inefficiencies in terms of pathogen activities, especially for fungal pathogens, which eventually leads to their death [64]. Induced systemic resistance is triggered by microbe associated molecular patterns (MAMPS), such as lipopolysaccharides, that plants recognise and respond to by turning on their defence systems [75]. Since MAMPS differ among PGPM, it is believed that microbial consortia made up of more than one microbe may induce stronger systemic resistance than single strains [72], although further research needs to be done for a clearer understanding of this potential. PGPM can not only mitigate crop, but also suppress crop pests, such as spidermites [76], moths [77], aphids [78], nematodes [67,79], leaffolder pest [80] and cutworms [81], which greatly contribute to losses incurred in crop production, right from planting to harvesting and storage, if not managed well. PGPR control pests through mechanisms such as production of volatile compounds, such as compounds such as β-ocimene and β-caryophyllene [76], that attract natural enemies of the pest in question. For example, a study by Pangesti et al. [77] showed an increase in the concentration of parasitoid *Microplitis mediator*, a natural predator of *Mamestra brassicae* following the inoculation of *Arabidopsis thaliana* roots with the rhizobacterium *Pseudomonas fluorescens* WCS417r. Other mechanisms through which PGPM mitigate the effects of pests include, increased activity of antioxidant enzymes and increased content of proteins and phenolics in plants, etc. In other cases, the biocontrol agent may not have an effect on the biotic antagonist, but will enhance plant yield in the presence of the antagonist [78]. This particular strategy seems very useful, especially in cases where biotic stress factors such as weeds and pests become resistant or unresponsive to other control strategies. It may also enhance/preserve species diversity, hence maintaining ecosystem functionality.

Some PGPM are efficient against pathogens as single strains, while others perform better as a consortium. Details of single strains vs. consortia are discussed later in this review. Table 1 lists PGPM with potential biocontrol activities against the pathogens of various crop species. With the increasing campaign against the use of chemicals, as a means of combating climate change, such strains are a promising substitute for chemicals that are currently prevalent in agricultural production. Currently, the global biocontrol market is approximately 2 billion USD [79], and is expected to grow further. More research on existing microbial species or microbe-produced compounds with biocontrol properties is still desirable, as is the identification of new ones.

Table 1. Biocontrol species against biotic stressors of different crop species.

PGPM	Biotic Stress	Host Plant	Reference
Bacillus amyloliquefaciens LY-1	*Peronophythora litchii*	Litchi (*Litchi chinensis* Sonn.)	[71]
Burkholderia cepacia	*Fusarium oxysporum*	*Solanum tuberosum*	[65]
Pseudomonas fluorescens	*Fusarium graminearum*	*Triticum aestivum*(wheat) cv. Tabuki	[66]
Pseudomonas fluorescens CHAO	*Gaeumannomyces graminis var. tritici*	*Triticum* sp.	[64]
Pseudomonas fluorescens CHAO	*Thielaviopsis basicola*	*Nicotiana tabacum*	[64]
Bacillus spp.	*Heterodera glycines*	*Glycine max.*	[82]
Serratia proteamaculans	*Meloidogyne incognita*	*Solanum lycopersicum* L.	[72]
Bacillus aryabhattai A08	*Meloidogyne incognita*	*Solanum lycopersicum* L.	[83]
Serratia plymuthica HRO-C48	*Botrytis cinerea*	–	[84]
Serratia plymuthica strain C-1, Chromobacterium sp. strain C-61 and *Lysobacter enzymogenes* strain C-3 consortium	*Phytophthora capsici*	*Cupsicum* spp.	[85]
Paenibacillus sp. 300 + *Streptomyces* sp. 385,	*Oxysporum f.* sp. *Cucumerinum*	*Cucumis sativus*	[86]
Pseudomonas fluorescens WCS 358	*Fusarium oxysporum f* sp. *Raphani*	*Raphanus sativus*	[87]
Pseudomonas fluorescens	*Macrophomina phseolina*	*Coleus forskohlii Briq.*	[68]
Pseudomonas aeruginosa 7NSK2	*Pythium splendens*	*Lycopersicon esculentum*	[88]
Pseudomonas fluorescens	*Pythium* spp.	*Triticum* sp.	[64]
Pseudomonas fluorescens	*Pythium ultimum*	*Gossypium* sp.	[64]
Bradyrhizobium japonicum NCIM 2746	*Rhizopus* sp. and, *Fusarium* sp.	*Glycine max* L.	[89]
Paenibacillus lentimorbus B30488	*Scelerotium rolfsii*	*Solunum lycopersicum* L.	[69]

Table 1. *Cont.*

PGPM	Biotic Stress	Host Plant	Reference
Pseudomonas putida UW4	*Agrobacterium tumefaciens*	*Solanum lycopersicum*	[90]
Burkholderia phytofirmans PsJN	*Agrobacterium tumefaciens*	*Solanum lycopersicum*	[90]
Bacillus cereus PX35, Bacillus subtilis SM21 and Serrati asp. XY2	*Meloidogyne incognito*	*S. lycopersicum*	[91]
Pseudomonas fluorescens strain S35	*Phytophthora infestans*	*Solanum tuberosum*	[73]
Pseudomonas frederiksbergensis strain 49 and Pseudomonas fluorescens strain 19 consortium	*Phytophthora infestans*	*Solanum tuberosum*	[73]
Pseudomonas putida strain R32	*Phytophthora infestans*	*Solanum tuberosum*	[73]
Pseudomonas chlororaphis spp. *strain R47*	*Phytophthora infestans*	*Solanum tuberosum*	[73]
Pseudomonas spp. *strain S49*	*Phytophthora infestans*	*Solanum tuberosum*	[73]
Bacillus and Pseudomonas spp. *consortium*	*Fusarium oxysporum U3 and Alternaria* sp. *U10*	*Nicotiana attenuata*	[92]
Chaetomium sp. *C72 and Oidodendron* sp. *Oi3 consortium*	*Fusarium oxysporum U3 and Alternaria* sp. *U10*	*Nicotiana attenuata*	[92]
Pseudomonas chlororaphis R47	*Phytophthora infestans*	*Solanum tuberosum*	[69,93]
Pseudomonas fluorescens strain LBUM 636	*Phytophthora infestans*	*Solanum tuberosum*	[94]
Agrobacterium radiobacter var radiobacter	*Crown gall*	*Solunum lycopersicon*	[95]
Tricoderma koningiopsis Th003 WP	*Fusarium oxysporum*	*Physalis peruviana*	[67]
Trichoderma harzianum Tr6 + Pseudomonas sp. *Ps14*	*Fusarium oxysporum f.* sp. *radicis cucumerinum*	*Cucumis sativus*	[96]
Pseudomonas sp. *Ps14*	*Botrytis cinerea*	*Arabidopsis thaliana*	[96]
Trichoderma harzianum Tr6	*Botrytis cinerea*	*Arabidopsis thaliana*	[96]
Pseudomonas putida	*Spodoptera litura*	*Solanum lycopersicum L.*	[96]
Pseudomnas flourescences Pf1, Bacillus subtilis Bs and Trichoderma viridae Tv consortium	*Lasiodiplodia theobromae*	*Polianthes tuberosa L.*	[97]
Pseudomonas sp. *23S*	*Clavibacter michiganensis*	*Solanum lycopersicum L.*	[98]
Peanibacillus lentimorbus B-30488	*cucumber mosaic virus*	*Nicotiana tabacum cv White burley*	[99]
Serratia liquefaciens MG1	*Alternaria alternate*	*Solanum lycopersicum*	[100]
Xanthomonas sp. *WCS2014-23, Stenotrophomonas* sp. *WCS2014-113 and Microbacterium* sp. *WCS2014-259*	*Hyaloperonospora arabidopsidis*	*Arabidopsis thaliana*	[74]
Lactobacillus plantarum SLG17 and Bacillus amyloliquefaciens FLN13	*Fusarium* spp.	*Triticum durum*	[101]
Fusarium oxysporum strain Fo162	*Aphis gossypii Glover*	*Cucurbita pepo*	[102]
Rhizobium etli strain G12	*Aphis gossypii Glover*	*Cucurbita pepo*	[102]
Bacillus subtilis strain BEB-DN	*Bemisia tabaci*	*Solanum lycopersicum*	[103]
Bacillus amyloliquefaciens (SN13)	*Rhizoctonia solani*	*Rice (Oryza sativa)*	[104]
Pseudomonas fluorescens Migula strains Pf1 and AH1	*Desmia funeralis*	*Oryza sativa*	[80]
Pseudomonas putida and Rothia sp.	*Spodoptera litura*	*Solanum lycopersicum*	[81]

4. PGPM and Abiotic Stress

With climate change, the occurrence of extreme abiotic stresses, such as floods, salinity, high temperature and drought are expected to increase [3,8–10,105]. In fact, much of this is already being experienced in some parts of the world. Winters are becoming warmer in some regions; rainfall is becoming scarcer and more erratic, causing droughts and desertification [1,2,104] in other regions. With less rainfall, salinity is more likely to occur, either through irrigation or natural causes [106–111].

All these factors affect crop production, and their management inputs are sufficiently costly that many farmers may not be able to afford them. Factors such as high temperatures can generally not be managed under field conditions. Therefore, there is the need for a strategy that is ecofriendly and manageable by the majority of crop producers. PGPM have been reported to mitigate effects of abiotic stress on plants, hence, allowing the plant to grow and yield relatively well under stress conditions [112–115]. Various researchers have reported the ability of a wide range of PGPM to enhance plant growth, in the presence of abiotic stressors, such as salinity [116,117], drought [114,118,119], heavy metals and acidity. In fact, the ability of some PGPMs to enhance plant growth is only triggered in the presence of stress [118]. They employ mechanisms such as the production of ROS scavenging compounds, possession of ACC deaminase (an enzyme that lowers ethylene concentration in plants exposed to stress), and the production of exopolysaccharides and osmolytes. For example, Akhtar et al. [120] observed an increase in the antioxidant activity of catalase (CAT) in the roots of drought stressed maize plants treated with *Bacillus licheniformis* (FMCH001). Treated plants also exhibited a higher dry weight and higher water use efficiency. Yang et al. [121] also reported the increased activity of catalase and dehydroascobate reductase enzymes in salinity stressed Quinoa plants treated with an endophytic bacterium known as *Burkholderia Phytofirmans* PsJN, compared to the untreated plants. The former also exhibited a higher shoot biomass, grain weight and grain yield compared to the latter. Some rhizobia spp. produce the compound rhizobitoxine, which inhibits the activity of ACC synthetase, hence lowering ethylene activity that would otherwise inhibit nitrogen fixation. A PGPM may possess one or more of these mechanisms, all of which act to help a plant thrive under stress conditions. Like plants, PGPM can also be affected by abiotic stress, such as salinity, high temperature and drought, which can lower their efficacy in promoting plant growth, or even death of the microbe, in cases of prolonged exposure to extremes of such conditions [122]. Therefore, it is essential that the strains chosen for use are tolerant to the abiotic stress, whose effect in plants they tend to mitigate. Strains isolated from areas affected by abiotic stress may have an edge over those isolated under normal conditions, although this may not always be the case. The use of microbial consortia may be helpful, especially in areas where more than one factor inhibits crop growth (which is almost always the case under field conditions). However, more research needs to be conducted, for the better deployment of PGPM technology. The exploitation of such microbes has a definite potential to maintain crop production amidst increasing abiotic stresses that are rendering some currently arable land unfit for crop production. Table 2, below, shows some PGPM strains that have been discovered and characterized by researchers, with the potential to mitigate the effects of abiotic stress on a range of plant species.

Table 2. Examples of PGPM that enable plants to withstand abiotic stress.

PGPM	Abiotic Stress	Host Plant	Reference
Pseudomonas putida MTCC5279	Drought	chickpea (*Cicer arietinum*)	[114]
Pseudomonas fluorescens REN1	Flooding	Rice (*Oryza sativa*)	[123]
Variovorax paradoxus 5C-2,	Salinity	Peas	[115]
Bacillus amyloliquefaciens SQR9	salinity	Maize	[112]
Dietzia natronolimnaea	Salinity	Wheat (*Triticum aestivum*)	[116]
Serratia nematodiphila	Low temperature	pepper (*Capsicum annum*)	[124]
Burkholderia phytofirmans PsJN	Low temperature	grapevine (*Vitis vinifera*)	[125]
Pseudomonas vancouverensis	Low temperature	Tomato (*Solanum lycopersicum*)	[113]
Pseudomonas sp. S1	drought	*Capsicum annum*	[118]

Table 2. *Cont.*

PGPM	Abiotic Stress	Host Plant	Reference
Pseudomonas sp. S1	drought	*Vitis vinifera*	[118]
Achromobacter xylosoxidans	Flooding stress	*Ocimumsanctum*	[126]
Pseudomonas sp. 54RB + *Rhizobium* sp. Thal-8	Salinity	*Zea mays cv. Agaiti 2002*	[127]
Pseudomonas putida KT2440, *Sphingomonas* sp. OF178, *Azospirillum brasilense* Sp7 and *Acinetobacter* sp. EMM02) consortium	drought	Zea mays	[119]
Achromobacter xylosoxidans	salinity	*Catharanthus roseus*	[128]
Burkholdera cepacia SE4	salinity	*Cucumis sativus* L.	[124]
Pseudomonas putida (W2)	salinity	*Triticum aestivum* L.	[56]
Pseudomonas fluorescens (W17)	salinity	*Triticum aestivum* L.	[56]
Kocuria flava AB402	Arsenic toxicity	*Oryza sativum*	[129]
Bacillus vietnamensis AB403	Arsenic toxicity	*Oryza sativum*	[129]
Trichoderma spp. strain, M-35	Arsenic toxicity	*Cicer arietinum*	[130]
Burkholderia cepacia and Penicillium chrysogenum consortium	waste motor oil toxicity	*Sorghum bicolor*	[131]
Bacillus safensis	High temperature	*Triticum aestivum* L.	[132]
Pseudomonas aeruginosa	Zn-induced oxidative stress	*Triticum aestivum* L.	[133]
Bacillus licheniformis (FMCH001)	oxidative stress Drought	*Zea mays* L. cv. Ronaldinho	[120]
Burkholderia phytofirmans PsJN	Salinity	*Chenopodium quinoa* Willd	[121]

5. Commercialisation of Microbial Inoculants

Making PGPM technology available for farmers is key to ensuring their adaptation as agricultural inputs. Commercialisation of promising strains is one way of making promising strains accessible by farmers. Although various strains that possess desirable properties under laboratory and greenhouse conditions may be isolated, developing a commercial product, effective under field conditions, is not an easy task, especially because numerous factors determine the efficiency of introduced species. Characteristics such as: possession of multiple mechanisms of enhancing plant growth, ability to compete favorably and establish populations in the rhizosphere, persistence in the rhizosphere over seasons, and ability to be cultured in artificial environments [15,61,89] are desired for potential PGPM strains. However, many plant and soil factors, such as plant species, soil temperature, composition and prevalence of native microbes, soil pH, etc., may work together against a strain which is otherwise excellent under controlled environment conditions. Even before introduction into the field, factors such as formulation play a major part concerning a product's efficacy. For instance, solid inoculant formulations are desired for their longer shelf life, however, the process of drying microbes often results in lower microbial cell counts, hence lowering their competitiveness, since number contributes greatly to their ability to compete with native microbes [134]. Exposing a potential PGPM to some level of stress before formulation may increase its survival rates during formulation and after field application [134]. Before introducing a potential PGPM inoculant into the market, a series of events, such as greenhouse and field trials, characterization, toxicology profiling, etc. occur, most of which are intended to increase strain survival and efficacy in the field.

Formulation of Microbial Inoculants for Commercial Purposes and Their Mode of Application

Microbial inoculants are usually a combination of microbial cells and/or their parts/compounds and a nonliving carrier that may be in form of a liquid or solid material [15,24,38]. Microbial cells may be either active or dormant; in the latter case, they have to be activated before or after inoculation [15]. They may also be pure cultures (single strains) or a combination of microbial strains (microbial consortia) [15,38].

Formulation is a major contributor to the variation in performance of inoculants observed in farmers' fields and at research stations. Formulation can shield the microbe from adverse environmental conditions, increase their shelf life and also supply their nutritional requirements, hence enhancing their chances of survival in the field [23,134]. Normally, a group of microbes are isolated from their natural habitat (soil or plant tissue), tested for their ability to promote plant growth under a range of conditions, and the superior strains are selected for commercialisation purposes. The strains are multiplied and formulated under controlled environment conditions, after which the efficiency of the inoculant is evaluated under field conditions [23].

The method of formulation ought to consider the target crop, target market and mode of application, the latter because the type of formulation often dictates the mode of application of the inoculant. For instance, solid formulations are mainly applied through seed dressing, or broadcasting onto the field, while liquid formulations have a wide range of application methods [15,24,38]. Liquid carriers are mostly water and/or organic solvents (other than microbial media), such as glycerol and carboxymethyl cellulose that are added to increase properties such as stickiness and dispersal abilities [23]. There are several types of solid carriers, such as clay, vermiculite, peat and charcoal [15]. Care should be taken, when selecting microbial carriers, to ensure they have no negative impact on the environment or the microbe itself [15,38].

Although they are easy to handle and work with, liquid carriers may require specialised storage conditions (cool conditions that necessitate a cooling mechanism) for a long shelf life [23], which makes their marketing and use in developing countries difficult, due to limited and unstable power supply on most farms. Solid formulations, on the other hand, are bulky and may require larger storage facilities, when compared to liquids. However, materials such as peat have an outstanding reputation as inoculant carriers, and are successfully used in both North and South America [23]. The formulation method opted for should ensure the affordability of the final product by the target market, since a very expensive product is likely to meaningfully increase production costs, which is undesirable. For instance, sterile carriers are preferred over nonsterile carriers [23], however the former are costlier than the latter, which may make them unaffordable to many farmers across the globe. The formulation method should also ensure the compatibility of the inoculant with agronomic practices, such as weed control methods, irrigation, etc.

Once a formulated product exhibits positive responses, in field and greenhouse trials, it is put on the market for accessibility by farmers. While the isolation and characterisation of microbial strains from their natural habitats is largely done by academic research institutions, the production of microbial inoculants for commercial purposes is dominated by registered companies, which obtain patents and rights over specific inoculants. Table 3 below shows such microbial based products on the market as plant growth stimulants.

Table 3. Examples of microbial inoculants currently available on the market, and their producing companies.

Inoculant	Country	Producer	Use	Reference
Bacillus megaterium	Sri Lanka	BioPowerLanka	Phosphorus solubilisation	[135]
Pseudomonas striata, B. Polymyxa and B.megaterium consortium	India	AgriLife	Phosphorus solubilisation	[135]
Acidithiobacillus ferrooxidans	India	AgriLife	Iron mobilization	[135]
Trichoderma and Bradyrhizobium spp. (Excalibre-SA) consortium	USA	ABM®	N fixation Growth stimulation	[18]
BIODOZ® (B. japonicum)	Denmark	Novozymes	Nitrogen fixation	[134]
Cell-Tech® (B.japonicum)	Belgium	Monsanto (Bayer)	Nitrogen fixation	[134]
Nitragin® *S. meliloti*	Belgium	Monsanto BioAg™ (Bayer)	Nitrogen fixation	[134]
Cedomon® *Pseudomonas chlororaphis*	Sweden	BioAgriAB	Biopesticide	[134]
Sheathguard™ *Pseudomonas fluorescens*	India	AgriLife	Biopesticide	[134]
Galltrol® -A *Agrobacterium radiobacter*	USA	AgBioChem	Biopesticide	[134]
HISTICK® *Bradyrhizobium japonicum*	Germany	BASF SE	Nitrogen fixation	[135]
Bacillus + Pseudomonas + Lactobacillus + Saccharomyces spp.	Canada	EVL Inc	Biostimulant	
Xen Tari (*Bacillus thuringiensis*)	USA	Valent USA	Biopesticide	[136]
VOTIVO FS seed treatment (*Bacillus firmus*)	USA	Bayer	Biopesticide	[136]
VectoLex FG (*Bacillus sphaericus*)	USA	Valent Biosciences	Biopesticide	[136]
Venerate XC (*Burkholderia rinojensis*)	USA	Marrone Bio Innovations	Biopesticide	[136]
Zequanox (*Pseudomonas fluorescens*)	USA	Marrone Bio Innovations	Biopesticide	[136]
BotaniGard ES/WP, Mycotrol (*Beauveria bassiana*)	USA	Lam International	Biopesticide	[136]
Naturalis L (*Beauveria bassiana*)	USA	Troy BioSciences	Biopesticide	[136]
BioCeres WP (*Beauveria bassiana*)	USA	BioSafe	Biopesticide	[136]
Met-52 EC and Met-52 G (*Metarhizium brunneum* (anisopliae s.L.)	USA	Novozymes	Biopesticide	[136]
MeloCon WG (*Purpureocillium lilacinum*)	USA	Bayer	Biopesticide	[136]
Cyd-X, Cyd-X HP (*Cydia pomonella* (CpGV)	USA	Certis USA	Biopesticide	[136]
FruitGuard (*Plodia interpunctella* GV	USA	Agrivir	Biopesticide	[136]
Serenade (*Bacillus subtilis* QST 713)		Agraquest	Biocontrol	[79]
Bacillus firmus I-1582 WP5 (*B. firmus* I-1582)		Bayer Crop Science	Biocontrol	[79]
Cedomon (*Pseudomonas chlororaphis* MA342)		Bioagri		[79]
Proradix (*Pseudomonas* sp. DSMZ 13134)		Sourcon–Padena Germany, Itary	Biocontrol	[79]
Novodor (*B. thuringiensis ssp. tenebrionis* NB 176)	USA	Valent Bioscience	Biocontrol	[79]

6. Limitations to Global Use of Microbial Inoculants

Although microbial inoculants are viewed as the most viable hope, with regard to sustainable agriculture in the face of climate change, their use and adoption globally are still wanting, due to a range of reasons, that vary between developed and developing countries. Adaptation to use of microbial inoculants is developing at a relatively faster pace [24] in the developed world than in developing areas, such as Africa, where their use is restricted by limited availability of resources and knowledge, among other factors. In the developed world, microbial use is slowed largely by inconsistencies in enhancing plant growth, in which case crop producers opt for chemicals, which generally provide stable results. There are many cases where the excellent performance of an inoculant observed during pre-commercialisation trials does not translate to efficiency on farmers' fields. Even when it does, sometimes the results are not consistent, which frustrates the farmers. Some of these inconsistences may be attributed to biotic and abiotic soil factors and plant factors which directly or indirectly affect the introduced microorganism(s) [23]. For instance, some inoculants are cultivar and species specific, in that applying them outside the target species will yield no results. Soil factors such as salinity and temperature are dynamic and affect the survival and effectiveness of the applied microbial strains. This implies that soil conditions should always be favorable for the introduced microbe, otherwise inconsistencies are bound to prevail. Therefore, there is a need to sensitise farmers regarding the proper use of microbial products to minimise such inconsistencies. Unless sensitisation is properly conducted, we cannot rule out inappropriate practices such as farmers applying rhizobial inoculants together with high doses of nitrogen fertilizer, expecting better results than the inoculant or fertilizer used alone. In fact, nitrogen fertilizer will inhibit biological nitrogen fixation. Similarly, applying a biocontrol to a soil or plant that lacks the pathogen it can antagonise/suppress may not yield results. It is also important to understand the status of the soil/plant as the application of microbial inoculants may inhibit plant growth where the soil/plant already contains optimal concentrations of the compound that the microbe produces to enhance plant growth. For instance, application of IAA producing PGPM on plants with an already optimal concentration of IAA may yield negative effects on the plant, due to excess IAA [43]. Understanding soil conditions will also guide the farmer regarding how often to apply the inoculant. Some require seasonal, annual or even twice in a season application, while after some time, application may not be necessary, especially where the microbe establishes reasonable populations in the soil. Successful microbial inoculants employ mechanisms that give them a competitive advantage over the native strains. For instance, rhizobia and mycorrhizal fungi have a signaling system with their host plants, which gives them an advantage over their competitors. Introduced microbes may also outcompete native microbes through the production of antimicrobials, which may kill or deter other microbes, as well as the production of siderophores that give them a competitive advantage over other microbes for iron resources in the soil, hence proliferating better, especially in iron limited soils [14]. Nevertheless, it is important to increase the competitive advantage of introduced microbes, by ensuring high microbial concentrations in the inoculant and use of adequate formulations [18]. With approaches such as metagenomics, the microbial population of the target environment can be studied, and potential PGPM studied for their ability to out compete the latter in field, greenhouse and laboratory conditions. However, this may not be an easy task, given that microbial populations in crop production fields may differ meaningfully due to a wide range of factors. Location and plant specific nature of some phytomicrobiome elements for inoculant production should also be prioritised, since such microbes, to a great extent, are more adapted to the environment and/or plant conditions, which may increase their chances of survival and persistence in the soil. The idea of using microbial consortia may also work to our benefit, as will be discussed below. This does not, however, disqualify single strain inoculants; their advantages are also discussed below.

In less developed countries, especially in sub-Saharan Africa, reasons for low adoption also vary between large- and small-scale farmers. For large-scale farmers, such as those in Zimbabwe, South Africa and Kenya, the ineffectiveness of many microbe-based products in the field contributes meaningfully to the low adoption of microbial inoculants [137,138]. For small scale farmers, costs and inadequate knowledge of such products are the major drivers. These two factors, especially costs, also limit the use of other agricultural inputs, such as high-quality seed. Exceptions can be made for a smaller group of small scale farmers, whose farms' researchers run experiments/field trials, because then, they can obtain access to the inputs from researchers largely free of charge, otherwise, they mostly depend on crop rotations (which are sometimes not properly done) and animal manures, while others just grow their preferred crops year in year out. The lack of knowledge can be attributed to the large gap between research and extension. Researchers achieve good findings, but due to poor funding and poor dissemination techniques, this knowledge never reaches the farmer [138]. Publications do not help much, because many small-scale farmers are illiterate, and even those who can read have limited access to technologies such as smart phones, computers and the internet. It should be noted that many small-scale farmers are also low-income earners, who struggle to meet their basic needs. In countries where governments are not directly involved in the distribution of agricultural inputs, dealers may not be willing to extend products to people who they well know cannot afford them, which leads to unavailability of and/or inaccessibility of the products by the farmers. In such cases, intervention strategies should definitely be at least a bit different and more vigorous. First and foremost, the knowledge of existence of PGPM technology needs to be spread to these largely small scale farmers. Projects like N2 Africa have done a good job in trying to spread the BNF technology, although more effort is still needed. Extension officers should be updated on new findings and products, and be properly facilitated to extend this knowledge to the farmers. Governments may consider subsidizing products and getting directly involved in their distribution to the farmers. Promiscuous soybean varieties are already a good strategy of eliminating the need for inoculation. It would be better to develop strategies that enable the use of farm-based PGPM inoculants, as many farmers have limited access to agro-input markets, in part due to poor transport networks. Locally made cooling facilities such as charcoal based refrigerators and unglazed clay pots may also be helpful. However, the former would be a contradictory measure, given that it would encourage deforestation. The whole sensitisation process should involve all stakeholders, such as governments, extension officers, agricultural schools, and private companies that contract small scale farmers to grow crops for them for use as raw materials. The latter, especially, provide the farmer with chemicals such as pesticides and fertilizers; therefore, their involvement cannot be ignored.

7. Microbial Consortia

In order to address issues associated with the use of single strains as inoculants, microbial consortia have gained popularity. This may be relevant, especially now that the prevalence of both biotic and abiotic stresses due to climate change are likely to increase. Microbial consortia technology involves the use of more than one microbial species in a single inoculant product. The microbes may have the same or different modes of action [18,70,89], and may be from different phyla, genera or even other groupings, for example, a combination of bacterial and fungal strains. Microbial consortia may have an advantage over single strains when the species synergistically interact and confer benefits to each other [70,71,89,117]. For instance, one strain may breakdown a substrate, unavailable to other species, converting it into forms that the other members of the consortium can utilise as a source of nutrients [14], or produce exopolysaccharides which offer protection against stress to all members of the consortium [134], produce compounds which are signals that activate plant growth promotion capability of other members of the consortium, through the production of plant growth stimulating compounds, that they would otherwise not produce, for instance, in pure culture. In cases where microbes with the same mode of action are used, members may have varying tolerance to different biotic and abiotic stresses, which enhances survival of at least a member that will confer intended benefits to the plant. In the case of different modes of action, these complement each other and confer a more effective benefit to the plant. It could also be that some members of the consortium are simply helpers of the strains meant to benefit the plant. Such helper strains, for instance mycorrhiza helper bacteria, should facilitate the target strain in plant colonisation, conferring benefits to the plant.

Researchers have reported inefficient strains that became efficient in a consortium. For example, Santhanam et al. [89] observed that the inclusion of two bacterial strains with insignificant effects on mortality of sudden wilt pathogens in tobacco, in a consortium with three other bacteria improved resistance of plants to the same pathogen, in comparison to the consortium of 3 used alone. Mycorrhizal fungi, in association with a helper bacterium, may have better established mycelia and plant root colonisation, if the bacterium produces substances that directly enhance the germination of fungal spores, or indirectly enhance the establishment of mycorrhiza through the production of antimicrobials that reduce competition from other microbes or minor pathogens [14].

Because of the interaction advantage, some microbes perform better in microbial consortia than when applied individually [70,89]. However, the reverse is true for some PGPM species, as reported by other researchers [70,92,100]. Therefore, the role that single strain inoculants play cannot be written off easily, especially because microbial consortia also have their shortcomings. Coming up with effective compatible combinations in which all members actively benefit the plant can be challenging, practically given that some members of the consortium may produce compounds lethal to other members [133]. Even if the produced compounds do not go to the extreme of killing other members, they may cause a shutdown of their plant growth promoting system, or interfere with their growth, as de Vrieze et al. [70] observed in a consortium of five *Pseudomonas* strains. In such cases, it is probable that only a subset of the consortium members will actively benefit the plant, the rest being "dormant" or dead. Difficulties concerning the formulation of microbial consortia may also be associated with the variations in optimal growth conditions. For more than one species, or even genus, creating conditions that will favour all members while retaining their ability to promote plant growth may not be easy. Finally, manufacturing consortia can be challenging, as very small changes at the outset can result in very different levels of consortium members in the final product, resulting in product inconsistencies.

8. Microbial Compounds as "Inoculants"

The use of microbial compounds as "inoculants" is slowly gaining popularity after successful trials [139–145]. To be a true inoculant, the material must contain living cells that colonize the plant. In this case, the technology may be the product of microbial growth and may be more valuable as a result of climate change were biotic and abiotic factors may lower or completely halt the effectiveness of microbial cell based inoculants. This practice involves the separation of cell free supernatant from microbial cells, and the subsequent separation and purification of the compound from the cell-free supernatant, mainly through high pressure liquid chromatography (HPLC). The pure compound is then tested for its ability to promote plant growth under greenhouse and field conditions, prior to commercialisation. Before commercialisation, other tests, such as the effect of the compound on non-target organisms and humans, as well as checks regarding legal regulations, are usually carried out. The effect of the compound on non-target organisms such as plants, humans and animals ought to be substantially understood too, as with studying the residual effects of the compound (how much of it remains in the edible parts of the plant, and in the soil, following application). Therefore, before any compound can be commercialised, its ability to be purified, and produced on a large scale, should be verified [143]. The compound should be identified and characterised based on its physiological and biological properties.

The efficacy and type of microbial bioactive compounds produced are influenced by microbial species and conditions to which the PGPM is exposed. Slight alterations in growth conditions may result in different compounds, produced at different levels, and with varying degrees of efficacy. For instance, varying the pH, a *Pseudomonas* species culture caused it to produce different phenazine compounds with varying efficacy against *Fusarium oxysporum* f. sp. radicis-lycopersici [143]. Sometimes, the PGPM has to be exposed to stressful conditions before it will produce bioactive compounds, as such compounds may only be produced to enhance the survival of the microorganism under stressful conditions. Therefore, it is important to have an adequate understanding of the conditions under which a certain PGPM will produce plant growth stimulating compounds.

So far, not many (compared to microbial strains) bioactive compounds have been identified for use in crop production. The Smith laboratory at McGill University has thuricin17 and lipochitooligosaccharide (LCO). Thuricin 17 is a bacteriocin secreted by *Bacillus thuringiensis*, a non-symbiotic endophytic bacterium. The compound is known to have anti-microbial properties, which gives *Bacillus thuringiensis* a competitive advantage over other bacteria of the same grouping [140]. After a series of experiments, thuricin 17 was discovered to have growth promoting properties for tomato, soybean, canola, arabidopsis, and rapeseed and switch grass [117,140–142,144,145]. More trials are on-going, and the technology has yet to be commercialised. Lipo-chitooligosaccharide, on the other hand, is produced by rhizobia, as a signal to its host plants [139]. Formerly extensively studied for its role in the nodulation process, the compound is currently patented and being marketed by Novozymes as a plant growth stimulant, where its effects are greatest under abiotic stress conditions. Other compounds such as phenazine-1-carboxylic acid (PCA) have also been commercialised [143,146–153]. Table 4 shows the various compounds with potential use as agro-inputs. Some of them are already commercialised.

Table 4. Microbial compounds of agricultural importance.

Compound	Producing Microbe	Function	Comment	Reference
LCO	*Bradyrhizobium japonicum*	Biostimulant	*Stimulates plant growth under stressed and non stressed conditions.*	[117,146]
Thuricin17	*Bacillus thuringiensis*	Biostimulant	*Enhances growth of different crops eg Soybean in stressed and non stressed conditions*	[141,142]
Anisomycin	*Streptomyces* sp.	herbicide	*Effective against Digitaria spp.*	[149]
Phenazine-1-carboxyamide (PCN)	*Pseudomonas* spp.	biocontrol	*It is effective against; Fusarium oxysporum f. sp. Radicis-lycopersici, Xanthomonas oryzae pv. Oryzae, Rhizoctonia solani, Botrytis cinerea*	[143,148,149,151]
Phenazine-1-carboxylic acid (PCA)	*Pseudomonas* spp.	biocontrol	*It is effective against Fusarium oxysp.orum f. sp.Radicis-lycopersici, Colletotrichum orbiculare, Gaeumannomyces graminis var. tritici, Phytophthora capsici*	[143,146,149,152, 153]
Pyocyanin (PYO)	*Pseudomonas* spp.	biocontrol	*Effective against: Sclerotium rolfsii, Macrophomina phaseolina*	[154–156]
Pyrrolnitrin	*Burkholderia pyrrocinia 2327*	biocontrol	*It has antifungal properties against; Ralstonia solani, Phytophthora capsici, and Fusarium oxysporum*	[157,158]
Phencomycin	*Burkholderia glumae 411gr-6*	biocontrol	*Effective against; Alternaria brassicicola, Aspergillus oryzae, Cladosporium cucumerinum, Colletotrichum gloeosporioides*	[159]
Ornibactin	*Burkholderia contaminans MS14*	biocontrol	*Siderophore with biocontrol activity against Erwinia amylovora, Ralstonia solanacearum, Pseudomonas syringae B301, Clavibacter michiganensis subsp. michiganensis*	[160]
Iturin A2	*Bacillus subtilis B47*	biocontrol	*Effective against fungi; Bipolaris maydis*	[161]
Mycosubtilin	*Bacillus subtilis*	biocontrol	*Has anti fungal properties, effective against; Bremia lactucae*	[162]
Herboxidiene	*Streptomyces* sp. A7847	herbecide	*Effective on a number of weed sp.*	[163]
Phosphinothricin	*Streptomyces hygroscopicus*	herbecide		

Table 4. *Cont.*

Compound	Producing Microbe	Function	Comment	Reference
Cyanobacterin	*Scytonema hofmanni*	*herbecide*	*Effective on cynobacteria, algae and higher plants*	[164]
Avermectin	*Streptomyces avermitilis*	*Insectide nematocide*	*Effective against Spider mites, Citrus red mite, horn worms, army worms, etc.*	[165]

9. Microbial Cells or Microbial Compounds?

Given the current understanding, a question would be, what should a crop producer adopt, given a choice between the microbial cells and microbial compound based products. The answer to such a question cannot be as definite as that specific factors may call for either of the two, or even the use of both simultaneously. Before one reaches the level of farmer preferences, soil and environmental factors as well as economic implications, intended use and handling may be major considerations. For instance, in the reclamation of areas heavily affected by abiotic stress, use of microbial cells may not be a good idea, if they are not able to survive some harsh conditions. Even if they did, the efficacy of their plant growth promotion capacity may be greatly affected. Compounds, on the other hand, are less affected by such abiotic stresses, and hence have a greater chance of being successful under such conditions. The use of compounds or both compound(s) and microbial cells may be desirable, especially when an abiotic stress such as drought interrupts signaling between plant and PGPM. In such a case, external application of the signal may rectify the disruption. Prudent et al. [142] observed a 17% increase in soybean biomass under drought conditions following co-inoculation with *Bradyrhizobium japonicum* and thuricin17, compared to inoculation with the rhizobial cells alone. The use of microbial compounds may also be a better choice in cases where the microbe is a facultative pathogen, such as *Agrobacterium* spp. [92]. In such cases, the pathogen effect of the microbe on plants is minimised. Application of microbial compounds may also benefit a wider range of crop species compared to microbial cells, given that many microbes can be at least somewhat species specific. A case would be that of lipochitooligosaccharides (LCOs), which can be utilised to enhance growth of legumes and non-leguminous crops [148], under stressed and non-stressed conditions [115,137], but to a greater extent, under stress conditions. For instance, LCOs enhanced fruit and flower production in tomato (*Lycopersicon esculentum*) plants [148], and stimulated the growth of soybean and corn plants [139]. The compound was also reported to enhance the germination of soybean seeds subjected to high NaCl concentrations [117] and canola [145]. Such benefits from LCO would not be provided to these crops had *Bradyrhizobium japonicum* been applied. Compounds are also less bulky and less costly, in most cases requiring small doses to be efficient. This relieves crop producers of storage and transportation concerns.

However, there are scenarios where the use of microbial cells is inevitable. For instance, the role that rhizobia play in nitrogen fixation, or mycorrhizae in P mobilisation and acquisition by plant roots could not be fulfilled by microbial compounds. Nitrogen fixing bacteria cannot be substituted by compounds in areas were N is limiting. Microbial cells have the potential to establish microbial populations in the rhizosphere, which may eliminate the need for further inoculation, a characteristic most farmers would desire, since it not only has positive financial implications, but also saves labour. Based on this, it is safe to assume that marketing companies would opt for compounds, since they guarantee continuous sales. However, the long and laborious process of isolating and purifying microbial compounds may also contribute to their scarcity and willingness of some researchers and companies to take that route.

10. Way Forward and Recommendations

With climate change conditions increasing, and the desperate need to come up with sustainable approaches of enhancing crop productivity to meet the food demand of the growing population, microbes are a prominent source of hope. However, a great deal still needs to be done to bridge the gap between their use in developed and developing countries. More research should be done to address issues of inconsistencies observed on crop producers' fields, following the use of microbial inoculants. It is obvious that single strains and consortia, or microbial cells and microbial compounds are issues that need to be evaluated on a case-by-case basis. Therefore, a better suggestion would be that more research be done to provide consumers with options that can address their unique needs, while being economically viable.

11. Conclusions

Lowering the effects of climate change on crop production, through reducing greenhouse gas emissions, is one of the major focuses of researchers in recent times. With proper manipulation, plant growth promoting microorganisms and compounds, they produce have potential to enhance growth and yield of plants exposed to biotic and biotic stress(es). This can complement other strategies, such as conservation farming and breeding for stress tolerant crop cultivars, to create an integrated approach of enhancing crop production in the face of climate change. Given that the prevalence of stress is predicted to increase with climate change, more research is needed to come up with better and more effective alternatives of utilising PGPM technology; not only to enhance plant growth, but also to reduce greenhouse gas emissions from the agricultural sector, which is a meaningful contributor.

Author Contributions: J.N. gathered all reading material and wrote the review; D.L.S. did all the editing and guidance on scientific knowledge. All authors have read and agreed to the published version of the manuscript.

References

1. Lott, F.C.; Christidis, N.; Stott, P.A. Can the 2011 East African drought be attributed to human-induced climate change? *Geophys. Res. Lett.* **2013**, *40*, 1177–1181. [CrossRef]
2. Rossi, F.; Olguın, E.J.; Diels, L.; de Philippis, R. Microbial fixation of CO_2 in waterbodies and in drylands to combat climate change, soil loss and desertification. *New Biotechnol.* **2015**, *32*, 109–120. [CrossRef]
3. Bradley, B.A.; Curtis, C.A.; Chambers, J.C. Bromus Response to Climate and Projected Changes with Climate Change. In *Exotic Brome-Grasses in Arid and Semiarid Ecosystems of the Western US*; Germino, M., Chambers, J., Brown, C., Eds.; Springer Series on Environmental Management; Springer: Cham, Switzerland, 2016; pp. 257–274. [CrossRef]
4. Richards, M.B.; Wollenberg, E.; van Vuuren, D. National contributions to climate change mitigation from agriculture: Allocating a global target. *Clim. Policy* **2018**, *18*, 1271–1285. [CrossRef]
5. Loboguerrero, A.M.; Campbell, B.M.; Cooper, P.J.M.; Hansen, J.W.; Rosenstock, T.; Wollenberg, E. Food and Earth Systems: Priorities for Climate Change Adaptation and Mitigation for Agriculture and Food Systems. *Sustainability* **2019**, *11*, 1372. [CrossRef]
6. IPCC. *2014: Climate Change 2014: Synthesis Report. Contribution of Working Groups I, II and III to the Fifth Assessment Report of the Intergovernmental Panel on Climate Change*; Pachauri, R.K., Meyer, L.A., Eds.; IPCC: Geneva, Switzerland, 2014; p. 151.
7. Porter, J.R.; Xie, L.; Challinor, A.J.; Cochrane, K.; Howden, S.M.; Iqbal, M.M.; Lobell, D.B.; Travasso, M.I. Food security and food production systems. In *Climate Change 2014: Impacts, Adaptation, and Vulnerability Part A: Global and Sectoral Aspects. Contribution of Working GroupII to the Fifth Assessment Report of the Intergovernmental Panel on Climate Change*; Field, C.B., Barros, V.R., Dokken, D.J., Mach, K.J., Mastrandrea, M.D., Bilir, T.E., Chatterjee, M., Ebi, K.L., Estrada, Y.O., Genova, R.C., et al., Eds.; Cambridge University Press: Cambridge, UK; New York, NY, USA, 2014; pp. 485–533.
8. Dawson, T.P.; Perryman, A.H.; Osborne, T.M. Modelling impacts of climate change on global food security. *Clim. Chang.* **2016**, *134*, 429–440. [CrossRef]

9. Bouwer, L.M.; Bubeck, P.; Aerts, J.C. Changes in future flood risk due to climate and development in a Dutch polder area. *Glob. Environ. Chang.* **2010**, *20*, 463–471. [CrossRef]

10. Mirza, M.M.Q. Climate change, flooding in South Asia and implications. *Reg. Environ. Chang.* **2011**, *11*, 95–107. [CrossRef]

11. Nam, W.; Hayes, M.J.; Svoboda, M.D.; Tadesse, T.; Wilhite, D.A. Drought hazard assessment in the context of climate change for South Korea. *Agric. Water Manag.* **2015**, *160*, 106–117. [CrossRef]

12. Barea, J.M. Future challenges and perspectives for applying microbial biotechnology in sustainable agriculture based on a better understanding of plant-microbiome interactions. *J. Soil Sci. Plant Nutr.* **2015**, *15*, 261–282. [CrossRef]

13. Gupta, G.; Parihar, S.S.; Ahirwar, N.K.; Snehi, S.K.; Singh, V. Plant Growth Promoting Rhizobacteria (PGPR): Current and Future Prospects for Development of Sustainable Agriculture. *Microb. Biochem. Technol.* **2015**, *7*, 96–102. [CrossRef]

14. Bender, S.F.; Wagg, C.; van der Heijden, M.G.A. An Underground Revolution: Biodiversity and Soil Ecological Engineering for Agricultural Sustainability. *Trends Ecol. Evol.* **2016**, *31*, 440–452. [CrossRef] [PubMed]

15. Babalola, O.O.; Glick, B.R. The use of microbial inoculants in African agriculture: Current practice and future prospects. *J. Food Agric. Environ.* **2012**, *10*, 540–549.

16. Smith, D.L.; Subramanian, S.; Lamont, J.R.; Bywater-Ekegärd, M. Signaling in the phytomicrobiome: Breadth and potential. *Front. Plant Sci.* **2015**, *6*, 709. [CrossRef] [PubMed]

17. Smith, D.L.; Gravel, V.; Yergeau, E. Signaling in the Phytomicrobiome. *Front. Plant Sci.* **2017**, *8*, 611. [CrossRef] [PubMed]

18. Backer, R.; Rokem, J.S.; Ilangumaran, G.; Lamont, J.; Praslickova, D.; Ricci, E.; Subramanian, S.; Smith, D.L. Plant Growth-Promoting Rhizobacteria: Context, Mechanisms of Action, and Roadmap to Commercialization of Bio stimulants for Sustainable Agriculture. *Front. Plant Sci.* **2018**, *9*, 1473. [CrossRef]

19. Hartmann, A.; Rothballer, M.; Hense, B.A.; Schröder, P. Bacterial quorum sensing compounds are important modulators of microbe-plant interactions. *Front. Plant Sci.* **2014**, *5*, 131. [CrossRef]

20. Sánchez-Cañizares, C.; Jorrín, B.; Poole, P.S.; Tkacz, A. Understanding the holobiont: The interdependence of plants and their microbiome. *Curr. Opin. Microbiol.* **2017**, *38*, 188–196. [CrossRef]

21. Compant, S.; Samad, A.; Faist, H.; Sessitsch, A. A review on the plant microbiome: Ecology, functions and emerging trends in microbial applications. *J. Adv. Res.* **2019**, *19*, 29–37. [CrossRef]

22. Berg, M.; Koskella, B. Nutrient and dose dependent protection against a plant pathogen. *Curr. Biol.* **2018**, *28*, 2487–2492. [CrossRef]

23. Bashan, Y.; de-Bashan, L.E.; Prabhu, S.R.; Hernandez, J. Advances in plant growth-promoting bacterial inoculant technology: Formulations and practical perspectives (1998–2013). *Plant Soil* **2014**, *378*, 1–33. [CrossRef]

24. Alori, E.T.; Dare, M.O.; Babalola, O.O. Microbial Inoculants for Soil Quality and Plant Health. In *Sustainable Agriculture Reviews*; Lichtfouse, E., Ed.; Springer: Cham, Switzerland, 2017; Volume 22, pp. 281–307. [CrossRef]

25. Malusa, E.; Sas-Paszt, L.; Ciesielska, J. Technologies for Beneficial Microorganisms Inocula Used as Biofertilizers. *Sci. World J.* **2012**, *2012*, 491206. [CrossRef]

26. Silva, E.C.; Nogueira, R.J.M.C.; Silva, M.A.; Albuquerque, M.B. Drought Stress and Plant Nutrition. *Plant Stress* **2011**, *5*, 32–41.

27. DaMatta, F.; Loos, R.A.; Silva, E.A.; Loureiro, M.E.; Ducatti, C. Effects of soil water déficit and nitrogen nutrition on water relations and photosynthesis of pot-grown *Coffea canephora* Pierra. *Trees* **2002**, *16*, 555–558. [CrossRef]

28. Tietema, A.; De Boer, W.; Riemer, L.; Verstraten, J.M. Nitrate production in nitrogen saturated acid forest soils: Vertical distributions and characteristics. *Soil Biol. Biochem.* **1992**, *24*, 235–240. [CrossRef]

29. Tripathi, D.K.; Singh, S.; Gaur, S.; Singh, S.; Yadav, V.; Liu, S.; Singh, V.P.; Sharma, S.; Srivastava, P.; Prasad, S.M.; et al. Acquisition and Homeostasis of Iron in Higher Plants and Their Probable Role in Abiotic Stress Tolerance. *Front. Environ. Sci.* **2018**, *5*, 86. [CrossRef]

30. Wang, W.; Zheng, Q.; Shen, Q.; Guo, S. The Critical Role of Potassium in Plant Stress Response. *Int. J. Mol. Sci.* **2013**, *14*, 7370–7390. [CrossRef]

31. Waraich, E.A.; Ahmad, R.; Halim, A.; Aziz, T. Alleviation of temperature stress by nutrient management in crop plants: A review. *J. Soil Sci. Plant Nutr.* **2012**, *12*, 221–244. [CrossRef]

32. Waraich, E.A.; Ahmad, R.; Ashraf, M.Y.; Saifullah; Ahmad, M. Improving agricultural water use efficiency by nutrient management in crop plants. Acta Agriculturae Scandinavica. *Sect. B Plant Soil Sci.* **2011**, *61*, 291–304. [CrossRef]

33. Karmakar, R.; Das, I.; Dutta, D.; Rakshit, A. Potential Effects of Climate Change on Soil Properties: A Review. *Sci. Int.* **2016**, *4*, 51–73. [CrossRef]

34. Wu, F.U.; Bao, W.; Li, F.L.; Wu, N. Effects of water stress and nitrogen supply on leaf gas exchange and fluorescence parameters of *Sophora davidii* seedlings. *Photosynthetica* **2008**, *46*, 40–48. [CrossRef]

35. Faye, I.; Diouf, O.; Guisse', A.; Se'ne, M.; Diallo, N. Characterizing Root Responses to Low Phosphorus in Pearl Millet [*Pennisetum glaucum* (L.) R. Br.]. *Agron. J.* **2006**, *98*, 1187–1194. [CrossRef]

36. Alori, E.T.; Fawole, O.B. Impact of chemical inputs on arbuscular mycorrhiza spores in soil: Response of AM Spores to fertilizer and herbicides. *Alban J. Agric. Sci.* **2017**, *16*, 10–13.

37. Lal, R. Restoring Soil Quality to Mitigate Soil Degradation. *Sustainability* **2015**, *7*, 5875–5895. [CrossRef]

38. Alori, E.T.; Babalola, O.O. Microbial Inoculants for Improving Crop Quality and Human Health in Africa. *Front. Microbiol.* **2018**, *9*, 2213. [CrossRef] [PubMed]

39. Naamala, J.; Jaiswal, S.K.; Dakora, F.D. Microsymbiont diversity and phylogeny of native Bradyrhizobia associated with soybean (*Glycine max* L. Merr.) nodulation in South African soils. *Syst. Appl. Microbiol.* **2016**, *39*, 336–344. [CrossRef]

40. Graham, J.H.; Leonard, R.T.; Menge, J.A. Membrane-Mediated Decrease in Root Exudation Responsible for Phorphorus Inhibition of Vesicular-Arbuscular Mycorrhiza Formation. *Plant. Physiol.* **1981**, *68*, 548–552. [CrossRef]

41. Sprent, J.I.; Stephens, J.H.; Rupela, O.P. Environmental effects on nitrogen fixation. In *World Crops: Cool Season Food Legumes*; Summerfield, R.J., Ed.; Current Plant Science and Biotechnology in Agriculture; Springer: Dordrech, The Netherlands, 1988; Volume 5, pp. 801–810. [CrossRef]

42. Marschner, H.; Dell, B. Nutrient uptake in mycorrhizal symbiosis. *Plant Soil* **1994**, *159*, 89–102. [CrossRef]

43. Glick, B.R. Plant growth promoting bacteria: Mechanisms and applications. *Scientifica* **2012**, *2012*, 1–15. [CrossRef]

44. Goldstein, A.H.; Krishnaraj, P.U. Phosphate solubilizing microorganisms vs. phosphate mobilizing microorganisms: What separates a phenotype from a trait? In *First International Meeting on Microbial Phosphate Solubilization. Developments in Plant and Soil Sciences*; Velázquez, E., Rodríguez-Barrueco, C., Eds.; Springer: Dordrecht, The Netherlands, 2007; Volume 102, pp. 203–213. [CrossRef]

45. Meding, S.M.; Zasoski, R.J. Hyphal-mediated transfer of nitrate, arsenic, cesium, rubidium, and strontium between arbuscular mycorrhizal forbs and grasses from a California oak woodland. *Soil Biol. Biochem.* **2008**, *40*, 126–134. [CrossRef]

46. Hayat, R.; Ali, S.; Amara, U.; Khalid, R.; Ahmed, I. Soil beneficial bacteria and their role in plant growth promotion: A review. *Ann. Microbiol.* **2010**, *60*, 579–598. [CrossRef]

47. Straker, C.J.; Mitchell, D.T. The activity and characterization of acid phosphatases in endomycorrhizal fungi of the Ericaceae. *New Phytol.* **1986**, *104*, 243–256. [CrossRef]

48. Khan, M.S.; Zaidi, A.; Ahemad, M.; Oves, M.; Wani, P.A. Plant growth promotion by phosphate solubilizing fungi—Current perspective. *Arch. Agron. Soil Sci.* **2010**, *56*, 73–98. [CrossRef]

49. Elias, F.; Woyessa, D.; Muleta, D. Phosphate Solubilisation Potential of Rhizosphere Fungi Isolated from Plants in Jimma Zone, Southwest Ethiopia. *Int. J. Microbiol.* **2016**, *2016*, 5472601. [CrossRef] [PubMed]

50. Daly, D.H.; Velivelli, S.L.S.; Prestwich, B.D. The Role of Soil Microbes in Crop Biofortification. In *Agriculturally Important Microbes for Sustainable Agriculture*; Meena, V., Mishra, P., Bisht, J., Pattanayak, A., Eds.; Springer: Singapore, 2017. [CrossRef]

51. Khan, A.; Singh, J.; Upadhayay, V.K.; Singh, A.V.; Shah, S. Microbial Biofortification: A Green Technology Through Plant Growth Promoting Microorganisms. In *Sustainable Green Technologies for Environmental Management*; Shah, S., Venkatramanan, V., Prasad, R., Eds.; Springer: Singapore, 2019. [CrossRef]

52. Bhatti, T.M.; Yawar, W. Bacterial solubilization of phosphorus from phosphate rock containing sulfur-mud. *Hydrometallurgy* **2010**, *103*, 54–59. [CrossRef]

53. Radzki, W.; Gutierrez Manero, F.J.; Algar, E.; Lucas Garcıa, J.A.; Garcıa-Villaraco, A.; Solano, B.R. Bacterial siderophores efficiently provide iron to iron-starved tomato plants in hydroponics culture. *Antonie Van Leeuwenhoek* **2013**, *104*, 321–330. [CrossRef]

54. Sharma, A.; Johri, B.N. Growth promoting influence of siderophore-producing *Pseudomonas* strains GRP3A and PRS9 in maize (*Zea mays* L.) under iron limiting conditions. *Microbiol. Res.* **2003**, *158*, 243–248. [CrossRef]

55. Yehuda, Z.; Shenker, M.; Romheld, V.; Marschner, H.; Hadar, Y.; Chen, Y. The Role of Ligand Exchange in the uptake of Iron from Microbial Siderophores by Cramineous Plants. *Plant Physiol.* **1996**, *112*, 1273–1280. [CrossRef]

56. Nadeem, S.M.; Zahir, Z.A.; Naveed, M.; Ashghar, H.N.; Arshad, M. Rhizobacteria capable of producing ACC deaminase may mitigate salt stress in wheat. *Soil Biol. Biochem.* **2010**, *74*, 533–542. [CrossRef]

57. Meena, V.S.; Maurya, B.R.; Prakash, J. Does a rhizospheric microorganism enhance K$^+$ availability in agricultural soils? *Microbiol. Res.* **2014**, *169*, 337–347. [CrossRef]

58. Cardoso, I.M.; Kuyper, T.W. Mycorrhizas and tropical soil fertility. *Agric. Ecosyst. Environ.* **2006**, *116*, 72–84. [CrossRef]

59. Jalili, F.; Khavazi, K.; Pazira, E.; Nejati, A.; Rahmani, H.A.; Sadaghiani, H.R.; Miransari, M. Isolation and characterization of ACC deaminase-producing fluorescent pseudomonads, to alleviate salinity stress on canola (*Brassica napus* L.) growth. *J. Plant Physiol.* **2008**, *166*, 667–674. [CrossRef] [PubMed]

60. Ali, S.; Charles, T.C.; Glick, B.R. Amelioration of high salinity stress damage by plant growth promoting bacterial endophytes that contain ACC deaminase. *Plant Physiol. Biochem.* **2014**, *80*, 160–167. [CrossRef] [PubMed]

61. Pérez-Montano, F.; Alías-Villegas, C.; Bellogín, R.A.; del Cerro, P.; Espuny, M.R.; Jiménez-Guerrero, I.; López-Baena, F.J.; Ollero, F.J.; Cubo, T. Plant growth promotion in cereal and leguminous agricultural important plants: From microorganism capacities to crop production. *Microbiol. Res.* **2014**, *169*, 325–336. [CrossRef] [PubMed]

62. Jha, C.K.; Saraf, M. Plant growth promoting Rhizobacteria (PGPR): A review. *E3 J. Agric. Res. Dev.* **2015**, *5*, 0108–0119.

63. Dodd, I.C.; Ruiz-Lozano, J.M. Microbial Enhancement of crop resource use efficiency. *Curr. Opin. Biotechnol.* **2012**, *23*, 236–242. [CrossRef]

64. Hassen, A.I.; Bopape, F.L.; Sanger, L.K. Microbial Inoculants as Agents of Growth Promotion and Abiotic Stress Tolerance in Plants. In *Microbial Inoculants in Sustainable Agricultural Productivity*; Singh, D., Singh, H., Prabha, R., Eds.; Springer: New Delhi, India, 2016; pp. 23–36. [CrossRef]

65. Recep, K.; Fikrettin, S.; Erkol, D.; Cafer, E. Biological control of the potato dry rot caused by *Fusarium* species using PGPR strains. *Biol. Control* **2009**, *50*, 194–198. [CrossRef]

66. Moussa, T.A.A.; Almaghrabi, O.A.; Abdel-Moneim, T.S. Biological control of the wheat root rot caused by *Fusarium graminearum* using some PGPR strains in Saudi Arabia. *Ann. Appl. Biol.* **2013**, *163*, 72–81. [CrossRef]

67. Díaz, A.; Smith, A.; Mesa, P.; Zapata, J.; Caviedes, D.; Cotes, A.M. *Control of Fusarium Wilt in Cape Gooseberry by Trichoderma koningiopsis and PGPR*; Pertot, I., Elad, Y., Barka, E.A., Clément, C., Eds.; Working Group Biological Control of Fungal and Bacterial Plant Pathogens; IOBC Bulletin: Dijon, France, 2013; Volume 86, pp. 89–94.

68. Vanitha, S.; Ramjegathesh, R. Bio Control Potential of *Pseudomonas fluorescens* against Coleus Root Rot Disease. *J. Plant Pathol. Microb.* **2014**, *5*, 216. [CrossRef]

69. Dixit, R.; Agrawal, L.; Gupta, S.; Kumar, M.; Yadav, S.; Chauhan, P.S.; Nautiyal, C.S. Southern blight disease of tomato control by 1-aminocyclopropane1-carboxylate (ACC) deaminase producing *Paenibacillus lentimorbus* B30488. *Plant Signal. Behav.* **2016**, *11*, e1113363. [CrossRef]

70. Li, X.L.; George, E.; Marschner, H. Extension of the phosphorus depletion zone in VA-mycorrhizal white clover in a calcareous soil. *Plant Soil* **1991**, *136*, 41–48. [CrossRef]

71. Wu, Y.; Lin, H.; Lin, Y.; Shi, J.; Xue, S.; Hung, Y.; Chen, Y.; Wang, H. Effects of biocontrol bacteria *Bacillus amyloliquefaciens* LY-1 culture broth on quality attributes and storability of harvested litchi fruit. *Postharvest Biol. Technol.* **2017**, *132*, 81–87. [CrossRef]

72. Zhao, D.; Zhao, H.; Zhao, D.; Zhua, X.; Wang, Y.; Duan, Y.; Xuan, Y.; Chen, L. Isolation and identification of bacteria from rhizosphere soil and their effect on plant growth promotion and root-knot nematode disease. *Biol. Control* **2018**, *119*, 12–19. [CrossRef]

73. de Vrieze, M.; Germanier, F.; Vuille, N.; Weisskopf, L. Combining Different Potato-Associated *Pseudomonas* Strains for Improved Biocontrol of *Phytophthora infestans*. *Front. Microbiol.* **2018**, *9*, 2573. [CrossRef] [PubMed]

74. Berendsen, R.L.; Vismans, G.; Yu, K.; Song, Y.; de Jonge, R.; Burgman, W.P.; Burmølle, M.; Herschend, J.; Bakker, P.A.; Pieterse, C.M. Disease-induced assemblage of a plant-beneficial bacterial consortium. *ISME J.* **2018**, *12*, 1496–1507. [CrossRef] [PubMed]

75. Gadhave, K.R.; Hourston, J.E.; Gange, A.C. Developing Soil Microbial Inoculants for Pest Management: Can One Have Too Much of a Good Thing? *J. Chem. Ecol.* **2016**, *42*, 348–356. [CrossRef]

76. Schausberger, P.; Peneder, S.; Juerschik, S.; Hoffmann, D. Mycorrhiza changes plant volatiles to attract spidermite enemies. *Funct. Ecol.* **2012**, *26*, 441–449. [CrossRef]

77. Pangesti, N.; Weldegergis, B.T.; Langendorf, B.; van Loon, J.J.; Dicke, M.; Pineda, A. Rhizobacterial colonization of roots modulates plant volatile emission and enhances the attraction of a parasitoid wasp. to host-infested plants. *Oecologia* **2015**, *178*, 1169–1180. [CrossRef]

78. Herman, M.A.B.; Nault, B.A.; Smart, C.D. Effects of plant growth–promoting rhizobacteria on bell pepper production and green peach aphid infestations in New York. *Crop Prot.* **2008**, *27*, 996–1002. [CrossRef]

79. Velivelli, S.L.S.; Sessitsch, A.; Prestwich, B.D. The Role of Microbial Inoculants in Integrated Crop Management Systems. *Potato Res.* **2014**, *57*, 291–309. [CrossRef]

80. Karthiba, L.; Saveetha, K.; Suresh, S.; Raguchander, T.; Saravanakumar, D.; Samiyappan, R. PGPR and entomopathogenic fungus bioformulation for the synchronous management of leaffolder pest and sheath blight disease of rice. *Pest. Manag. Sci.* **2010**, *66*, 555–564. [CrossRef]

81. Bano, A.; Muqarab, R. Plant defence induced by PGPR against *Spodoptera litura* in tomato (*Solanum lycopersicum* L.). *Plant Biol.* **2017**, *19*, 406–412. [CrossRef] [PubMed]

82. Xiang, N.; Lawrence, K.S.; Kloepper, J.W.; Donald, P.A.; McInroy, J.A. Biological control of *Heterodera glycines* by spore-forming plant growth-promoting rhizobacteria (PGPR) on soybean. *PLoS ONE* **2017**, *12*, e0181201. [CrossRef] [PubMed]

83. Viljoen, J.F.; Labuschagne, N.; Fourie, H.; Sikora, R.A. Biological control of the root-knot nematode *Meloidogyne incognita* on tomatoes and carrots by plant growth-promoting rhizobacteria. *Trop Plant Pathol.* **2019**, *44*, 284–291. [CrossRef]

84. Frankowski, J.; Lorito, M.; Scala, F.; Schmid, R.; Berg, G.; Bahl, H. Purification and properties of two chitinolytic enzymes of *Serratia plymuthica* HRO-C48. *Arch. Microbiol.* **2001**, *176*, 421–426. [CrossRef]

85. Kim, Y.C.; Jung, H.; Kim, K.Y.; Park, S.K. An effective biocontrol bioformulation against *Phytophthora* blight of pepper using growth mixtures of combined chitinolytic bacteria under different field conditions. *Eur. J. Plant Pathol.* **2008**, *120*, 373–382. [CrossRef]

86. Singh, P.P.; Shin, Y.C.; Park, C.S.; Chung, Y.R. Biological control of Fusarium wilt of cucumber by chitinolytic bacteria. *Phytopathology* **1999**, *89*, 92–99. [CrossRef]

87. Leeman, M.; Ouder, F.M.D.; Pelt, J.A.V.; Dirk, F.P.M.; Steij, H.; Bakker, P.A.; Schippers, B. Iron availability affects induction of systemic resistance to Fusarium wilt of radishes by *Pseudomonas fluorescens*. *Phytopathology* **1996**, *86*, 149–155. [CrossRef]

88. Buysens, S.; Heungens, K.; Poppe, J.; Höftee, M. Involvement of pyochelin and pyoverdin in suppression of Pythium-induced damping-off of tomato by *Pseudomonas aeruginosa* 7NSK2. *Appl. Environ. Microbiol.* **1996**, *62*, 865–871. [CrossRef]

89. Khandelwal, S.R.; Manwar, A.V.; Chaudhari, B.L.; Chincholkar, S.B. Siderophorogenic bradyrhizobia boost yield of soybean. *Appl. Biochem. Biotechnol.* **2002**, *102*, 155–168. [CrossRef]

90. Toklikishvili, N.; Dandurishvili, M.; Tediashvili, N.; Lurie, G.S.; Szegedi, E.; Glick, B.R.; Chermin, L.N. Inhibitory effect of ACC deaminase-producing bacteria on crown gall formation in tomato plants infected by *Agrobacterium tumefaciens* or A. vitis. *Plant Pathol.* **2010**, *59*, 1023–1030. [CrossRef]

91. Niu, D.D.; Zheng, Y.; Zheng, L.; Jiang, C.H.; Zhou, D.M.; Guo, J.H. Application of PSX biocontrol preparation confers root-knot nematode management and increased fruit quality in tomato under field conditions. *Biocontrol. Sci. Technol.* **2016**, *26*, 174–180. [CrossRef]

92. Santhanam, S.; Luu, V.T.; Weinhold, A.; Goldberg, J.; Oh, Y.; Baldwin, I.T. Native root-associated bacteria rescue a plant from a sudden-wilt disease that emerged during continuous cropping. *Proc. Natl. Acad. Sci. USA* **2015**, *112*, E5013–E5020. [CrossRef] [PubMed]

93. Hunziker, L.; Bönisch, D.; Groenhagen, U.; Bailly, A.; Schulz, S.; Weisskopf, L. *Pseudomonas* strains naturally associated with potato plants produce volatiles with high potential for inhibition of *Phytophthora infestans*. *Appl. Environ. Microbiol.* **2015**, *81*, 821–830. [CrossRef] [PubMed]

94. Guyer, A.; de Vrieze, M.; Bönisch, D.; Gloor, R.; Musa, T.; Bodenhausen, N.; Bailly, A.; Weisskopf, L. The Anti-Phytophthora Effect of Selected Potato-Associated Pseudomonas Strains: From the Laboratory to the Field. *Front. Microbiol.* **2015**, *6*, 1309. [CrossRef] [PubMed]

95. New, P.B.; Kerr, A. Biological Control of Crown Gall: Field Measurements and Glasshouse Experiments. *J. Appl. Buct.* **1972**, *35*, 279–287. [CrossRef]

96. Allizadeh, H.; Behboudi, K.; Masoud, A.; Javan-Nikkhah, M.; Zamioudis, C.; Corné, M.J.P.; Bakker, A.H.M. Induced systemic resistance in cucumber and *Arabidopsis thaliana* by the combination of *Trichoderma harzianum* Tr6 and *Pseudomonas* sp. Ps14. *Biol. Control* **2013**, *65*, 14–23. [CrossRef]

97. Durgadevi, D.; Srivignesh, S.; Sankaralingam, A. Effect of Consortia Bioformulation of Rhizobacteria on Induction of Systemic Resistance in Tuberose against Peduncle Blight Disease. *Int. J. Bio Resour. Stress Manag.* **2018**, *9*, 510–517. [CrossRef]

98. Takishita, Y.; Charron, J.B.; Smith, D.L. Biocontrol Rhizobacterium *Pseudomonas* sp. 23S Induces Systemic Resistance in Tomato (*Solanum lycopersicum* L.) Against Bacterial Canker *Clavibacter michiganensis* subsp. michiganensis. *Front. Microbiol.* **2018**, *9*, 2119. [CrossRef]

99. Kumar, S.; Chauhan, P.S.; Agrawal, L.; Raj, R.; Srivastava, A.; Gupta, S.; Mishra, S.K.; Yadav, S.; Singh, P.C.; Raj, S.K.; et al. *Paenibacillus lentimorbus* Inoculation Enhances Tobacco Growth and Extenuates the Virulence of Cucumber mosaic virus. *PLoS ONE* **2016**, *11*, e0149980. [CrossRef]

100. Schuhegger, R.; Ihring, A.; Gantner, S.; Bahnweg, G.; Knappe, C.; Vogg, G.; Hutzler, P.; Schmid, M.; Breusegem, F.V.; Eberl, L.; et al. Induction of systemic resistance in tomato by N-acyl-L-homoserine lactone-producing rhizosphere bacteria. *Plant Cell Environ.* **2006**, *29*, 909–918. [CrossRef]

101. Baffoni, L.; Gaggia, F.; Dalanaj, N.; Prodi, A.; Nipoti, P.; Pisi, A.; Biavati, B.; Gioia, D.D. Microbial inoculants for the biocontrol of *Fusarium* spp. in durum wheat. *BMC Microbiol.* **2015**, *15*, 242. [CrossRef] [PubMed]

102. Martinuz, A.; Schouten, A.; Menjivar, R.; Sikora, R. Effectiveness of systemic resistance toward *Aphis gossypii* (Hom., Aphididae) as induced by combined applications of the endophytes *Fusarium oxysporum* Fo162 and *Rhizobium etli* G12. *Biol. Control* **2012**, *62*, 206–212. [CrossRef]

103. Valenzuela-Soto, J.H.; Estrada-Hernandez, M.G.; Ibarra-Laclette, E.; Delano-Frier, J.P. Inoculation of tomato plants (*Solanum lycopersicum*) with growth-promoting *Bacillus subtilis* retards whitefly *Bemisia tabaci* development. *Planta* **2010**, *231*, 397–410. [CrossRef] [PubMed]

104. Srivastava, S.; Bist, V.; Srivastava, S.; Singh, P.C.; Trivedi, P.K.; Asif, M.H.; Chauhan, P.S.; Nautiyal, C.S. Unraveling aspects of *Bacillus amyloliquefaciens* mediated enhanced production of rice under biotic stress of *Rhizoctonia solani*. *Front. Plant Sci.* **2016**, *7*, 587. [CrossRef] [PubMed]

105. Collins, M.; Knutti, R.; Arblaster, J.; Dufresne, L.; Fichefet, T.; Friedlingstein, P.; Gao, X.; Gutowski, W.J.; Johns, T.; Krinner, G.; et al. Long-term Climate Change: Projections, Commitments and Irreversibility. In *Climate Change 2013: The Physical Science Basis. Contribution of Working Group I to the Fifth Assessment Report of the Intergovernmental Panel on Climate Change*; Stocker, T.F., Qin, D., Plattner, G.-K., Tignor, M., Allen, S.K., Boschung, J., Nauels, A., Xia, Y., Bex, V., Midgley, P.M., Eds.; Cambridge University Press: New York, NY, USA, 2013; pp. 1029–1136.

106. Xu, D.Y.; Kang, X.W.; Zhuang, D.F.; Pan, J.J. Multi-scale quantitative assessment of the relative roles of climate change and human activities in desertification—A case study of the Ordos Plateau, China. *J. Arid Environ.* **2010**, *74*, 498–507. [CrossRef]

107. Tank, N.; Saraf, M. Salinity-resistant plant growth promoting rhizobacteria ameliorates sodium chloride stress on tomato plants. *J. Plant Interact.* **2010**, *5*, 51–58. [CrossRef]

108. Rousk, J.; Elyaagubi, F.K.; Jones, D.L.; Godbold, D.L. Bacterial salt tolerance is unrelated to soil salinity across an arid agroecosystem salinity gradient. *Soil Biol. Biochem.* **2011**, *43*, 1881–1887. [CrossRef]

109. Egamberdieva, D.; Lugtenberg, B. Use of Plant Growth-Promoting Rhizobacteria to Alleviate Salinity Stress in Plants. In *Use of Microbes for the Alleviation of Soil Stresses*; Miransari, M., Ed.; Springer Science + Business Media: New York, NY, USA, 2014; pp. 73–96. [CrossRef]

110. Shrivastava, P.; Kumar, R. Soil salinity: A serious environmental issue and plant growth promoting bacteria as one of the tools for its alleviation. *Saudi J. Biol. Sci.* **2015**, *22*, 123–131. [CrossRef]

111. Yan, N.; Marschner, P.; Cao, W.; Zuo, C.; Qin, W. Influence of salinity and water content on soil microorganisms. *Int. Soil Water Conserv. Res.* **2015**, *3*, 316–323. [CrossRef]

112. Chen, L.; Liu, Y.; Wu, G.; Njeri, K.V.; Shen, Q.; Zhang, N.; Zhang, R. Induced maize salt tolerance by rhizosphere inoculation of *Bacillus amyloliquefaciens* SQR9. *Physiol. Plant* **2016**, *158*, 34–44. [CrossRef]

113. Subramanian, P.; Mageswari, A.; Kim, K.; Lee, Y.; Sa, T. Psychrotolerant endophytic *Pseudomonas* sp. strains OB155 and OS261 induced chilling resistance in tomato plants (*Solanum lycopersicum* mill.) by activation of their antioxidant capacity. *Mol. Plant Microbe Interact.* **2015**, *28*, 1073–1081. [CrossRef] [PubMed]

114. Tiwari, S.; Lata, C.; Chauhan, S.P.; Chandra Shekhar Nautiyal, C.P. *Pseudomonas putida* attunes morphophysiological, biochemical and molecular responses in *Cicer arietinum* L. during drought stress and recovery. *Plant Physiol. Biochem.* **2016**, *99*, 108–117. [CrossRef] [PubMed]

115. Wang, Q.; Dodd, I.C.; Belimov, A.A.; Jiang, F. Rhizosphere bacteria containing 1-aminocyclopropane-1-carboxylate deaminase increase growth and photosynthesis of pea plants under salt stress by limiting Na$^+$ accumulation. *Funct. Plant Biol.* **2016**, *43*, 161–172. [CrossRef] [PubMed]

116. Bhartirt, N.; Pandey, S.S.; Barnawal, D.; Patel, V.K.; Kalra, A. Plant growth promoting rhizobacteria *Dietzia natronolimnaea* modulates the expression of stress responsive genes providing protection of wheat from salinity stress. *Sci. Rep.* **2016**, *6*, 34768. [CrossRef]

117. Subramanian, S.; Ricci, E.; Souleimanov, A.; Smith, D.L. A proteomic approach to lip-chitooligosaccharide and thuricin 17 effects on soybean germination under salt stress. *PLoS ONE* **2016**, *11*, e0160660. [CrossRef] [PubMed]

118. Rolli, E.; Marasco, R.; Vigani, G.; Ettoumi, B.; Mapelli, F.; Deangelis, M.L.; Gandolfi, C.; Casati, F.; Previtali, F.; Gerbino, R.; et al. Improved plant resistance to drought is promoted by the root-associated microbiome as a water stress-dependent trait. *Environ. Microbiol.* **2015**, *17*, 316–331. [CrossRef]

119. Molina-Romero, D.; Baez, A.; Quintero-Hernández, V.; Castañeda-Lucio, M.; Fuentes-Ramírez, L.E.; Bustillos-Cristales, M.D.R.; Rodríguez-Andrade, O.; Morales-García, Y.E.; Munive, A.; Muñoz-Rojas, J. Compatible bacterial mixture, tolerant to desiccation, improves maize plant growth. *PLoS ONE* **2017**, *12*, e0187913. [CrossRef]

120. Akhtar, S.S.; Amby, D.B.; Hegelund, J.N.; Fimognari, L.; Großkinsky, D.K.; Westergaard, J.C.; Muller, R.; Melba, L.; Liu, F.; Roitsch, T. *Bacillus licheniformis* FMCH001 increases water use efficiency via growth stimulation in both normal and drought conditions. *Front. Plant Sci.* **2020**, *11*, 297. [CrossRef]

121. Yang, A.; Akhtar, S.S.; Fu, Q.; Naveed, M.; Iqbal, S.; Roitsch, T.; Jacobsen, S.E. *Burkholderia Phytofirmans* PsJN Stimulate Growth and Yield of Quinoa under Salinity Stress. *Plants* **2020**, *9*, 672. [CrossRef]

122. Zahran, H.H. Rhizobium-Legume Symbiosis and Nitrogen Fixation under Severe Conditions and in an Arid Climate. *Microbiol. Mol. Biol. Rev.* **1999**, *63*, 968–989. [CrossRef]

123. Etesami, H.; Mirseyedhosseini, H.; Alikhani, H.A. Bacterial biosynthesis of 1-aminocyclopropane-1-caboxylate (ACC) deaminase, a useful trait to elongation and endophytic colonization of the roots of rice under constant flooded conditions. *Physiol. Mol. Biol. Plants* **2014**, *20*, 425–434. [CrossRef] [PubMed]

124. Kang, S.M.; Khan, A.L.; Waqs, M.; You, Y.H.; Hamayun, M.; Joo, G.; Shahzad, R.; Choi, K.; Lee, I.J. Gibberellin-producing *Serratia nematodiphila* PEJ1011 ameliorates low temperature stress in *Capsicum annuum* L. *Eur. J. Soil Biol.* **2015**, *68*, 85–93. [CrossRef]

125. Fernandez, O.; Theocharis, A.; Bordiec, S.; Feil, R.; Jacquens, L.; Clément, C.; Fontaine, F.; Barka, E.A. *Burkholderia phytofirmans* PsJN Acclimates Grapevine to Cold by Modulating Carbohydrate Metabolism. *Mol. Plant Microbe Interact.* **2012**, *25*, 496–504. [CrossRef]

126. Barnawal, D.; Bharti, N.; Maji, D.; Chanotiya, C.S.; Kalra, A. 1-Aminocyclopropane-1-carboxylic acid (ACC) deaminase containing rhizobacteria protect *Ocimum sanctum* plants during water logging stress via reduced ethylene generation. *Plant Physiol. Biochem.* **2012**, *58*, 227–235. [CrossRef] [PubMed]

127. Bano, A.; Fatima, M. Salt tolerance in *Zea mays* (L.) following inoculation with Rhizobium and Pseudomonas. *Biol. Fertil. Soils* **2009**, *45*, 405–413. [CrossRef]

128. Karthikeyan, B.; Joe, M.M.; Islam, M.D.R.; Sa, T. ACC deaminase containing diazotrophic endophytic bacteria ameliorate salt stress in *Catharanthus roseus* through reduced ethylene levels and induction of antioxidative defense systems. *Symbiosis* **2012**, *56*, 77–86. [CrossRef]

129. Mallick, I.; Bhattacharyya, C.; Mukherji, S.; Sarkar, S.C.; Mukhopadhyay, U.K.; Ghosh, A. Effective rhizoinoculation and biofilm formation by arsenic immobilizing halophilic plant growth promoting bacteria (PGPB) isolated from mangrove rhizosphere: A step towards arsenic rhizoremediation. *Sci. Total Environ.* **2018**, *610*, 1239–1250. [CrossRef] [PubMed]

130. Tripathi, P.; Singh, P.C.; Mishra, A.; Srivastava, S.; Chauhan, R.; Awasthi, S.; Mishra, S.; Dwivedi, S.; Tripathi, P.; Kalra, A.; et al. Arsenic tolerant *Trichoderma* sp. reduces arsenic induced stress in chickpea (*Cicer arietinum*). *Environ. Pollut.* **2017**, *223*, 137–145. [CrossRef]

131. Sánchez-Yáñez, J.M.; Alonso-Bravo, J.N.; Dasgupta-Schuber, N.; Márquez-Benavides, L. Bioremediation of soil contaminated by waste motor oil in 55,000 and 65,000 and phytoremediation by *Sorghum bicolor* inoculated with *Burkholderia cepacia* and *Penicillium chrysogenum*. *J. Selva Andina Biosph.* **2015**, *3*, 86–94.

132. Sarkar, J.; Chakraborty, B.; Chakraborty, U. Plant Growth Promoting Rhizobacteria Protect Wheat Plants Against Temperature Stress Through Antioxidant Signalling and Reducing Chloroplast and Membrane Injury. *J. Plant Growth Regul.* **2018**, *37*, 1396–1412. [CrossRef]

133. Islam, F.; Yasmeen, T.; Ali, O.; Ali, S.; Arif, S.M.; Sabir Hussain, S.; Rizv, H. Influence of *Pseudomonas aeruginosa* as PGPR on oxidative stress tolerance in wheat under Zn stress. *Ecotoxicol. Environ. Saf.* **2014**, *104*, 285–293. [CrossRef] [PubMed]

134. Berninger, T.; Lopez, O.G.; Bejarano, A.; Preininger, C.; Sessitsch, A. Maintenance and assessment of cell viability in formulation of non-sporulating bacterial inoculants. *Microb. Biotechnol.* **2018**, *11*, 277–301. [CrossRef]

135. Mehnaz, S. An Overview of Globally Available Bioformulations. In *Bioformulations: For Sustainable Agriculture*; Arora, N., Mehnaz, S., Balestrini, R., Eds.; Springer: New Delhi, India, 2016; pp. 267–281. [CrossRef]

136. Arthur, S.; Dara, S.K. Microbial biopesticides for invertebrate pests and their markets in the United States. *J. Invertebr. Pathol.* **2018**, *165*, 13–21. [CrossRef]

137. Babalola, O.O.; Glick, B.R. Indigenous African agriculture and plant associated microbes: Current practice and future transgenic prospects. *Sci. Res. Essays* **2012**, *7*, 2431–2439.

138. Aremu, B.R.; Alori, E.T.; Kutu, R.F.; Babalola, O.O. Potentials of Microbial Inoculants in Soil Productivity: An Outlook on African Legumes. In *Microorganisms for Green Revolution. Microorganisms for Sustainability*; Panpatte, D., Jhala, Y., Vyas, R., Shelat, H., Eds.; Springer: Singapore, 2017; Volume 6. [CrossRef]

139. Souleimanov, A.; Prithiviraj, B.; Smith, D. The major Nod factor of *Bradyrhizobium japonicum* promotes early growth of soybean and corn. *J. Exp. Bot.* **2002**, *53*, 1929–1934. [CrossRef] [PubMed]

140. Gray, E.J.; Lee, K.D.; Souleimanov, A.M.; Di Falco, M.R.; Zhou, X.; Ly, A.; Charles, T.C.; Driscoll, B.T.; Smith, D.L. A novel bacteriocin, thuricin 17, produced by plant growth promoting rhizobacteria strain *Bacillus thuringiensis* NEB17: Isolation and classification. *J. Appl. Microbiol.* **2006**, *100*, 545–554. [CrossRef]

141. Subramanian, S.; Souleimanov, A.; Smith, D.L. Proteomic Studies on the Effects of Lipochitooligosaccharide and Thuricin 17 under Unstressed and Salt Stressed Conditions in *Arabidopsis thaliana*. *Front. Plant Sci.* **2016**, *7*, 1314. [CrossRef]

142. Arunachalam, S.; Schwinghamer, T.; Dutilleul, P.; Smith, D.L. Multi-Year Effects of Biochar, Lipo-Chitooligosaccharide, Thuricin 17, and Experimental Bio-Fertilizer for Switchgrass. *Agron. J.* **2018**, *110*, 77–84. [CrossRef]

143. Navarro, M.O.P.; Piva, A.C.M.; Simionato, A.S.; Spago, F.R.; Modolon, F.; Emiliano, J.; Azul, A.M.; Chryssafidis, A.L.; Andrade, G. Bioactive Compounds Produced by Biocontrol Agents Driving Plant Health. In *Microbiome in Plant Health and Disease*; Kumar, V., Prasad, R., Kumar, M., Choudhary, D., Eds.; Springer: Singapore, 2019; pp. 337–374. [CrossRef]

144. Prudent, M.; Salon, C.; Souleimanov, S.A.; Emery, R.J.N.; Smith, D.L. Soybean is less impacted by water stress using *Bradyrhizobium japonicum* and thuricin-17 from *Bacillus thuringiensis*. *Agron. Sustain. Dev.* **2015**, *35*, 749–757. [CrossRef]

145. Schwinghamer, T.; Souleimanov, A.; Dutilleul, P.; Smith, D.L. Supplementation with solutions of lipo-chitooligosaccharide Nod Bj V (C18:1, MeFuc) and thuricin 17 regulates leaf arrangement, biomass, and root development of canola (*Brassica napus* [L.]). *Plant Growth Regul.* **2015**, *78*, 31–41. [CrossRef]

146. Yuan, L.; Li, Y.; Wang, Y.; Zhang, X.; Xu, Y. Optimization of critical medium components using response surface methodology for phenazine-1-carboxylic acid production by *Pseudomonas* sp. M-18Q. *J. Biosci. Bioeng.* **2008**, *3*, 232–237. [CrossRef] [PubMed]

147. Xu, S.; Pan, X.; Luo, J.; Wu, J.; Zhou, Z.; Liang, X.; He, Y.; Zhou, M. Effects of phenazine-1-carboxylic acid on the biology of the plant-pathogenic bacterium *Xanthomonas oryzae* pv. Oryzae. *Pestic. Biochem. Physiol.* **2015**, *117*, 39–46. [CrossRef] [PubMed]

148. Chen, C.; Mciver, J.; Yang, Y.; Bai, Y.; Schultz, B.; Mciver, A. Foliar application of lipochitooligosaccharides (Nod factors) to tomato (*Lycopersicon esculentum*) enhances flowering and fruit production. *Can. J. Plant Sci.* **2007**, *87*, 365–372. [CrossRef]

149. Duke, S.O.; Lydon, J. Herbicides from Natural Compounds. *Weed Technol.* **1987**, *1*, 122–128. [CrossRef]

150. Zhang, Z.K.; Huber, D.J.; Qu, H.X.; Yun, Z.; Wang, H.; Huang, Z.H.; Huang, H.; Jiang, Y.M. Enzymatic browning and antioxidant activities in harvested litchi fruit as influenced by apple polyphenols. *Food Chem.* **2015**, *171*, 191–199. [CrossRef]

151. Shanmugaiah, V.; Mathivanan, N.; Varghese, B. Purification, crystal structure and antimicrobial activity of phenazine-1-carboxamide produced by a growth-promoting biocontrol bacterium, *Pseudomonas aeruginosa* MML2212. *J. Appl. Microbiol.* **2010**, *108*, 703–711. [CrossRef]

152. Huang, H.; Sun, L.; Bi, K.; Zhong, G.; Hu, M. The effect of phenazine-1-carboxylic acid on the morphological, physiological, and molecular characteristics of *Phellinus noxius*. *Molecules* **2016**, *21*, 613. [CrossRef]

153. Puopolo, G.; Masi, M.; Raio, A.; Andolfi, A.; Zoina, A.; Cimmino, A.; Evidente, A. Insights on the susceptibility of plant pathogenic fungi to phenazine-1-carboxylic acid and its chemical derivatives. *Nat. Prod. Res.* **2013**, *27*, 956–966. [CrossRef]

154. Gheorghe, I.; Popa, M.; Marutescu, L.; Saviuc, C.; Lazar, V.; Chifiriuc, M.C. Lessons from interregn communication for development of novel, ecofriendly pesticides. In *New Pesticides and Soil Sensors*; Grumezescu, A.M., Ed.; Academic Press: London, UK, 2017; pp. 1–46. [CrossRef]

155. Rane, M.R.; Sarode, P.D.; Chaudhari, B.L.; Chincholkar, S.B. Exploring antagonistic metabolites of established biocontrol agent of marine origin. *Appl. Biochem. Biotechnol.* **2008**, *151*, 665–675. [CrossRef]

156. Kare, E.; Arora, N.K. Dual activity of pyocyanin from *Pseudomonas aeruginosa*—Antibiotic against phytopathogen and signal molecule for biofilm development by rhizobia. *Can. J. Microbiol.* **2011**, *57*, 708–713. [CrossRef]

157. Jung, B.K.; Hong, S.J.; Park, G.S.; Kim, M.C.; Shin, J.H. Isolation of *Burkholderia cepacia* JBK9 with plant growth–promoting activity while producing pyrrolnitrin antagonistic to plant fungal diseases. *Appl. Biol. Chem.* **2018**, *61*, 173–180. [CrossRef]

158. Okada, A.; Banno, S.; Ichiishi, A.; Kimura, M.; Yamaguchi, I.; Fujimura, M. Pyrrolnitrin interferes with osmotic signal transduction in *Neurospora crassa*. *J. Pestic. Sci.* **2005**, *30*, 378–383. [CrossRef]

159. Han, J.W.; Kim, J.D.; Lee, J.M.; Ham, J.H.; Lee, D.; Kim, B.S. Structural elucidation and antimicrobial activity of new phencomycin derivatives isolated from *Burkholderia glumae* strain 411gr6. *J. Antibiot.* **2014**, *67*, 721. [CrossRef]

160. Deng, P.; Foxfire, A.; Xu, J.; Baird, S.M.; Jia, J.; Delgado, K.H.; Shin, R.; Smith, L.; Lu, S.E. Siderophore product ornibactin is required for the bactericidal activity of Burkholderia contaminans MS14. *Appl. Environ. Microbiol.* **2017**, *83*, e00051-17. [CrossRef]

161. Ye, Y.; Li, Q.; Fu, G.; Yuan, G.; Miao, J.; Lin, W. Identification of antifungal substance (Iturin A2) produced by *Bacillus subtilis* B47 and its effect on southern corn leaf blight. *J. Integr. Agric.* **2012**, *11*, 90–99. [CrossRef]

162. Deravel, J.; Lemière, S.; Coutte, F.; Krier, F.; Hese, N.V.; Béchet, M.; Sourdeau, N.; Höfte, M.; Leprêtre, A.; Jacques, P. Mycosubtilin and surfactin are efficient, low ecotoxicity molecules for the biocontrol of lettuce downy mildew. *Appl. Microbiol. Biotechnol.* **2014**, *98*, 6255–6264. [CrossRef]

163. Isaac, B.G.; Ayer, S.W.; Elliott, R.C.; Stonard, R.J. Herboxidiene: A potent phytotoxic polyketide from Streptomyces sp. A7847. *J. Org. Chem.* **1992**, *57*, 7220–7226. [CrossRef]

164. Saxena, S.; Pandey, A.K. Microbial metabolites as eco-friendly agrochemicals for the next millennium. *Appl. Microbiol. Biotechnol.* **2001**, *55*, 395–403. [CrossRef]

165. Tanaka, Y.; Omura, S. Agroactive compounds of microbial origin. *Annu. Rev. Microbiol.* **1993**, *47*, 57–87. [CrossRef]

PERMISSIONS

The contributors of this book come from diverse backgrounds, making this book a truly international effort. This book will bring forth new frontiers with its revolutionizing research information and detailed analysis of the nascent developments around the world.

We would like to thank all the contributing authors for lending their expertise to make the book truly unique. They have played a crucial role in the development of this book. Without their invaluable contributions this book wouldn't have been possible. They have made vital efforts to compile up to date information on the varied aspects of this subject to make this book a valuable addition to the collection of many professionals and students.

This book was conceptualized with the vision of imparting up-to-date information and advanced data in this field. To ensure the same, a matchless editorial board was set up. Every individual on the board went through rigorous rounds of assessment to prove their worth. After which they invested a large part of their time researching and compiling the most relevant data for our readers.

The editorial board has been involved in producing this book since its inception. They have spent rigorous hours researching and exploring the diverse topics which have resulted in the successful publishing of this book. They have passed on their knowledge of decades through this book. To expedite this challenging task, the publisher supported the team at every step. A small team of assistant editors was also appointed to further simplify the editing procedure and attain best results for the readers.

Apart from the editorial board, the designing team has also invested a significant amount of their time in understanding the subject and creating the most relevant covers. They scrutinized every image to scout for the most suitable representation of the subject and create an appropriate cover for the book.

The publishing team has been an ardent support to the editorial, designing and production team. Their endless efforts to recruit the best for this project, has resulted in the accomplishment of this book. They are a veteran in the field of academics and their pool of knowledge is as vast as their experience in printing. Their expertise and guidance has proved useful at every step. Their uncompromising quality standards have made this book an exceptional effort. Their encouragement from time to time has been an inspiration for everyone.

The publisher and the editorial board hope that this book will prove to be a valuable piece of knowledge for researchers, students, practitioners and scholars across the globe.

LIST OF CONTRIBUTORS

Alejandro Jiménez-Gómez and Zaki Saati-Santamaría
Microbiology and Genetics Department, University of Salamanca, 37007 Salamanca, Spain
Spanish-Portuguese Institute for Agricultural Research (CIALE), Villamayor, 37185 Salamanca, Spain

Martin Kostovcik
Department of Genetics and Microbiology, Faculty of Science, Charles University, 12844 Prague
BIOCEV, Institute of Microbiology, the Czech Academy of Sciences, 25242 Vestec, Czech Republic

Raúl Rivas, Encarna Velázquez, Pedro F. Mateos and Paula García-Fraile
Microbiology and Genetics Department, University of Salamanca, 37007 Salamanca, Spain
Spanish-Portuguese Institute for Agricultural Research (CIALE), Villamayor, 37185 Salamanca, Spain
Associated R&D Unit, USAL-CSIC (IRNASA), Villamayor, 37185 Salamanca, Spain

Esther Menéndez
Microbiology and Genetics Department, University of Salamanca, 37007 Salamanca, Spain
Spanish-Portuguese Institute for Agricultural Research (CIALE), Villamayor, 37185 Salamanca, Spain
Associated R&D Unit, USAL-CSIC (IRNASA), Villamayor, 37185 Salamanca, Spain
MED—Mediterranean Institute for Agriculture, Environment and Development, Institute for Advanced Studies and Research (IIFA), Universidade de Évora, Pólo da Mitra, Ap. 94, 7006-554 Évora, Portugal

Maqshoof Ahmad, Muhammad Luqman and Azhar Hussain
Department of Soil Science, The Islamia University of Bahawalpur, Bahawalpur 63100, Pakistan

Xiukang Wang
College of Life Sciences, Yan'an University, Yan'an 716000, China

Thomas H. Hilger
Hans-Ruthenberg Institute, University of Hohenheim, 70593 Stuttgart, Germany

Farheen Nazli
Pesticide Quality Control Laboratory, Punjab Agriculture Department, Government of Punjab, Bahawalpur 63100, Pakistan

Zahir Ahmad Zahir
Institute of Soil and Environmental Sciences, University of Agriculture, Faisalabad 38000, Pakistan

Muhammad Latif
Department of Agronomy, The Islamia University of Bahawalpur, Bahawalpur 63100, Pakistan

Qudsia Saeed
College of Natural Resources and Environment, Northwest Agriculture and Forestry University, Xianyang 712100, China

Hina Ahmed Malik
Institute of Food and Agriculture Sciences, University of Florida, Gainesville, FL 110690, USA

Adnan Mustafa
National Engineering Laboratory for Improving Quality of Arable Land, Institute of Agricultural Resources and Regional Planning, Chinese Academy of Agricultural Sciences, Beijing 100081, China

Tasawar Abbas, Zahir Ahmad Zahir and Muhammad Naveed
Soil Microbiology and Biochemistry Laboratory, Institute of Soil and Environmental Sciences, University of Agriculture, Faisalabad 38040, Pakistan

Sana Abbas
Department of Chemistry, Government College Women University Faisalabad, Faisalabad 38040, Pakistan

Mona S. Alwahibi and Mohamed Soliman Elshikh
Department of Botany and Microbiology, College of Science, King Saud University, Riyadh 11451, Saudi Arabia

Waleed Asghar, Shiho Kondo, Riho Iguchi, Ahmad Mahmood and Ryota Kataoka
Department of Environmental Sciences, Faculty of Life and Environmental Sciences, University of Yamanashi, Kofu, Yamanashi 400-8510, Japan

Abdujalil Narimanov
Laboratory of Medicinal Plants Genetics and Biotechnology, Institute of Genetics and Plant Experimental Biology, Uzbekistan Academy of Sciences, Tashkent Region, Kibray 111208, Uzbekistan

Dilfuza Jabborova
Laboratory of Medicinal Plants Genetics and Biotechnology, Institute of Genetics and Plant Experimental Biology, Uzbekistan Academy of Sciences, Tashkent Region, Kibray 111208, Uzbekistan
Division of Microbiology, ICAR-Indian Agricultural Research Institute, Pusa, New Delhi 110012, India
Leibniz Centre for Agricultural Landscape Research (ZALF), D-15374 Müncheberg, Germany

Annapurna Kannepalli
Division of Microbiology, ICAR-Indian Agricultural Research Institute, Pusa, New Delhi 110012, India

Stephan Wirth
Leibniz Centre for Agricultural Landscape Research (ZALF), D-15374 Müncheberg, Germany

Said Desouky
Botany and Microbiology Department, Faculty of Science, Al-Azhar University, Cairo 11651, Egypt

Kakhramon Davranov
Institute of Microbiology, Academy of Sciences of Uzbekistan, Tashkent 100128, Uzbekistan

Riyaz Z. Sayyed
Department of Microbiology, PSGVP Mandal's, Arts, Science & Commerce College, Shahada 425409, Maharashtra, India

Roslinda Abd Malek
Institute of Bioproduct Development (IBD), Universiti Teknologi, Malaysia (UTM), Skudai, Johor Bahru 81310, Malaysia

Asad Syed and Ali H. Bahkali
Department of Botany and Microbiology, College of Science, King Saud University, Riyadh 11451, Saudi Arabia

Lorena del Rosario Cappellari, Julieta Chiappero, Tamara Belén Palermo, Walter Giordano and Erika Banchio
INBIAS Instituto de Biotecnología Ambiental y Salud (CONICET – Universidad Nacional de Río Cuarto), Campus Universitario, 5800 Río Cuarto, Argentina

Muhammad Nawaz, Hasnain Ishaq and Naeem Iqbal
Department of Botany, Government College University Faisalabad, Lahore 54000, Pakistan

Sabtain Ishaq
Department of Botany, University of Agriculture Faisalabad, Faisalabad 38000, Pakistan

Shafaqat Ali
Department of Environmental Sciences and Engineering, Government College University, Lahore 54000, Pakistan
Department of Biological Sciences and Technology, China Medical University, Taichung 40402, Taiwan
Department of Environmental Sciences and Engineering, Government College University, Allama Iqbal Road, Faisalabad 38000, Pakistan

Muhammad Rizwan
Department of Environmental Sciences and Engineering, Government College University, Lahore 54000, Pakistan
Department of Environmental Sciences and Engineering, Government College University, Allama Iqbal Road, Faisalabad 38000, Pakistan

Abdulaziz Abdullah Alsahli and Mohammed Nasser Alyemeni
Botany and Microbiology Department, College of Science, King Saud University, Riyadh 11451, Saudi Arabia

Qasim Ali, Muhammad Tariq Javed, Muhammad Zulqurnain Haider and Noman Habib
Department of Botany, Government College University, New Campus, Jhang Road, Faisalabad 38000, Pakistan

Rashida Perveen
Department of Physics, University of Agriculture, Faisalabad 38040, Pakistan

Mohammed Nasser Alyemeni
Botany and Microbiology Department, College of Science, King Saud University, Riyadh l1451, Saudi Arabia

Hamed A. El-Serehy and Fahad A. Al-Misned
Department of Zoology, College of Science, King Saud University, Riyadh l1451, Saudi Arabia

Noshin Ilyas and Roomina Mazhar
Department of Botany, PMAS-Arid Agriculture University, Rawalpindi 46300, Pakistan

Humaira Yasmin
Department of Bio-Sciences, COMSATS University, Islamabad 45550, Pakistan

Wajiha Khan
Department of Biotechnology, COMSATS University Islamabad, Abbottabad Campus, Abbottabad 22010, Pakistan

Sumera Iqbal
Department of Botany, Lahore College for Women University, Lahore 54000, Pakistan

Hesham El Enshasy
Institute of Bioproduct Development (IBD), Universiti Teknologi Malaysia (UTM), Skudai, Johor 81310, Malaysia
School of Chemical and Energy Engineering, Faculty of Engineering, Universiti Teknologi Malaysia (UTM), Skudai, Johor 81310, Malaysia
City of Scientific Research and Technology Applications (SRTA), New Burg Al Arab, Alexandria 21934, Egypt

Daniel Joe Dailin
Institute of Bioproduct Development (IBD), Universiti Teknologi Malaysia (UTM), Skudai, Johor 81310, Malaysia
School of Chemical and Energy Engineering, Faculty of Engineering, Universiti Teknologi Malaysia (UTM), Skudai, Johor 81310, Malaysia

Bani Kousar
Department of Plant Sciences Faculty of Biological Sciences, Quaid-i-Azam University, Islamabad 45320, Pakistan

Asghari Bano
Department of Biosciences, University of Wah, Wah Cantt 47040, Pakistan

Naeem Khan
Department of Agronomy, Institute of Food and Agricultural Sciences, University of Florida, Gainesville, FL 32608, USA

Qiong He, Dongya Wang, Bingxue Li and Haiyan Wu
Guangxi Key Laboratory of Agric-Environment and Agric-Products Safety, Agricultural College of Guangxi University, Nanning 530004, China

Ambreen Maqsood
Guangxi Key Laboratory of Agric-Environment and Agric-Products Safety, Agricultural College of Guangxi University, Nanning 530004, China
Department of Plant Pathology, Faculty of Agriculture and Environmental Sciences, The Islamia University of Bahawalpur, Bahawalpur 63100, Pakistan

Judith Naamala and Donald L. Smith
Department of Plant Science, McGill University, Lakeshore Road, Ste. Anne de Bellevue, 21111, Montreal, QC H9X3V9, Canada

Index

A

Abiotic Stress, 23, 29, 32, 36, 53, 101-103, 108-114, 116, 118, 127-128, 153, 155, 170, 191, 214-217, 221-223, 229-230, 232, 234

Acacia Biochar, 37, 39-40, 42, 45-46, 48-51

Agricultural Lands, 6, 38

Agricultural System, 38

Allelopathic Bacteria, 56-57, 59-62, 64, 66, 68-69, 71

Ammonia Production, 39, 51

Antimetabolites, 56, 68

Ascorbic Acid, 54, 88, 104, 106-107, 117, 119-120, 122-123, 125-127, 129-130, 132, 136, 143, 153, 155-156

B

Bacteria, 1-2, 4-6, 21, 23-24, 29-36, 38-39, 51-53, 56-62, 64, 66, 68-69, 71-74, 76-77, 79, 81-85, 87, 98-103, 112-115, 158, 165, 170, 172, 174-175, 178, 194-195, 199, 209, 213, 215-218, 227-230, 233-237

Bacterial Community Composition, 72, 81, 173

Bacterial Culture, 87, 103

Biochar Application, 37-38, 42, 45, 50-52, 55, 86, 92-94, 97

Biochar Treatments, 37, 40, 42, 48, 93

Biofertilizers, 2, 29-31, 53, 72, 112, 217, 219, 232

Biological Control, 2, 35, 56-57, 70-72, 113, 173, 198-199, 210, 212-213, 234-236

Biomass Carbon, 37, 41, 43, 47

Biomass Reduction, 117, 125

Biotic Stresses, 117

Bradyrhizobium Japonicum, 86-87, 98-99, 220, 225, 230, 238

Brassica Rapa, 72, 76, 80-81, 85

C

Chitin Decomposition, 39, 51

Crop Production, 29, 37, 39, 55, 57, 70, 72-73, 153-154, 214-215, 217, 220, 222, 226, 229, 231, 234

Cyanide Production, 39, 51, 58, 62, 68, 71, 160

D

Desertification, 38, 221, 231, 236

Drought, 36, 53, 82, 85, 109-116, 118, 124, 128-133, 137, 141-142, 144-147, 149-157, 159, 171, 173-176, 193-196, 214, 216, 221-223, 230-232, 237

E

Effective Strategy, 37, 158

Egyptian Acacia, 37, 39-40, 42, 45-46, 48-51

Environmental Stresses, 117-118, 125, 157, 178

F

Fertilizers, 1-2, 30-31, 35, 37-39, 52-53, 72-73, 81-83, 85, 87, 103, 118, 196, 214, 219, 227

Food Waste Materials, 72-73, 80

G

Germination, 4, 59-60, 63-66, 68, 82, 85, 87-90, 95, 102, 118, 126, 128, 134, 154, 160, 165, 171-172, 174, 176, 195-196, 198-199, 201, 207-208, 210-211, 228, 230, 237

Glandular Trichomes, 101, 108, 112, 115

Glycinebetaine, 116-117, 155-156

Growth Regulators, 50, 85, 117, 128, 155, 193

H

High Temperature, 37, 50, 156, 216, 221-223

I

Irrigation, 40-41, 54-56, 100, 116, 118, 133-134, 137, 161, 221, 224

L

Liquid Food Waste, 72-74, 80, 82

Local Effective Microorganisms, 73, 83

Low Rainfall, 37, 50

M

Manure Biochar, 37, 39, 51, 100

Mentha Piperita, 101, 105-108, 113, 115-116, 130

Metal Stress, 98-99, 112, 117

Microorganism-plant Interactions, 101-102

Microorganisms, 2, 31-32, 38, 57, 71-73, 76, 80, 83, 99, 102, 112, 130, 154, 175, 195-197, 214-219, 231-233, 236, 238

N

Nitrate Nitrogen, 37, 47, 82

Nitrogen Fixation, 4, 6, 21, 32, 53, 72, 75, 98, 171, 216-217, 219, 222, 225-226, 230, 233, 237

Nitrogen Mineralization, 52, 73, 83

Nutrient Availability, 37-38, 51, 54-55, 72, 95, 97, 151, 173

Nutrient-solubilizing, 37-40

O

Operational Taxonomic Units, 5, 72, 75

Organic Compounds, 72, 87, 101-103, 113, 133, 149

Organic Fertilizers, 52, 72-73, 85

Organic Matter, 4, 37-43, 46-47, 50-51, 54, 73, 87, 99, 134

Organic Phosphate, 72

P

Phenolic Compound, 101, 103

Phosphate Solubilizing, 33-34, 36, 38, 53, 99, 233

Phytotoxic Metabolites, 56, 68

Plant Growth, 1-2, 4-6, 21, 23-24, 29-32, 34-35, 37-39, 50-55, 71-73, 75-79, 81-91, 95-104, 111-118, 120, 124-125, 128-129, 149, 153, 155-156, 158-160, 171-178, 182, 191-197, 212, 214-219, 222-224, 226-239

Plant Growth-promoting Microbes, 72-73

Plant Nutrients, 86-89, 92, 95, 217

Plant Species, 36, 68-69, 101, 110, 118, 125, 149, 177, 191, 194, 217-218, 222-223

Plant-growth Promoting Rhizobacteria, 86-87

Post-harvest Soil, 37, 41

Pseudomonas Putida, 61, 84, 86-87, 98-99, 221-223, 237

R

Rhizobacteria, 31-32, 34, 37-38, 53-63, 68-69, 71, 83, 85-89, 91-94, 97, 99-103, 112-116, 128-130, 153, 158-159, 171, 173-178, 192, 194-197, 212, 215, 232, 234-238

S

Salicylic Acid, 109, 112, 114, 116-117, 125, 128-130, 154-155, 177, 181, 187-189, 192, 194-197

Salinization, 38, 172

Semi-arid Climate, 37, 48

Semi-arid Regions, 37, 39, 50-51, 117

Siderophores Production, 6, 39, 51

Soil Carbon Contents, 38, 50

Soil Enzymes, 86-87, 89, 93-94, 96-98

Soil Fertility, 31, 39, 50, 52, 72, 86-87, 96, 175, 216-217, 234

Soil Health, 37-39, 50, 72, 176

Soil Nutrients, 86-87, 93, 96-97, 178, 217

Soil Organic Matter, 37, 39, 42, 50-51, 87

Soil Productivity, 38, 50, 238

Soil Quality, 38, 72-73, 97, 101, 232-233

Sustainable Agriculture, 31, 35, 53, 56, 72, 103, 194, 226, 232-233, 238

U

Unused Resources, 72, 80

V

Volatile Organic Compounds, 101-103, 113

W

Weed Invasion, 56, 61, 69

Wheat Straw Biochar, 37-40, 42, 45-46, 48-51

Printed in the USA
CPSIA information can be obtained
at www.ICGtesting.com
JSHW051409091023
49903JS00006B/354